Metallurgical Slags

Environmental Geochemistry and Resource Potential

Chemistry in the Environment Series

Titles in the series:
1: Graphene-based 3D Macrostructures for Clean Energy and Environmental Applications
2: Metallurgical Slags: Environmental Geochemistry and Resource Potential

How to obtain future titles on publication:
A standing order plan is available for this series. A standing order will bring delivery of each new volume immediately on publication.

For further information please contact:
Book Sales Department, Royal Society of Chemistry, Thomas Graham House, Science Park, Milton Road, Cambridge, CB4 0WF, UK
Telephone: +44 (0)1223 420066, Fax: +44 (0)1223 420247
Email: booksales@rsc.org
Visit our website at www.rsc.org/books

Metallurgical Slags
Environmental Geochemistry and Resource Potential

By

Nadine M. Piatak
United States Geological Survey, USA
Email: npiatak@usgs.gov

and

Vojtěch Ettler
Charles University, Czech Republic
Email: vojtech.ettler@natur.cuni.cz

ROYAL SOCIETY OF **CHEMISTRY**

Chemistry in the Environment Series No. 2

Print ISBN: 978-1-78801-887-6
PDF ISBN: 978-1-83916-457-6
EPUB ISBN: 978-1-83916-458-3
Print ISSN: 2516-2624
Electronic ISSN: 2516-2632

A catalogue record for this book is available from the British Library

The Royal Society of Chemistry is a charity, registered in England and Wales, Number 207890, and a company incorporated in England by Royal Charter (Registered No. RC000524), registered office: Burlington House, Piccadilly, London W1J 0BA, UK, Telephone: +44 (0) 20 7437 8656.

For further information see our website at www.rsc.org

Printed in the United Kingdom by CPI Group (UK) Ltd, Croydon, CR0 4YY, UK

Preface

The aim of this book is to summarize the current state of knowledge on the environmental geochemistry and resource potential of metallurgical slags. Hundreds of millions of tonnes of slag, a by-product of pyrometallurgical processing of ferrous and non-ferrous ores or recyclable materials, are generated annually worldwide. These slags are either landfilled, reprocessed, or repurposed. Limited space, as the global population explodes exponentially; degradation of our water, air, and soil due to human activities; and global concern over our planet's finite natural resources inspire us to better understand our industrial footprint and minimize it, and find value in by-products and waste with the ultimate goal of a circular economy. Currently, research on the environmental impacts and utilization of slags is mainly presented as case studies in numerous specialized journals. Several valuable review manuscripts have been published in journals, mostly in the last 5 years, and several books have been dedicated to particular aspects of slag use, such as in construction, or evaluating the ecosystem impacts of all types of mining waste. However, a book providing a compilation and detailed discussions of metallurgical slags' uses and impacts does not exist. Here, we provide a comprehensive overview of how metallurgical slags are generated, their fate in the environment, and their numerous beneficial uses, unlike any other book published to date.

The first chapter, *Metallurgical Slag – Environmental Liability or Valuable Resource?* by Nadine Piatak and Vojtěch Ettler, provides an introduction to our current knowledge of slag's potential as a contaminant and resource. The second chapter, *Metallurgical Overview and Production of Slags* by Elias Matinde and Joalet Steenkamp, discusses the smelting conditions under which slag is generated as a by-product. The furnace operating conditions and chemistries of their charge, and formation of slag during ironmaking,

Chemistry in the Environment Series No. 2
Metallurgical Slags: Environmental Geochemistry and Resource Potential
Edited by Nadine M. Piatak and Vojtěch Ettler

Published by the Royal Society of Chemistry, www.rsc.org

steelmaking, and the production of ferroalloys and non-ferrous metals (*e.g.*, aluminium, lead, zinc, copper, nickel, and platinum-group metals) are examined. Next, the third chapter, *Geochemical and Mineralogical Characterization of Slags* by Nadine Piatak, Vojtěch Ettler, and Darryl Hoppe, provides a summary of the analytical methods commonly used to characterize slag, and then details the chemical and mineralogical variations among slag types, including blast furnace slags, steel slags, ferroalloy slags, and base metal slags.

The fourth chapter, *Weathering of Slags* by Jakub Kierczak, Anna Pietranik, and Nadine Piatak, is the first of several chapters focusing on the environmental impacts of slags. This chapter explains the processes that affect slags over time in the environment and the approaches used to understand this weathering, and it includes a discussion of the mineralogical controls and the formation of secondary phases. The fifth chapter, *Slag Leaching Properties and Release of Contaminants* by Vojtěch Ettler and Martina Vítková, focuses on factors that influence the release of metal(loid)s into the environment, particularly from base metal slags. A thorough discussion of leaching scenarios is presented and how their results combined with speciation modeling and mineralogy can be used to understand geochemical weathering processes. The sixth and final chapter on environmental impacts is *Environmental Impact of Slag Particulates* by Vojtěch Ettler and Jakub Kierczak. This chapter highlights the mineralogy and degradation of slag particulates emitted within flue dust by smelter stacks or eroded from disposal sites with a focus on mainly base metal slags. Slag particulate transmission and its ultimate fate in soils and vegetation, and potential impacts to ecosystems and humans are examined.

The final three chapters of the book focus on the use of slag and its potential to host valuable recoverable metals. Chapter seven, *Diverse Applications of Slag in the Construction Industry* by Ivanka Netinger Grubeša and Ivana Barišić, details the current prolific use of slags, mostly ferrous slags, in concrete, mortar, road construction, cement, and other construction products. The specific and unique properties that slags impart to various materials are covered. Next, chapter eight, *Environmental Applications of Slags* by Helena Gomes, William Mayes, and Rebecca Ferrari, reveals the diverse environmental applications of slags, with an emphasis on ferrous slags, which include water and wastewater treatment where slag filters or slag-constructed wetlands may be used to remove excess nutrients, pathogens, or metal and/or organic contaminants. In addition, slag's use to mitigate greenhouse gas emissions, sequester carbon, condition soils, and aid in the restoration of grassland or marine habitats is discussed. Lastly, chapter nine, *Secondary Metal Recovery from Slags* by Anna Potysz and Eric van Hullebusch, provides a detailed overview of methods used to recover metals from slags, including currently utilized physical separation, flotation, and roasting techniques. This chapter also highlights both biotic and abiotic hydrometallurgical leaching as prospective recovery techniques to be applied in the future.

We would like to thank the authors, who have worked diligently to contribute to this book, which we hope will become an important reference for environmental geochemists, metallurgists, mining and civil engineers, waste and resource managers, and university students. We are also grateful to the reviewers that have taken the time to critically evaluate each chapter and, through their comments and suggestions, greatly improve the overall context of the book. Reviewers include Cristian Barca (Aix-Marseille University, France), Harvey Belkin (US Geological Survey), Alexandra Courtin-Nomade (Paris-Saclay University, France), Jan Elsen (KU Leuven, Belgium), Jakub Kierczak (Wrocław University, Poland), Carlo Mapelli (Polytechnic University of Milan, Italy), Elias Matinde (MINTEK, South Africa), Tony Morrison (Macquarie University, Australia), Maxim Muravyov (Winogradsky Institute of Microbiology, Russian Academy of Sciences), Michael Parsons (Natural Resources Canada), Yiannis Pontikes (KU Leuven, Belgium), and Bob Seal (US Geological Survey). Nadine Piatak would like to particularly thank Bob Seal for his incredible mentorship and support over the many years. Lastly, we appreciate The Royal Society of Chemistry (especially Helen Armes, Connor Sheppard, Lewis Pearce, and the production staff) for guiding us through the publication process. The institutional funding from the Center for Geosphere Dynamics of Charles University (UNCE/SCI/006) and the US Geological Survey are greatly acknowledged for supporting our work on this book.

Reston and Prague

Nadine M. Piatak (US Geological Survey)
Vojtěch Ettler (Charles University, Czech Republic)

Contents

Chemistry in the Environment Series No. 2
Metallurgical Slags: Environmental Geochemistry and Resource Potential
Edited by Nadine M. Piatak and Vojtěch Ettler

Published by the Royal Society of Chemistry, www.rsc.org

CHAPTER 1

Introduction: Metallurgical Slags – Environmental Liability or Valuable Resource?

NADINE M. PIATAK*[a] AND VOJTĚCH ETTLER[b]

[a]US Geological Survey, 954 National Center, Reston, Virginia 20192, USA; [b]Institute of Geochemistry, Mineralogy and Mineral Resources, Faculty of Science, Charles University, Albertov 6, 128 00 Prague 2, Czech Republic
*E-mail: npiatak@usgs.gov

1.1 Slag Production

Slags are by-products generated by the pyrometallurgical processing of both ferrous and non-ferrous metal ores and secondary resources (recycling). They are formed during smelting from the reaction of gangue minerals in the furnace feed and fluxes and, possibly, reductants intentionally added during the smelting process. Fluxes combine with the impurities to form a slag melt that is immiscible with the metal and/or alloy phase targeted for recovery. Slags can be crystalline or glass-like depending on their composition and cooling rates and are usually silica-rich. Several major categories of slags exist, including: ferrous slags generated from the manufacture of iron (Fe) and steel products; ferroalloy slags formed from the production of bulk ferroalloys (*e.g.*, FeCr, FeMn, FeNi, and FeSi); and base metal slags formed from the recovery of metals such as Cu, Ni, Pb, and Zn, commonly from sulfidic ores. Other less commonly reported slags, some that belong in the

Chemistry in the Environment Series No. 2
Metallurgical Slags: Environmental Geochemistry and Resource Potential
Edited by Nadine M. Piatak and Vojtěch Ettler

Published by the Royal Society of Chemistry, www.rsc.org

base metal slag category, include slags generated from the extraction of Mn, platinum-group metals, and Sn from natural ores, or the recovery of metals from recyclable products, such as Pb–acid batteries, Al scrap, and other post-consumer waste, including batteries, magnets, catalytic converters, and computer circuit boards (see Chapter 2 for technological details).

Hundreds of millions of tonnes (t) of slags are produced annually worldwide in numerous pyrometallurgical processes in our industrialized society. Iron blast furnaces (BF) generate approximately 0.25 to 0.30 t of slag per t of crude iron for typical ore grades, with higher amounts generated for low ore grades (up to 1.0 to 1.2 t of slag per t of crude iron).[1] Steel furnaces commonly produce approximately 0.2 t of slag per t of steel, with some being returned to the furnace to recover entrained metal.[1] Based on iron and steel production, van Oss[1] estimated that between 295 and 354 Mt of BF slag and 169 to 254 Mt of steel slag were generated in 2017; the USA contributed 6 to 7 Mt of BF slag and 8 to 12 Mt of steel slag to global slag generation that year.[1] Similarly, Tripathy *et al.*[2] stated that approximately 430 Mt of ferrous slag from BFs were produced worldwide in 2016, with 87% being produced in Asia-Pacific, 7% in Europe, 3% each in North America and Latin America, and 1% in the Middle East and Africa, collectively. The most recent European statistics for 2018 indicate that 19.2 Mt of BF slag and 15.7 Mt of slag from steel production were produced in Europe.[3] For non-ferrous metal smelting industries, the production of slag per amount of metal produced is rather variable, dependent on the technology used. For example, Goonan[4] compiled data on material flows in 30 large-scale Cu smelters around the world and found that the amount of slag produced was in the range 0.65–5.14 t per 1 t of produced Cu, with a weighted average of 1.91 t (Figure 1.1).

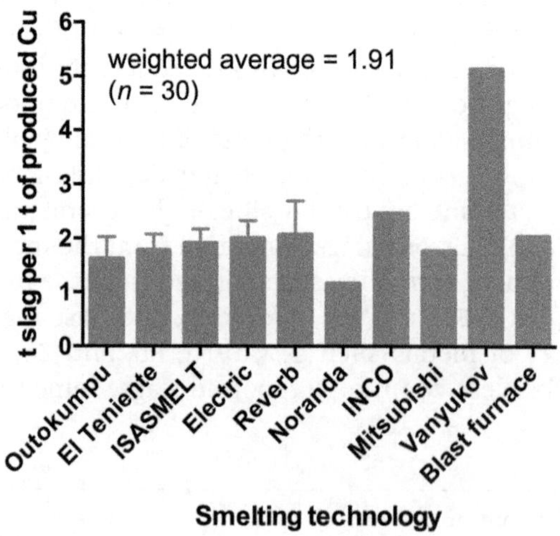

Figure 1.1 Copper slag production by different smelting technologies (tonnes of slag per 1 tonne of produced Cu) (compiled data for 30 large-scale smelters from ref. 4).

Given the fact that Cu production accounted for 20.4 Mt in 2018,[5] it can be estimated that approximately 39 Mt of Cu slag were generated in that year. Pan *et al.*[6] suggested that Pb slag produced in 2016 likely exceeded 5.5 Mt based on global Pb production of 11.1 Mt from both primary Pb ores and secondary resources. The amounts of slags generated from the production of base metals and other metals and alloys vary depending on the smelting conditions and how much metal is in production each year. However, overall, it is clear that ironmaking and steelmaking generate the largest volumes of slags, in the order of hundreds of millions of tonnes per year, whereas the amount of base metal slags generated, although not trivial in quantity, is an order of a magnitude lower than that for ferrous slags.

Depending on their chemical and mineralogical compositions, slags are either considered as waste, and many slags resulting from non-ferrous metallurgy contain appreciable amounts of potentially toxic elements [PTEs] (see Chapters 3, 4, 5, and 6 of this book), or as raw materials, which can find applications in civil engineering and environmental technologies (mainly ferrous slags, see Chapters 7 and 8) and/or can be used for metal recovery (mainly non-ferrous slags, see Chapter 9).

1.2 Literature Summary and Perspective

Scientific interest in slags has been increasing steadily over the last few decades with thousands of abstracts and manuscripts published annually over the last few years on slag research. The abundant scientific literature on slags can generally be divided into research on the environmental effects of slags or on the reuse, potential reuse, and value of slags. Several valuable review manuscripts have been published, typically focusing on one slag type and from one or limited perspectives. For example, Potysz *et al.*[7] provided an excellent review of Cu slags, highlighting their chemical and phase composition, environmental issues related to their disposal, and approaches for recovering valuable residual metals. Additionally, other insightful reviews include Pan *et al.*[6] on Pb slag generation, characteristics, and utilization; Naidu *et al.*[8] on uses of basic oxygen furnace slag, a type of steel slag; and Yildirim and Prezzi[9] on chemical, mineralogical, and morphological characteristics of all steel slag types. Special issues dedicated to slag research include "Metallurgical slags" in the journal *Minerals*[10] and an upcoming issue in *Applied Science* on "Utilization of steel furnace slag in cementitious composites." Currently, conferences and workshops on slags draw international attention and provide a venue for the newest research to be shared and advanced. Many books have also focused on principles of extractive metallurgy and contain some information about metallurgical slags.[11–15] Several books or chapters in books are dedicated to particular aspects of slag use, such as in construction,[16–18] or to evaluating the ecosystem impacts of mining waste, including slags.[19,20] Some reviews on slags have attempted a more holistic approach by including various aspects of metallurgical slags, two of which, by Piatak *et al.*[21] and Piatak,[22] included discussions of the characteristics of ferrous and base metals slags, their environmental impacts, and utilization potential. These reviews provided motivation for the creation of this book, which provides a comprehensive and detailed overview of how metallurgical slags are generated, their fate in the environment, and their numerous beneficial uses – all captured in one central publication.

Interestingly, published studies on base metal slags and ferrous slags generally have very different focuses, which influenced the perspectives of the chapters included in this book. The majority of research on ferrous slags has been on the physical and mechanical properties owing to their wide use in construction and cement, as well as their environmental applications. A search for the types of ferrous slag [*e.g.*, iron slag, steel slag, basic oxygen furnace (BOF) slag, electric arc furnace slag, BF slag, ladle slag] in the titles of abstracts and citations in the Scopus database[23] gives a query result of over 5000 documents for 2010 to 2020. A small fraction (around one in ten) of those documents are in the subject areas of Environmental Science or Earth and Planetary Science with the vast majority being in the Engineering and Material Science fields. These results imply that there is a major research focus on the physical and mechanical properties of this material for reuse, as highlighted in Chapters 7 and 8, and less of a focus on the environmental consequences of its disposal. As for the subject of weathering studies (discussed in Chapter 4), steel slags are intentionally weathered or activated before use in construction to hydrate the lime (CaO) so it does not expand and create instability or swelling when exposed to the atmosphere in the final product. The generation of secondary phases, also discussed in Chapter 4, such as portlandite ($Ca(OH)_2$) and calcite ($CaCO_3$), is intentional and studies have focused on how to facilitate and expedite it. Additionally, environmental applications of ferrous slags, such as in treating nutrient-rich or metal-rich wastewater or agricultural runoff, neutralizing acid mine drainage, or sequestering carbon dioxide gas, cause the dissolution of phases and precipitation of secondary phases; laboratory and field application studies abound on these applications, particularly for nutrient removal, as discussed in Chapter 8. Studies of ferrous slags have typically focused on the breakdown of the Ca and Mg silicates and oxides, which influence the efficacy of the material to be used in the various applications. To note, base metal slags are also used for construction purposes but the literature on ferrous slag use is overwhelmingly more abundant in comparison, which is partially attributed to the larger volume of ferrous slags generated annually compared to base metal slags, as previously discussed.

In contrast to ferrous slags, a large knowledge base exists on the environmental behavior of base metal slags enriched in PTEs. A Scopus search, similar to the one done for ferrous slag types, showed that studies on base metal slag types have more of an emphasis on Environmental Science and less on Engineering and Materials Science compared to ferrous slags. All over the world, there are examples of the leaching of deleterious metal(oid)s from base metal slags disposed of in waste piles (discussed in Chapter 5) or used as aggregate and in building materials, causing human and ecosystem impacts. Research spanning many years has focused on determining which phases host the PTEs, such as, Cd, Pb, and Zn, and understanding their susceptibility to and rates of weathering. Studies on the geochemical processes that facilitate the dissolution of phases found in base metal slags have revealed the importance of the weathering of sulfide phases, intermetallic compounds, and glass to metal(oid) release. The precipitation of secondary phases, their ability to sequester metal(oid)s, and their potential to later release those metal(oid)s upon further weathering have all been examined (see Chapter 4). Furthermore, slag particulates emitted within flue dust from smelter stacks and eroded from disposal sites may have deleterious effects on ecosystems and human health (presented in Chapter 6). The environmental behavior of base metal slags has been documented in numerous laboratory and field studies. In contrast to base metal slags, the weathering of Ca and Mg silicates and oxides is generally the source of environmental issues related to ferrous slag weathering, predominantly by the generation of alkaline or hyperalkaline runoff but also the release of PTEs, such as Cr and V. A focus on silicates and oxides in ferrous slag is in stark contrast to the sulfide phases, intermetallic compounds, and glass phases that draw attention in base metal slag studies due to them hosting PTEs. These phases in base metal slags also are of interest for reprocessing to recover residual metals (the focus of Chapter 9). In contrast, the sulfides (*i.e.*, CaS, MnS) that occur in ferrous slag commonly do not tend to host PTEs. In addition, the metallic component of ferrous slag is Fe, unlike the complex intermetallic phases reported in base metal slags; studies on metallic Fe focus on reprocessing to recovery the metal (also in Chapter 9) instead of environmental issues associated with its oxidation.

Overall, the different perspectives and research foci motivating studies on base metal slags compared to ferrous slags have led to unique and different knowledge bases. This ultimately influences which types of slag are highlighted in the various chapters of this book. For example, Chapters 4, 5, and 6 on environmental impacts of slags predominantly discuss base metal slags, whereas Chapters 7 and 8 on the use of slags in construction and environmental applications mostly refer to ferrous slags, and Chapter 9 on metal recovery includes both ferrous and non-ferrous case studies. In addition, Chapter 2 on metallurgy and Chapter 3 on the chemical and mineralogical characteristics of slags include discussions of ferrous, ferroalloy, and non-ferrous slags.

1.3 Slag as a Source of Information About the Metallurgical Process

Slags contain the "DNA" of the metallurgical processes under which they were generated; their mineralogical and chemical compositions can be used to reconstruct the raw materials, conditions of operation of the smelting furnace, and metal extraction efficiency. This information is especially useful for archaeologists and archaeometallurgists, who investigate (pre)historic metal production sites.[24,25] Chemical compositions can be used for the estimation of viscosity indices; they describe the slag melt viscosity, which in turn reflects the efficiency of the removal of the metallic fraction during the smelting process.[26] The use of ternary diagrams aids in the estimation of the minimum temperatures in the smelting furnace[21,26–34] and can be enhanced when combined with modeling using FACT Sage[26] or MELTS[34] metallurgical or petrological software. Alternatively, for slags with specific compositions, geothermometry, traditionally used in geology, has also been tested for evaluating conditions for slag crystallization.[31,34] The slag textures, and shapes and sizes of crystals can be used for estimating the slag melt cooling regime.[26] All of these methods can be used to unravel the history at archaeological smelting sites and at current legacy slag disposal sites, where the historical documents are missing.

1.4 Slag as an Environmental Threat

Owing to high levels of PTEs (metals and metalloids) in some types of slag, certain slags are considered as potentially hazardous wastes despite the fact that the standardized leaching tests used for the waste classification often indicate that contaminant leaching is low over the short timeframe that these tests are administered.[35] However, a 12 year study of the leaching of Pb–Zn slags indicated that metals and metalloids, mainly from the dissolution of sulfides and intermetallic compounds, may be released to a greater extent in the long term.[36] When Cu slags are used as a construction material (*e.g.*, as a barrier material in waterways and coast protection), they may represent a significant diffuse source of pollution.[37,38] However, the incorporation of Cu slags in some materials, such as in ceramic tiles, may result in the immobilization of PTEs, such as As and Pb.[39] For ferrous slags, alkalinity generated from the hydration and subsequent dissociation of Ca and Mg oxides and silicates, and the concurrent release of oxyanions (*e.g.*, As, Cr, Mo, Se, V), may cause ecosystem impacts.[40] A combination of chemical and mineralogical investigations, leaching tests, weathering experiments, field and onsite investigations, and geochemical modeling can be used to assess the environmental impacts of slags and outline directions for their future utilization.[41,42]

Uncontrolled and unfenced slag disposal sites can represent a critical environmental risk for populations living and/or working in their vicinity, especially in developing countries.[43] Recent investigations in the former Pb–Zn mining and smelting district in Kabwe (Zambia) indicated that fine dust from the granulated slag heaps might have direct human health implications. When incidental ingestion of the slag dust was assumed as the major exposure scenario for the local population, extractions in simulated gastric fluids indicated particularly high amounts of bioaccessible toxic Pb.[44]

1.5 Metal Recovery from Slag

It is common to have incomplete recovery of the targeted metal during pyrometallurgical extraction, resulting in an enriched slag. In the making of iron and steel, residual Fe metal or ferroalloys may end up in the slag. Blast furnace slag may be recycled into the steel plant as a secondary source of Fe and as a pre-melted flux.[16] Furthermore, BOF slag can be recycled into the sinter and blast furnace process to recover Fe, and possible Mn, as well as being a source of fluxing compounds.[16] Additionally, Cu slag may contain a weight percent or more of Cu; this may be higher than the concentration of Cu present in the original ore.[22] This Cu enrichment is usually due to entrapment of matte in the slag and is especially common when slag melt is quenched by granulation, which limits the gravity separation of metal-bearing sulfide droplets. The reprocessing of Cu slags is a routine part of some extraction circuits and is generally based on comminution in a mill followed by flotation to recover residual sulfides, which are fed back into the smelter.[45–47] The resulting slag tailings obtained after flotation are then deposited in the tailing ponds and can be chemically stabilized using carbonate and Fe oxide-bearing soils, which was found to be an efficient way of immobilizing potentially leachable metal(loid) contaminants (Figure 1.2).[45]

Besides incomplete recovery of the targeted metals, the partitioning of co-occurring or byproduct elements into the slag phase, as compared to the metal or gas phases, during smelting warrants further exploration as slag may be a promising source of some technologically critical and valuable elements. Figure 1.3 is an element radar chart that graphically represents the tendencies of elements to partition into the slag, gas, or metal phase during

Figure 1.2 Tailing impoundment where slag tailing after milling and re-flotation (for recovery of Cu-bearing sulfides) is co-disposed with carbonate-rich mine tailing and Fe oxide-bearing lateritic soil (Tsumeb, Namibia). This stabilization technology helps to minimize contaminant release from the slag (photo: Vojtěch Ettler). Reproduced from ref. 45 with permission from Elsevier, Copyright 2017.

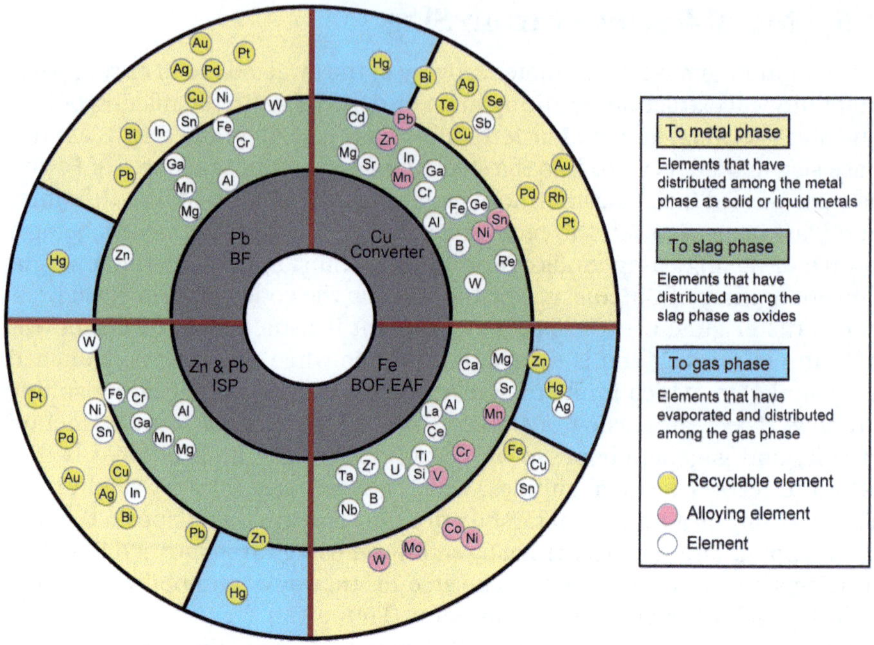

Figure 1.3 Element radar chart showing distributions of elements among the metal, slag, and gas phases during pyrometallurgical reprocessing. BOF, basic oxygen furnace, EAF, electric arc furnace, BF, blast furnace, ISP, imperial smelting process. Modified from Nakajima *et al.*[48] Reproduced from ref. 22 with permission from Elsevier, Copyright 2018.

smelting. This figure is based on thermodynamic analyses for metal recyclability by Nakajima *et al.*[48] and Nakajima *et al.*[49] but is also illustrative of element enrichment of slags during pyrometallurgical processing of natural ores. These types of studies can guide us to identify certain elements that could be recoverable from slag. For example, Ga, a critical element used in high technology, partitions into the slag phase during Cu, Pb, and Zn smelting (Figure 1.3).

Chemical and biological leaching approaches, originally developed and used for low-quality metal ores, have already been tested on slags. As shown in Figure 1.3, steel slags may contain elements such as Al, Cr, and V. Laboratory leaching studies have shown these metals can be recovered from steel slags.[50,51] Abhilash *et al.*[52] explored acid leaching of rare earth elements from blast furnace slag, elements that may also partition into the slag (Figure 1.3). Copper slags, as already discussed, contain residual Cu but may also contain Zn and Co, among other metals; these metals have also been shown to be amenable to recovery by leaching.[53–56] For bioleaching, Potysz *et al.*[57] reviewed advantages and limitations in metal leaching and subsequent recovery and showed that detailed slag characterization and appropriate choices of pH, slag particle size, liquid-to-solid (L/S) ratio, microorganism species, and nutrient supply based on laboratory testing are needed before pilot-scale and site-scale bioleaching applications (see also Chapter 9). Another less explored approach to recover metals is carbothermal reduction of slags; according to Sarfo *et al.*,[58] Cu and Fe can be recovered as metals from Cu slag leaving a secondary glass slag that can be used in the ceramic or glass industries. Unlike the routine use of flotation to recover Cu from some Cu slags, these more novel approaches (*i.e.*, bioleaching, carbothermal reduction) have not yet been implemented on an industrial scale. This is an area that warrants additional research to progress to full-scale implementation.

1.6 Slag Use Applications

Slags from iron and steel production have already found numerous applications in the construction industry.[2,17,59] Iron and steel companies typically provide their slag free of charge to slag-processing companies. The slag melt is cooled in a specified way to impart the desired physical properties and may be then crushed and sieved to sort by size, depending on the end use. Magnetic separators can also be used to recover entrained metal, which can be sold back to the furnace operator. The slag is then sold by the slag-processing companies on the open market for use in concrete, road bases, cement, or raw building materials.[1] Although commonly marketed for use in construction, the applications of ferrous slags are broadening with research showing that the material may also have use in water and wastewater treatment, where slag filters or slag-constructed wetlands can be used to remove excess nutrients, pathogens, or metal and/or organic contaminants.[40] Ferrous slags are also being examined for aiding in the mitigation of greenhouse gas emissions, sequestering carbon, conditioning soils, and contributing to the restoration of grassland or marine habitats.[40]

Slags originating from non-ferrous metal production are also used in construction applications; however, the quantities generated annually are significantly less than those for ferrous slags. Therefore, there are smaller amounts of non-ferrous slags to be reused and they also may contain PTEs. Nonetheless, Cu slags have found use as construction materials, abrasive tools, and ballast,[60,61] and Pb slags have been used as fine aggregate in cement mortar, which limits leaching of PTEs,[62] among other uses of non-ferrous slags.

1.7 Outlook

The aim of this book is to summarize the current state of knowledge on metallurgical slags, with a special focus on the environmental behavior of slags and their current applications, as well as to explore future directions, especially in view of their reprocessing for residual base and critical metal recovery. Since 2015, the circular economy is being discussed with increasing intensity, particularly in Europe;[63] it intimately concerns mineral industries and metallurgy.[64] Reuter[65] even stated that metallurgy is at the heart of a circular economy as metals all have intrinsic recycling potentials. However, van Schalkwyk *et al.*[66] drew attention to the inevitable losses that occur during smelting operations, which represent a true limitation to circularity. The policy documents on the circular economy do not operate with technological parameters, which are often overlooked and/or neglected in the material flows related to metal production.[66] In a series of excellent papers devoted to the cyclic use of metals and life-cycle assessment in the metallurgical industry,[65,67,68] the complexity of the system (*e.g.*, primary ores, metal production circuits, waste generation and treatment) has been demonstrated with respect to the aims of the circular economy in the mineral industry and metal production. Many technologically important minor elements are directly linked to base metal production (*e.g.*, Bi in Pb ores, Te in Cu ores) and production of some of these elements is so interwoven that they can hardly be considered separately.[67] These interconnected processes in individual metallurgical operations are often not fully known because they generally do not operate at thermodynamic equilibrium conditions, are based on multiphase reactions at high temperature gradients, and are unpredictable by mathematical models.[67] The metallurgical industry is based on experience acquired over the years in a "learning-by-doing" mode, called "tacit" knowledge by many authors.[65,67] Preservation of this old metallurgical knowledge and its coupling with modern metallurgical expertise and novel tools, such as digitalization, process simulations, and modeling in multi-metal metallurgical processing, currently applied for the environmental assessment of metal production (including waste slag generation), are of key importance for optimization of the industrial extractive process towards a circular economy for metals.[65]

Metallurgical slags are one of the key segments in the complex "puzzle" of the metal extraction and recycling industries, as we clearly demonstrate in this book. For this reason, studies that focus on maximizing the recovery of residual metals from slags while generating a more environmentally friendly and useable byproduct are important for establishing a sustainable metallurgical industry and moving towards a circular economy. However, if industry is to adapt methods to recover metals from slags in their operations, they need to be profitable, which is not always the case considering current extraction costs and metal prices.[69]

Acknowledgements

Vojtěch Ettler acknowledges the Czech Science Foundation (GAČR) project no. 19-18513S and the institutional funding from the Centre for Geosphere Dynamics (UNCE/SCI/006). The valuable comments from Harvey Belkin (US Geological Survey) and Robert (Bob) Seal (US Geological Survey) helped to significantly improve the original version of the manuscript. Any use of trade, firm, or product names is for descriptive purposes only and does not imply endorsement by the US Government.

References

1. H. G. van Oss, *2017 Minerals Yearbook*, US Geological Survey, 2020, p. 69.1.
2. S. K. Tripathy, J. Dasu, Y. R. Murthy, G. Kapure, A. R. Pal and L. O. Filippov, *J. Cleaner Prod.*, 2020, **262**, 121354.
3. EUROSLAG, Statistics, 2018, https://www.euroslag.com/products/statistics/statistics-2018/, accessed March 2021.
4. T. G. Goonan, *Flows of Selected Materials Associated with World Copper Smelting*, US Geological Survey Open-file Report 2004-1395, US Geological Survey, Reston, Virginia, 2005.
5. US Geological Survey, *Mineral Commodity Summaries 2020*, US Geological Survey, 2020, DOI: 10.3133/mcs2020.
6. D. Pan, L. Li, X. Tian, Y. Wu, N. Cheng and H. Yu, *Resour., Conserv. Recycl.*, 2019, **146**, 140.
7. A. Potysz, E. D. van Hullebusch, J. Kierczak, M. Grybos, P. N. L. Lens and G. Guibaud, *Crit. Rev. Environ. Sci. Technol.*, 2015, **45**, 2424.
8. T. S. Naidu, C. M. Sheridan and L. D. van Dyk, *Miner. Eng.*, 2020, **149**, 106234.
9. I. Z. Yildirim and M. Prezzi, *Adv. Civ. Eng. Mater.*, 2011, **2011**, 463638.
10. J. Kierczak and V. Ettler, *Minerals*, 2020, **11**, 24.
11. M. E. Schlesinger, K. C. Sole and W. G. Davenport, *Extractive Metallurgy of Copper*, Elsevier, Oxford, UK, 5 edn, 2011.
12. T. Rosenqvist, *Principles of Extractive Metallurgy*, Tapir Academic Press, Trondheim, 2 edn, 2004.
13. F. K. Crundwell, M. S. Moats, V. Ramachandran, T. G. Robinson and W. G. Davenport, *Extractive Metallurgy of Nickel, Cobalt and Platinum-group Metals*, Elsevier, Amsterdam, 2011.
14. R. J. Sinclair, *The Extractive Metallurgy of Zinc*, The Australasian Institute of Mining and Metallurgy (AusIMM), Carlton Victoria, 2005.
15. M. Geerdes, R. Chaigneau and I. Kurunov, *Modern Blast Furnace Ironmaking: an Introduction* IOS Press, 3 edn, 2015.
16. S. Ndlovu, G. S. Simate and E. Matinde, *Waste Production and Utilization in the Metal Extraction Industry*, CRC Press, Boca Raton, Florida, 2017.
17. I. Netinger Grubeša, I. Barišić, A. Fucic and S. S. Bansode, *Characteristics and Uses of Steel Slag in Buidling and Construction*, Elsevier, Duxford, UK, 2016.
18. G. C. Wang, *The Utilization of Slag in Civil Infrastructure Construction*, Woodhead Publishing, Duxford, UK, 2016.
19. *Environmental Geochemistry – Site Characterization, Data Analysis, and Case Histories*, ed. B. De Vivo, H. E. Belkin and A. Lima, Elsevier, Amsterdam, Netherlands, 2 edn, 2018.
20. *Sampling and Monitoring for the Mine Life Cycle*, ed. V. T. McLemore, K. S. Smith and C. C. Russell, Society for Mining, Metallurgy, and Exploration, Englewood, Colorado, USA, 2014.
21. N. M. Piatak, M. B. Parsons and R. R. Seal II, *Appl. Geochem.*, 2015, **57**, 236.

22. N. M. Piatak, in *Environmental Geochemistry: Site Characterization, Data Analysis and Case Histories*, ed. B. De Vivo, H. E. Belkin and A. Lima, Elsevier, 2 edn, 2018, ch. 19, pp. 487–519.
23. Elsevier, Scopus, https://www.scopus.com/search/form.uri?display=basic&zone=header&origin=#basic, accessed January 2021.
24. M. F. Charlton, P. Crew, T. Rehren and S. J. Shennan, *J. Anthropol. Archaeol.*, 2010, **29**, 352.
25. S. Chirikure and F. Bandama, *Archaeometry*, 2014, **56**, 296.
26. V. Ettler, R. Červinka and Z. Johan, *Archaeometry*, 2009, **51**, 987.
27. N. M. Piatak and R. R. Seal II, *Appl. Geochem.*, 2010, **25**, 302.
28. N. M. Piatak, R. R. Seal II and J. M. Hammarstrom, *Appl. Geochem.*, 2004, **19**, 1039.
29. A. Vartiainen and M. Kytö, *Scand. J. Metall.*, 2002, **31**, 298.
30. R. Warchulski, A. Gawęda, M. Kądziołka-Gaweł and K. Szopa, *Mineral. Mag.*, 2015, **79**, 459.
31. R. Warchulski, *Eur. J. Mineral.*, 2016, **28**, 375.
32. M. Vítková, V. Ettler, Z. Johan, B. Kříbek, O. Šebek and M. Mihaljevič, *Mineral. Mag.*, 2010, **74**, 581.
33. J. Kierczak and A. Pietranik, *Can. Mineral.*, 2011, **49**, 1281.
34. K. Kądziołka, A. Pietranik, J. Kierczak, A. Potysz and T. Stolarczyk, *J. Archaeol. Sci.*, 2020, 105142.
35. V. Ettler, Z. Johan, B. Kříbek, O. Šebek and M. Mihaljevič, *Appl. Geochem.*, 2009, **24**, 1.
36. V. Ettler and Z. Johan, *Appl. Geochem.*, 2014, **40**, 97.
37. L. Duester, C. Brinkmann, T. A. Ternes and P. Heininger, *Crit. Rev. Environ. Sci. Technol.*, 2016, **46**, 434.
38. A. Schmukat, L. Duester, D. Ecker, H. Schmid, C. Heil, P. Heininger and T. A. Ternes, *J. Hazard. Mater.*, 2012, **227–228**, 257.
39. M. M. Jordán, M. A. Montero and F. Pardo-Fabregat, *J. Mater. Cycles Waste Manage.*, 2021, 707–716.
40. H. I. Gomes, W. M. Mayes, M. Rogerson, D. I. Stewart and I. T. Burke, *J. Cleaner Prod.*, 2016, **112**, 3571.
41. A. Lassin, P. Piantone, A. Burnol, F. Bodenan, L. Chateau, C. Lerouge, C. Crouzet, D. Guyonnet and L. Bailly, *J. Hazard. Mater.*, 2007, **139**, 430.
42. H. I. Gomes, W. M. Mayes, H. A. Baxter, A. P. Jarvis, I. T. Burke, D. I. Stewart and M. Rogerson, *J. Cleaner Prod.*, 2018, **202**, 401.
43. V. Ettler, M. Cihlová, A. Jarošíková, M. Mihaljevič, P. Drahota, B. Kříbek, A. Vaněk, V. Penížek, O. Sracek, M. Klementová, Z. Engel, F. Kamona and B. Mapani, *Environ. Int.*, 2019, **124**, 205.
44. V. Ettler, D. Štěpánek, M. Mihaljevič, P. Drahota, R. Jedlicka, B. Kříbek, A. Vaněk, V. Penížek, O. Sracek and I. Nyambe, *Chemosphere*, 2020, 127642.
45. A. Jarošíková, V. Ettler, M. Mihaljevič, B. Kříbek and B. Mapani, *J. Environ. Manage.*, 2017, **187**, 178.
46. A. Rozendaal and R. Horn, *Miner. Eng.*, 2013, **52**, 184.
47. A. Sarrafi, B. Rahmati, H. R. Hassani and H. H. A. Shirazi, *Miner. Eng.*, 2004, **17**, 457.

48. K. Nakajima, O. Takeda, T. Miki, K. Matsubae, S. Nakamura and T. Nagasaka, *Environ. Sci. Technol.*, 2010, **44**, 5594.
49. K. Nakajima, O. Takeda, T. Miki and T. Nagasaka, *Mater. Trans.*, 2009, **50**, 453.
50. J. Xiang, Q. Huang, X. Lv and C. Bai, *J. Cleaner Prod.*, 2018, **170**, 1089.
51. H. I. Gomes, V. Funari, W. M. Mayes, M. Rogerson and T. J. Prior, *J. Environ. Manage.*, 2018, **222**, 30.
52. A. Abhilash, P. Meshram, S. Sarkar and T. Venugopalan, *Miner. Metall. Process.*, 2017, **34**, 178.
53. A. Potysz and J. Kierczak, *Minerals*, 2019, **9**, 542.
54. A. H. Kaksonen, L. Lavonen, M. Kuusenaho, A. Kolli, H. Närhi, E. Vestola, J. A. Puhakka and O. H. Tuovinen, *Miner. Eng.*, 2011, **24**, 1113.
55. A. H. Kaksonen, S. Särkijärvi, J. A. Puhakka, E. Peuraniemi, S. Junnikkala and O. H. Tuovinen, *Miner. Eng.*, 2017, **106**, 97.
56. A. Potysz, A. Pędziwiatr, S. Hedwig and M. Lenz, *J. Environ. Chem. Eng.*, 2020, **8**, 104450.
57. A. Potysz, E. D. van Hullebusch and J. Kierczak, *J. Environ. Manage.*, 2018, **219**, 138.
58. P. Sarfo, G. Wyss, G. Ma, A. Das and C. Young, *Miner. Eng.*, 2017, **107**, 8.
59. T. A. Branca, V. Colla, D. Algermissen, H. Granbom, U. Martini, A. Morillon, R. Pietruck and S. Rosendahl, *Metals*, 2020, **10**, 345.
60. J. Vijayaraghavan, A. B. Jude and J. Thivya, *Resour. Policy*, 2017, **53**, 219.
61. K. Murari, R. Siddique and K. K. Jain, *J. Mater. Cycles Waste Manage.*, 2015, **17**, 13.
62. N. Saikia, G. Cornelis, G. Mertens, J. Elsen, K. Van Balen, T. Van Gerven and C. Vandecasteele, *J. Hazard. Mater.*, 2008, **154**, 766.
63. Communication from the Commission to the European Parliament, The Council and The European Economic and Social Commitee and the Commitee of the Regions, *Closing the loop – An EU action plan for the Circular Economy*, Report COM/2015/0614, Brussels, 2015.
64. J. Spooren, K. Binnemans, J. Björkmalm, K. Breemersch, Y. Dams, K. Folens, M. González-Moya, L. Horckmans, K. Komnitsas, W. Kurylak, M. Lopez, J. Mäkinen, S. Onisei, K. Oorts, A. Peys, G. Pietek, Y. Pontikes, R. Snellings, M. Tripiana, J. Varia, K. Willquist, L. Yurramendi and P. Kinnunen, *Resour., Conserv. Recycl.*, 2020, **160**, 104919.
65. M. A. Reuter, *Metall. Mater. Trans. B*, 2016, **47**, 3194.
66. R. F. van Schalkwyk, M. A. Reuter, J. Gutzmer and M. Stelter, *J. Cleaner Prod.*, 2018, **186**, 585.
67. E. V. Verhoef, G. P. J. Dijkema and M. A. Reuter, *J. Ind. Ecol.*, 2004, **8**, 23.
68. M. A. Reuter, A. van Schaik and J. Gediga, *Int. J. Life Cycle Assess.*, 2015, **20**, 671.
69. V. Ettler, A. Jarošíková, M. Mihaljevič, B. Kříbek, I. Nyambe, F. Kamona and B. Mapani, *J. Geochem. Explor.*, 2020, **218**, 106631.

CHAPTER 2

Metallurgical Overview and Production of Slags

ELIAS MATINDE*[a,b] AND JOALET D. STEENKAMP[a,b]

[a]Pyrometallurgy Division, Mintek, Praegville, Johannesburg, 2125, South Africa; [b]School of Chemical and Metallurgical Engineering, University of Witwatersrand, Johannesburg, 2050, South Africa
*E-mail: eliasm@mintek.co.za

2.1 Introduction

Metallurgical slags are complex crystalline or amorphous materials produced as by-products in the smelting and refining of metals, mattes, and alloys.[1] They are formed from the reaction between fluxes, intentionally added as part of the furnace charge, and gangue minerals in ores and/or reductants during the smelting process, or from the reaction between fluxes and dissolved impurities during refining.[1,2] Fluxing produces a low liquidus and complex slag melt that is immiscible with the metal and/or alloy phases and is thus easily separated during tapping.[2] Metallurgical slags play crucial roles in controlling the efficiency of smelting and refining processes through controlled slag–metal–gas reactions. The important metallurgical properties of slags, such as liquidus temperature, viscosity, surface tension, diffusivity, oxidation potential (pO_2), and foaming properties, are largely controlled by the composition and structure of slags.[1,2] Therefore, careful control of the physical and chemical properties of slags is indispensable for the efficient operation of pyrometallurgical processes.[1,2]

Chemistry in the Environment Series No. 2
Metallurgical Slags: Environmental Geochemistry and Resource Potential
Edited by Nadine M. Piatak and Vojtěch Ettler

Published by the Royal Society of Chemistry, www.rsc.org

Metallurgical slags are classified according to the type of metal/alloy and the unit process producing them. Broadly, metallurgical slags can be categorized into ferrous, ferroalloy, and non-ferrous slags.[3] Ferrous slags are produced in the primary production and refining of iron (Fe) and steel products, and predominantly belong to the CaO–SiO_2–Al_2O_3–MgO–FeO thermodynamic system (Section 2.2).[1–4] Ferroalloy slags are produced in the production of bulk and minor ferroalloys and predominantly consist of SiO_2, MgO, CaO, and Al_2O_3 as the main oxides (Section 2.3).[3,5] Similarly, non-ferrous slags are produced from the smelting and refining of non-ferrous metals from their natural ores and/or secondary materials (*e.g.*, scrap and process wastes) and belong to the FeO–SiO_2–Al_2O_3–CaO system (Section 2.4).[3,6]

The physicochemical properties of slags are strongly dependent on a number of factors, namely, the process in which they are generated, the quality of the raw materials, and the type of fluxes.[2–4] Since solid crystal formation is a function of both the chemical composition of the melt and its cooling rate, the mineralogical properties of solidified slags are thus strongly dependent on the slag chemistry, slag cooling rates, and cooling methods used after tapping from the respective smelting and refining furnaces.[1,3,4]

This chapter provides an overview of the unit processes used in the production of common metals and alloys. It also provides a detailed discussion on the generation and physicochemical properties of common types of slags produced in the production of metals and alloys from ores and secondary resources. Finally, this chapter provides fundamental knowledge that links the environmental and geochemical behavior of slags to the unit process, chemistry, and/or cooling or crystallization behaviors of slags.

2.2 Ironmaking and Steelmaking

Steel is one of the most ubiquitous engineering materials known to man. It is a key driver for industrialization and global economic growth and has irreplaceable applications in the construction, mining, transport, agriculture, and energy sectors. Most of the steel consumed globally is produced *via* two main routes, that is, the integrated blast furnace (BF)–basic oxygen furnace (BOF) and the electric arc furnace (EAF) routes. In the BF–BOF route, the blast furnace is the major source (about 75%) of the hot metal charged into the BOF, with the remaining 25% coming from direct charging of steel scrap and other forms of auxiliary metallic charges.[4,7,8] In contrast, the EAF route predominantly utilizes solid steel scrap and other forms of auxiliary iron charges (*e.g.*, direct reduced iron (DRI) and reverts) as raw materials.[7] The integrated BF–BOF route currently accounts for approximately 60% of global crude steel production, with the EAF and other process routes, such as open-hearth furnace and other smelting reduction processes, accounting for approximately 30% of global steel production.[7] The crude steel from either the BF–BOF or scrap-based EAF processes is further refined in secondary or ladle refining processes to produce various grades of steels. Section 2.2.1

briefly discusses the BF ironmaking process and the properties of ironmaking slags produced therein. The production of value-added steel products and the properties of steelmaking slags are briefly discussed in Section 2.2.2.

2.2.1 Ironmaking

2.2.1.1 Overview

The BF is a counter-current gas–solid–liquid reactor in which cold Fe-bearing materials and fluxes are converted at high temperatures in the presence of coke to produce hot metal and slag.[7] The process involves complex physicochemical reactions that include the reduction of Fe oxides, combustion and gasification of C, melting and separation of slag and metal, slag–coke reactions (*i.e.*, liquid state reduction reactions of metalloids), and the refining of hot metal through slag–metal–gas reactions.[4,7,8] Iron-bearing raw materials, charged in various proportions, include Fe ore lumps, sinter, and pellets, and, to some extent, pre-reduced charge (*e.g.*, DRI) and recycled Fe-bearing wastes.[4,7] The properties of the ferrous charge, such as gangue content, reducibility, and softening and melting behavior, significantly affect the operation and productivity of the BF.[7]

Slag formers, such as limestone ($CaCO_3$), dolomite ($Ca,Mg(CO_3)_2$), olivine ($(Mg,Fe)_2SiO_4$), and quartz (SiO_2), are added to balance the slag chemistry, liquidus temperature, and volume of slag produced.[4,7,8] Self-fluxing sinter and/or pellets are preferred as they are already designed to contain the right amount of slag formers needed to attain the desired slag properties. Typically, self-fluxing charges have high basicity (CaO to SiO_2 ratio >1.2) such that, together with coke ash, the final slag has a basicity of approximately 1.1 for the CaO to SiO_2 ratio.[4,7,8]

Metallurgical coke (containing approximately 88 wt% C, 0.2–0.5 wt% volatile matter, 0.6–0.8 wt% S, and 8–12 wt% ash on a dry basis), pulverized coal, and other auxiliary fuels provide the thermal energy required to drive the endothermic reduction reactions and melt the furnace charge.[4,9,10] The carbonaceous additives also provide the chemical potential required for the reduction of Fe oxides as well as the C required for the direct reduction of metalloids and carburization of hot metal. Modern BFs operate at approximately 170–200 kg of pulverized coal injection (PCI) and between 286 and 320 kg of coke per tonne of hot metal.[4,9,10] Preheated oxygen-enriched blast reacts with the coke and PCI in the raceway to generate high temperatures (in the range of 1900–2300 °C) and the $CO–H_2–N_2$ gas mixtures required for chemical reduction and heat transfer in the furnace.[8,10] The counter-current flow of gases, liquids, and solids accounts for the efficient heat transfer, mass transfer, and exchange of chemical enthalpies experienced in the BF.[11]

The BF process produces C-saturated hot metal (approximately 94 wt% Fe, 4.5 wt% C, 0.4–0.5 wt% Si, 0.5 wt% Mn, 0.03–0.06 wt% S, and <0.1 wt% P; temperature 1400–1500 °C), liquid slag (approximately 32–42 wt% CaO, 6–12 wt% MgO, 28–38 wt% SiO_2, and 8–20 wt% Al_2O_3; temperature 1500–1550 °C), flue gas (containing approximately 22 vol% CO, 22 vol% CO_2, 4 vol% H_2, and 50 vol% N_2), and flue dust (containing approximately 100–150 g of zinc per tonne of hot metal).[7] The supernatant slags are usually superheated to approximately 50 °C above their melting points and are periodically tapped into slag launders or slag pots. A modern BF is usually equipped with two to four tap-holes which are operated either intermittently or continuously depending on the size and design of the BF.

The hot metal from the BF is either subjected to pre-treatment or is sent directly to the steelmaking plant. The broad objective of hot metal pre-treatment is to decrease the impurity content in the hot metal prior to the decarburization treatment in the oxygen steelmaking process. The hot metal pre-treatment ultimately reducing the costs of downstream steelmaking steps.[12–15] In addition, some impurities are difficult to remove in the oxygen steelmaking process.[12,13] In particular, desulfurization is only possible under reducing conditions and cannot be achieved in the oxygen steelmaking process. Although the sulfide capacity of BF slag is usually high, the hot metal still contains 0.03–0.06 wt% S.[12–14] Hot metal desulfurization is carried out by deep injection of powdered reagents, such as CaO, CaC_2, Na_2CO_3, and/or mixtures thereof, in the tapping ladles or torpedo cars.[12–14] The sulfides formed in the supernatant slag layer, approximately 2–5 kg t^{-1} hot metal, are then skimmed off to permanently remove the S.[14] Similarly, desiliconization to approximately less than 0.2 wt% dissolved Si by injection of oxidants (*e.g.*, mill scale, Fe_2O_3 powder or $O_{2(g)}$) and basic fluxes (*e.g.*, CaO or MgO) is carried out to reduce the amount of CaO required and, ultimately, the volume of slag generated in the oxygen steelmaking stage.[12,13] Dephosphorization treatment is conducted to reduce the dissolved phosphorus (P) to as low as thermodynamically and practically possible, thereby allowing for the production of steels with lower P contents than otherwise would be practically achievable without penalties to the converter process.[12,13,15]

2.2.1.2 *Ironmaking Slags*

Ironmaking slags belong to the CaO–SiO_2–Al_2O_3–MgO slag system, with minor components of MnO, FeO, K_2O, Na_2O, TiO_2, S, and P. These slags are formed from the reaction between fluxes intentionally added as part of BF charge and the silicates, aluminosilicates, and sulfide gangue constituents in the ore, ash of coke and PCI.[4,7,16] The formation of BF slags occurs *via* two stages: primary and final slag. Primary slag is predominantly a mixture of FeO and oxides such as CaO, Al_2O_3, MgO, and SiO_2 present in fluxes and gangue, and is formed during the initial melting process of the reduced charge and fluxes in the softening and melting or cohesive zone prior to the final dissolution of gangue in ore and coke ash components.[7] The liquidus temperature of primary slag decreases with an increase in the amount of FeO present in the primary slag. For example, Geerdes *et al.*[17] propose that the liquidus temperature of BF slag decreases from 1540 °C in the absence of FeO, through 1370 °C at 20 wt% FeO, to 1220 °C at 40 wt% FeO when the CaO to SiO_2 ratio is fixed at 0.9.

The final slag is formed after the final reduction of FeO from the primary slag and the dissolution of silicates and other oxides from the ash components of coke and PCI as the slag descends through the bosh and combustion zone to the hearth. Typically, the final BF slag consists of 40 wt% CaO, 36 wt% SiO_2, 10 wt% MgO, and 10 wt% Al_2O_3.[7,17] The physical and chemical properties of the final slag, such as liquidus temperature (1300–1450 °C), slag basicity ($B_4 = 1.1 < \frac{CaO + MgO}{SiO_2 + Al_2O_3} < 1.3$), slag viscosity (<0.2 Pascal-second), sulfide capacity ($L_S = \frac{(\%S)}{[\%S]} > 90\%$) and slag rate (<300 kg t^{-1}) significantly affect the hot metal quality and the overall productivity of the BF.[7,17]

A modern BF produces approximately 200–300 kg of slag per tonne of hot metal depending on the burden composition and degree of PCI.[7] Approximately 400 million tonnes of BF slags are produced globally each year.[4,7] Depending on the cooling method used, solidified BF slags can be classified into pelletized, granulated, or slow-cooled slags.[3,4,18] Pelletized or expanded slags have a vesicular texture similar to volcanic rock and are most commonly used as lightweight aggregates. Granulated slags consist of glassy or amorphous phases as a result of vitreous solidification, and have chemical and hydraulic properties similar to Portland cement. Slow-cooled slags predominantly consist of crystalline Ca–Al–Mg silicate phases, existing mainly as melilite, a solid solution of åkermanite ($Ca_2MgSi_2O_7$) and gehlenite ($Ca_2Al_2SiO_7$), and merwinite ($Ca_3MgSi_2O_8$).[18,19] Other crystalline phases include rankinite ($Ca_3Si_2O_7$), wollastonite ($CaSiO_3$), forsterite (Mg_2SiO_4), pyroxene ($CaAl_2SiO_6$), and dicalcium silicate ($(Mg,Mn,Fe,Ca)_2SiO_4$).[18] Slow-cooled slags are hard and dense, with properties similar to crushed basalt stone, and are thus especially suitable for use as construction aggregates.[3,4,18,19] Further details on the mineralogical properties and applications of BF ironmaking slags are given in Chapters 3 and 7, respectively.

2.2.2 Steelmaking

2.2.2.1 Overview

Steelmaking refers to a series of chemical reactions and refining steps, occurring either sequentially or simultaneously, in the conversion of hot metal or crude steel to a specific composition of liquid steel. Steelmaking processes can be divided into two categories: primary and secondary steelmaking. Primary steelmaking involves the refining of hot metal in the oxygen steelmaking units and/or the melting of scrap steel, DRI, and other metallic additives in EAFs to produce crude steel. Secondary steelmaking involves the further refinement of crude steel in the argon–oxygen decarburization (AOD) process and ladle furnaces (LFs) before continuous casting.[20–24]

2.2.2.2 Basic Oxygen Steelmaking

Basic oxygen steelmaking (BOS) refers to primary converter processes that use magnesia (MgO) refractories, basic slags (slags rich in CaO and/or MgO), and pure oxygen to convert a charge of hot metal (approximately 80–90% by volume) and ambient scrap (approximately 10–20% by volume) into low-carbon (C) and low-P steel.[20–26] Basic oxygen steelmaking is also known as the basic oxygen process (BOP), the Linz–Donawitz (LD) process, or simply the BOF process, and is currently the most dominant primary steelmaking technology, accounting for over 60% of the global steel output. The broad objective of oxygen steelmaking is to oxidize the dissolved C (to less than 1 wt% C) and dissolved P (to less than 0.08 wt% P) from the crude steel as well as to adjust the bath temperature to appropriate levels required for further treatment of steel.[21]

The BOF is a batch operation that involves a series of sequential steps: (1) charging of scrap, hot metal, iron ore fines, and other ferrous charges, (2) addition of basic fluxes to form slag, (3) blowing oxygen to oxidize the dissolved impurities, and (4) tapping of crude steel and slag.[20–26] Figure 2.1 shows the schematic and typical material flow for a combined blown BOF.[4,26] The quantities of hot metal, scrap, oxygen, and fluxes vary depending on the starting composition and temperatures, and on the desired chemistry and temperatures of the steel to be tapped.[24]

The kinetics of the oxidation reactions (Table 2.1) and the formation and stability of the slag–metal emulsions in the furnace are controlled by parameters such as the oxygen lance design, lance height and oxygen blowing velocity.[20,21] Slag formation is facilitated by the simultaneous dissolution of calcined CaO and/or MgO fluxes introduced in the form

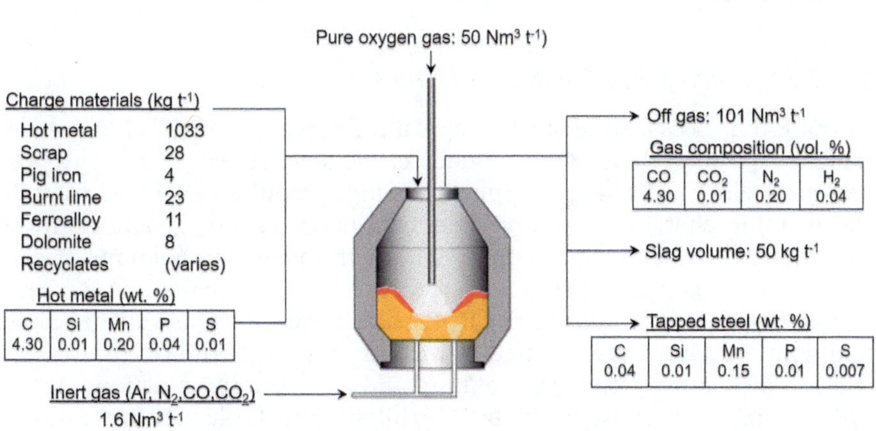

Figure 2.1 Schematic of combined blown basic oxygen furnace (BOF) charged with pre-treated hot metal. Adapted from ref. 26 with permission from the Association for Iron and Steel Technology, Copyright 2011.

Table 2.1 Reactions taking place in a standard basic oxygen furnace (BOF) operation. Reproduced from ref. 4 with permission from Taylor and Francis, Copyright 2017.[a]

	Chemical reactions	By-product
Primary oxidation reactions	$2[C] + O_{2(g)} \rightarrow 2CO_{(g)}$	Gas
	$CO_{(g)} + 0.5O_{2(g)} \rightarrow CO_{2(g)}$	
	$[C] + [O] \rightarrow CO_{(g)}$	
	$2[Fe] + O_{2(g)} \rightarrow 2(FeO)$	Slag
	$[Si] + O_{2(g)} \rightarrow (SiO_2)$	
Secondary oxidation and slag metal reactions	$[Fe] + [O] \rightarrow (FeO)$	
	$[Mn] + [O] \rightarrow (MnO)$	
	$(FeO) + [Mn] \rightarrow [Fe] + (MnO)$	
	$[Si] + 2(FeO) + 2(CaO) \rightarrow 2[Fe] + (2CaO{\cdot}SiO_2)$	
	$2[P] + 5(FeO) + 3(CaO) \rightarrow 5[Fe] + (3CaO{\cdot}P_2O_5)$	
	$3(CaO) + (SiO_2) \rightarrow (3CaO{\cdot}SiO_2)$	

[a]Notation: [X] = alloy; (X) = slag.

of burnt lime or dolomite and the subsequent reaction of the dissolved fluxes with the oxidation products (such as SiO_2, P_2O_5, MnO, FeO) to form the final BOF slag. The final slag has high basicity (CaO to SiO_2 ratio > 3) and the FeO content can be as high as 35–38 wt% FeO due to the strongly oxidizing conditions experienced in the furnace.[25,27] The amount of fluxes, and ultimately the volume of BOF slag, depends on the amount of dissolved silicon (Si), manganese (Mn), and P in the charge materials. Hot metal pre-treatment has significantly reduced the volume of steelmaking slags, from 100–110 kg slag per tonne of steel to as low as 40–50 kg per tonne of steel.[12–14,21,26]

2.2.2.3 *Electric Arc Furnace Steelmaking*

In the EAF process, solid steel scrap, DRI, fluxes, and other additives are melted at around 1600 °C to produce crude steel (Figure 2.2).[4] The process uses electrical energy supplied through graphite electrodes to melt the metallic charges. Auxiliary chemical energy is also generated from oxidation reactions during refining and/or from the combustions of carbonaceous fuels injected through specially designed oxy-fuel burners and oxygen lances.[28,29] Similar to the BOF, basic fluxes, such as calcined lime and dolomitic lime, are injected in the refining stage to form basic slags (CaO to SiO_2 ratio > 1.5). The refining and slag forming reactions in the EAF are similar to those in the BOF highlighted in Table 2.1. Intermittent de-slagging is often carried out to remove impurities from the furnace and prevent the reversion of impurities. Depending on the desired alloy chemistry and refining regime, the slag rate in the EAF process is in the range of 60–270 kg per tonne of steel.[4]

2.2.2.4 Secondary Steelmaking

The crude steel from the EAF and/or BOF is fed directly to secondary steelmaking furnaces.[4,20–23,30] Secondary steelmaking refers to a series of refining and alloying steps, carried out under carefully controlled melt conditions, used in the refinement of different grades of stainless steel (Figure 2.3).[4,20–23] Most commonly, secondary steelmaking includes unit steps such as deoxidation, desulfurization, decarburization to low- or ultra-low-carbon contents, degassing, removal of non-metallic inclusions, and, additionally, alloying and temperature adjustment.[22,30] The broad objectives of secondary refining include chemistry and temperature homogenization, deoxidation and degassing to control the level of dissolved impurities, and finally, to enhance the overall cleanliness of steel through the removal of dissolved impurities and non-metallic inclusions.[4,22]

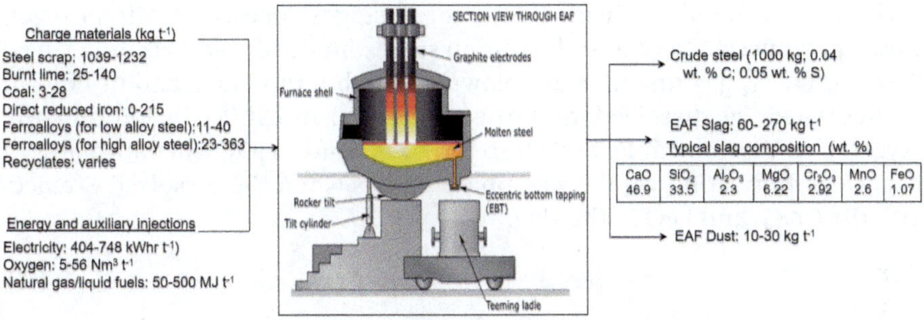

Figure 2.2 Section view and material flow in an electric arc furnace (EAF) process. Adapted from ref. 4 with permission from Taylor and Francis, Copyright 2013.

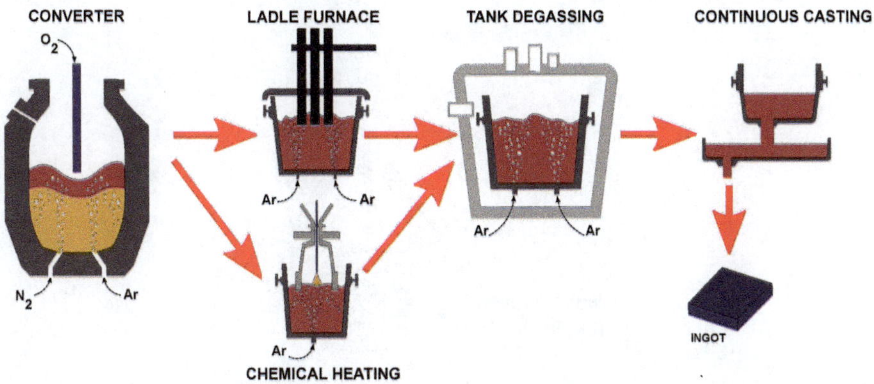

Figure 2.3 Overview of secondary metallurgy processes for refining of steel. Reproduced from ref. 22 with permission from Elsevier, Copyright 2014.

Stainless steels contain relatively high amounts of chromium (approximately 12–30 wt% Cr) and low amounts of C (approximately 0.03 wt% C). Varying amounts of alloying elements, such as nickel (Ni), molybdenum (Mo), titanium (Ti), and Mn, may also be added to obtain final products with the desired mechanical properties and performance.[30,31] The key to the cost-effective refining of stainless steel therefore lies in the ability to effectively oxidize the dissolved impurities without compromising the retention of alloying elements. The AOD process constitutes one of the most preferred methods of refining stainless steels and other high alloy steels, accounting for over 70% of the current global stainless steel production.[22,30–32]

The AOD consists of a top lance for oxygen or oxygen–inert gas (argon (Ar) or nitrogen (N_2)) injection and several bottom tuyères for blowing inert gas during the refining of stainless steel (Figure 2.4).[31,32] The combined blowing capabilities in the AOD process provide the high mass transfer conditions required for high decarburization rates, while the dilution with Ar or N_2 minimizes the unwanted oxidation of valuable alloying elements.[22,31,32]

The AOD process is divided into three stages: decarburization/oxygen blowing, reduction, and desulfurization stages. In the decarburization stage, O_2–Ar or O_2–N_2 gas mixtures are blown through a top lance and/or bottom tuyères to oxidize dissolved impurities, such as C and Si.[24,30] In addition, the oxygen also reacts with Fe and Cr to form FeO and Cr_2O_3, which dissolve in slag. Decarburization *via* the slag phase occurs when the dissolved C reacts with the Cr_2O_3 and FeO in the slag.[30]

$$(\mathrm{FeO}) + [\mathrm{C}] \rightarrow [\mathrm{Fe}] + \mathrm{CO} \tag{2.1}$$

$$(\mathrm{Cr_2O_3})_{\mathrm{in\ slag}} + 3[\mathrm{C}] \rightarrow 2[\mathrm{Cr}]_{\mathrm{in\ Fe}} + 3\mathrm{CO} \tag{2.2}$$

For a standard stainless steel bath containing 18 wt% Cr, for example, the slag at the end of oxygen blowing contains approximately 20–30 wt% Cr_2O_3 (Table 2.2).[24,30] The basicity of the top slag is controlled to between 1.6 and 1.8 by adding CaO or MgO fluxes as part of the charge material.[24] The chemistry and

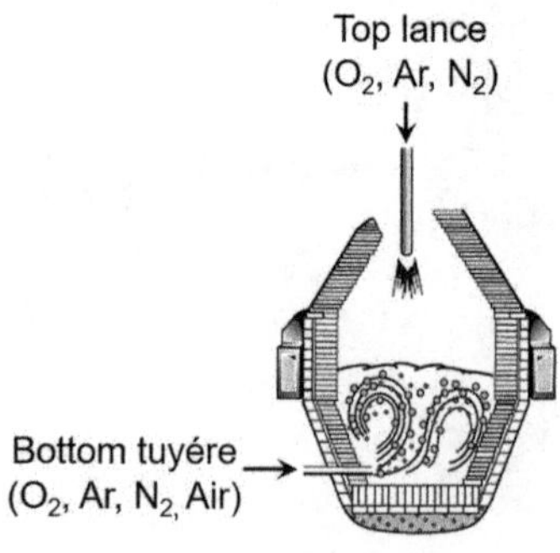

Figure 2.4 Schematic of a combined blown argon–oxygen decarburization (AOD) reactor. Reproduced from ref. 24 with permission from the Association for Iron and Steel Technology, Copyright 1998.

Table 2.2 Changes in the composition of argon–oxygen decarburization (AOD) slag after decarburization and after silicon reduction for 18 wt% Cr stainless steel. Reproduced from ref. 24 with permission from Association for Iron and Steel Technology, Copyright 1998.

	Slag composition (wt%)	
	After decarburization	After silicon reduction
FeO	4–6	1–2
MnO	4–8	1–3
SiO_2	12–18	30–40
Al_2O_3	18–22	3–8
CaO	8–15	33–43
MgO	7–15	10–20
Cr_2O_3	20–30	1–3

temperature of top slag are important parameters for controlling the viscosity of slag and, ultimately, minimizing the loss of Cr during oxidation.[22,30]

In order to attain an optimum yield of Cr, a reducing agent, usually Si in the form of ferrosilicon (FeSi) alloy, is added in the reduction stage to reduce the Cr_2O_3 in slag.[21,24,30]

$$2(Cr_2O_3)_{\text{in slag}} + 3[Si]_{\text{in Fe}} \rightarrow 4[Cr]_{\text{in Fe}} + 3(SiO_2) \quad (2.3)$$

The recovery of Cr is favored by high slag basicity, low SiO_2 activity in slag, high temperatures (approximately 1700 °C), and inert gas blowing.[22,24,30] De-slagging of the top slag is often conducted at the end of the reduction stage before the desulfurization treatment by injecting basic fluxes (CaO and CaF_2) into the bath followed by gentle stirring with Ar gas to enhance the contact between the steel and the slag. The reducing conditions in the slag favor deep desulfurization to as low as 0.003 wt% S.[21,30]

2.2.2.5 *Steelmaking Slags*

Steelmaking slags play a crucial role in the removal of impurities and nonmetallic inclusions. Steelmaking slags also provide chemical insulation from re-oxidation by atmospheric O_2 and dissolution of atmospheric N and H, as well as providing thermal insulation against heat losses.[22,24,30,33] The slags also have other unique functions in some processes. For example, slag foaming is an important practice in the modern EAF process since a good foaming slag captures the electrical arc heat, thereby minimizing heat losses, and shields the furnace refractories and water-cooled panels from thermal radiation. This in turn increases the energy efficiency and allows for higher power input and productivity.[33,34] Broadly, the physicochemical properties of steelmaking slags are controlled by the chemical (*e.g.*, basicity and oxidation potentials) and physical properties (*e.g.*, density, liquidus temperature, and viscosity).[22,30–34]

The chemistry of the steelmaking slags is unique to the nature of impurities and refining process being considered. Due to the highly basic slags and oxidizing conditions, BOF slags tend to have high CaO content (>35 wt% CaO, with up to 12 wt% free CaO), up to 10–35 wt% FeO, 7–18 wt% SiO_2, 0.5–4 wt% Al_2O_3, and 0.4–14 wt% MgO.[25,27,35] The chemical composition of EAF slags also varies greatly depending on the type of alloy produced and is typically in the range of 22–60 wt% CaO, 10–40 wt% FeO, 6–34 wt% SiO_2, 3–14 wt% Al_2O_3, and 3–13 wt% MgO with variable amounts of free CaO and MgO.[1,4,25] Rao[36] also proposed that EAF slags broadly consist of 6.8–23.4 wt% FeO, 39.9–52.2 wt% CaO, 4.6–12.8 wt% SiO_2, 1.3–20.5 wt% MgO, and 1.9–9.3 wt% MnO. The huge compositional variability of EAF slags as a result of the different alloy chemistry and technology requirements invariably leads to significant heterogeneities in the phase properties and geochemical behaviors of the solidified slags.[1,4,25,35,36]

The chemical composition of AOD slags also varies greatly depending on the grade of alloy being produced. AOD slags are formed from the secondary refining of the crude steel produced in the primary steelmaking processes. Characteristic of AOD slags is the presence of CaO, MgO, and SiO_2, as shown in Table 2.2, and the high content of alloying elements such as Cr (1.5–4.3 wt%) and Ni (0.2–0.45 wt%).[35,37] Similarly, ladle furnace (LF) slags originate from the final refinement of the steel produced in the AOD refining step, as shown in Figure 2.3. LF slags are designed to possess unique thermal and refining functions, such as low viscosity and high ability to absorb impurities and non-metallic inclusions.[22,24,30,35,38] Typically, LF slags contain approximately 10 wt% SiO_2 (as carry-over from the AOD), 10 wt% MgO (intentionally added to protect the slag line), 55 wt% CaO (added to adjust the slag basicity), and 10 wt% Al_2O_3 (formed as a deoxidation product).[22,24,30] Similar to AOD slags, LF slags are also strongly associated with elevated amounts of alloying elements such as Cr, Ni, and Mn.[1,4,25,39,40]

The refining of steel is a batch process. At the end of each refining cycle, steelmaking slags are tapped from the refining furnaces and solidified, either by slow-cooling in slag dumps or by rapid cooling using water jets.[4,25] Solidified slags tend to exist as complex and heterogeneous mixtures of silicates, aluminosilicates, spinels and solid solutions.[1,4,25] As highlighted earlier, the chemical and phase properties of steelmaking slags are inherently complex, varying greatly depending on the composition of starting materials, grade of steel produced, unit process producing the slag, and the cooling method applied.[25,35] Consequently, the knowledge of chemical, phase chemical and morphological properties is solidified steelmaking slags is crucial, mainly because their environmental performance and valorization potential is strongly dependent on these properties. Chapter 3 provides further details on the chemical and geochemical properties of solidified steelmaking slags.

2.3 Production of Ferroalloys

2.3.1 Overview

Various alloying elements, including Mn, Si, Cr, and Ni, are added to steel and other alloys to improve their properties. Often being technically and economically most feasible, the required elements are added as an alloy with Fe in the form of ferroalloys.[41,42]

Ferroalloys can be classified as *major* or bulk ferroalloys, which are produced in large quantities, and *minor* ferroalloys produced in smaller quantities but of high importance.[41] The major ferroalloys are ferromanganese (FeMn), silicomanganese (SiMn), ferrochromium (FeCr), ferrosilicon (FeSi), and ferronickel (FeNi).[43,44] The minor ferroalloys are ferrotungsten (FeW), ferromolybdenum (FeMo), ferrovanadium (FeV), ferroniobium (FeNb), ferrotitanium (FeTi), ferrozirconium (FeZr), ferroboron (FeB), ferroalloys containing rare earth elements (REEs), and ferroalloys containing alkaline earth metals, namely calcium (Ca), magnesium (Mg), strontium (Sr), and barium (Ba).[43,44]

Major ferroalloys are only used in steelmaking and in steel or Fe foundries, the exception being FeNi.[43,44] Minor ferroalloys have several applications. Steelmaking consumes around 85% to 90% of all ferroalloys and the remaining consumers are non-ferrous alloys (mainly Ni- or Ti-based) and the chemicals industry.[41]

The history of ferroalloy production is closely related to that of steelmaking.[42] Sir Henry Bessemer, often seen as the father of modern steelmaking, developed the eponymous converter-based steelmaking process for which he was awarded an English patent in 1855.[45] He initially added spiegeleisen (8 wt% Mn, 4 wt% C, and 86 wt% Fe) after oxygen blowing, in order to remove oxygen from the steel and to add Mn and C. From 1862 onwards, he added what he called *ferromanganese* (20–25 wt% Mn) and coined the term.[45] His inventions paved the way not only for modern steelmaking but also for the modern production of Mn ferroalloys.[46]

Commercial production of FeMn containing 65 wt% Mn started in France in 1875 using BF technology.[47] In 1877, the Terre Noire Company in France produced FeMn containing 80 wt% Mn and 6–7 wt% carbon.[48] FeSi as well as low amounts of FeTi and FeV could also be produced in a BF, but not FeCr owing to the high melting point of the slag formed.

The introduction of EAF technology for the production of ferroalloys toward the end of the 19th century allowed also for FeCr production.[42] The ability to produce FeCr and FeNi on an industrial scale led to the development of stainless steels. The advent of commercial stainless steel production is associated with the patent for austenitic Cr–Ni stainless steel by the company Krupp in 1912. It was followed by ferritic and martensitic stainless steels, created by others, and acid-resistant Mo-containing stainless steels later in the 1920s.[48] Today, the bulk ferroalloys are produced almost exclusively in EAFs, with only a few producers of FeMn in China, Poland, and Russia applying BF technology.[42,44,46]

The utilization of ferroalloys in steelmaking was also dependent on the development of analytical and characterization techniques. The measurement of steel composition became possible as the understanding of basic chemistry, including analytical chemistry, progressed in the 19th century. Simultaneously, advancements were made in materials investigation and characterization. H. C. Sorby, a contemporary of Bessemer in England, pioneered studies examining steel microstructures using optical microscopy and correlating it to steel composition and properties. By 1885, the value of metallography was appreciated in Europe. In the 1930s, electron microscopy was developed and became a common tool for materials scientists studying the structures of steel and the phenomena influencing its properties.[48] Table 2.3 summarizes the effects of the common alloying elements, added as major or minor ferroalloys, on the structure and enhanced properties of steel, as well as their typical applications.[42]

For the period 2012–2016, major ferroalloys accounted for between 57% and 94% (depending on how unspecified ferroalloys are classified) of the 57 to 67 million tonnes of ferroalloys produced per annum worldwide.[44] During this period, the production of FeCr and SiMn each accounted for approximately 30% of the total major ferroalloy production. FeMn and FeNi each accounted for approximately 10% and FeSi approximately 20%.[44] The production of major ferroalloys is therefore the main source of slags produced by the ferroalloy industry and the remainder of this section will focus on the production of FeMn, SiMn, FeSi, FeCr, and FeNi alloys.

Table 2.3 Common alloying elements, added as ferroalloys to steel, and their effects on the structure and properties of the steel as well as their typical applications during the steelmaking process. Adapted with permission from ref. 42.

Element	Structure	Properties	Applications
Mn	Austenite forming, Substitutional atom, Carbide stabilizer	Hardenability, Strength, Toughness, Hot workability	Deoxidation (milder than Si, Al), MnS formation, Low alloyed steels: ≤1.6 wt% Mn, Hadfield steel: 13 wt% Mn, Stainless steel: 5–10 wt% Mn
Si	Ferrite forming, Substitutional atom	Ferrite hardening	Deoxidation, Slag reduction, Mild steel: 0.3–0.6 wt% Si, Spring steel: 1.7–2.2 wt% Si, Electrical Si-steels: 1–4 wt% Si
Cr	Ferrite forming, Substitutional atom	Hardenability, Strength, Corrosion resistance	Low alloyed steels: ≤2 wt% Cr, Stainless steels: 11–25 wt% Cr, Heat-resistant steels, Tool steels
Ni	Austenite forming, Substitutional atom	Strength, Toughness, Corrosion resistance	Case hardening steel: ≤3.5 wt% Ni, Stainless steel: 8–20 wt% Ni, Heat resistant alloys: ±25 wt% Ni
Cr, Mo, Nb, Ti, V, W	Carbide forming	Hardness, Strength, Wear resistance	Tool steels, Wear resistant steels
B, Nb, Ti, V	Nitride forming, Grain refining	Strength, Ductility	Fine-grained steels, High strength low alloyed (HSLA) steels

2.3.2 Production of Ferromanganese Alloys

FeMn alloys produced commercially can be divided into high-carbon ferromanganese (HCFeMn), medium-carbon ferromanganese (MCFeMn) with or without nitrogen (N), and low-carbon ferromanganese (LCFeMn). The typical chemical compositions of these alloys are indicated in Table 2.4.[49] This section focusses on the production routes for HCFeMn and MCFeMn only.

HCFeMn is produced by the carbothermic reduction of Mn-bearing ores in a BF or alternating current (AC) submerged arc furnace (SAF).[46] The main source of energy in the BF is the combustion of coke by oxygen-enriched air and in the SAF, it is electricity.

Prior to World War II, HCFeMn was mostly produced in the BF.[46,51] Since then, BF production of HCFeMn has been replaced by SAF production owing to the scarcity of coke, the lower capital cost required when building a SAF, the availability of low-cost electricity, and the flexibility of the SAF in terms of commodities produced. For the period 2012–2016, 7–11% of the world production of HCFeMn was by BFs and the remainder by SAFs.[44] BF technology is still preferred by operations based on low-grade ore and where the cost of electricity is high.[51,52] As an example, in the 1980s in Japan two SAFs were converted to a single BF due to significant increases in electricity tariffs.[52] In Figure 2.5, the process conditions when producing HCFeMn in a BF are compared against those for pig iron production. The solid/gas reduction reactions of Mn ores, sinter, and pellets by CO are exothermic.[53] This results in higher temperatures in the upper part of the BF, which is addressed by modifications to the furnace design, resulting in higher off-gas temperatures.[51,52] Owing to the complexity of the mineralogy of the ores, Mn in the feed can be present in the di-, tri-, or tetra-valent state. Solid-state reduction only progresses to the divalent state. The final reduction of MnO to Mn occurs through the liquid slag reaction with coke, a highly endothermic process, which occurs in the lower part of the furnace.

The SAFs utilized in HCFeMn production are rated between 20 and 90 megavolt-ampere (MVA).[50] In the SAF, the reduction process can also be divided into solid–gas reactions in the burden and liquid–solid reactions in

Table 2.4 Categories of FeMn and their typical chemical compositions in wt% (balance Fe).[49,50] Adapted from ref. 50 with permission from Elsevier, Copyright 2013.

	Composition (wt%)					
Product	Mn	C	Si	P	S	N
HCFeMn	78.0–82.0	≤7.5	1–1.2	0.2–0.35	≤0.50	NR[a]
MCFeMn	80.0–85.0	1–1.5	0.35–1.5	≤0.30	≤0.20	NR
MCFeMn, high N	75–80	≤1.5	≤1.5	≤0.3	≤0.20	≥4
LCFeMn	80.0–90.0	≤0.75	2.0–7.0	0.20–0.30	≤0.20	NR

[a]NR: Not reported.

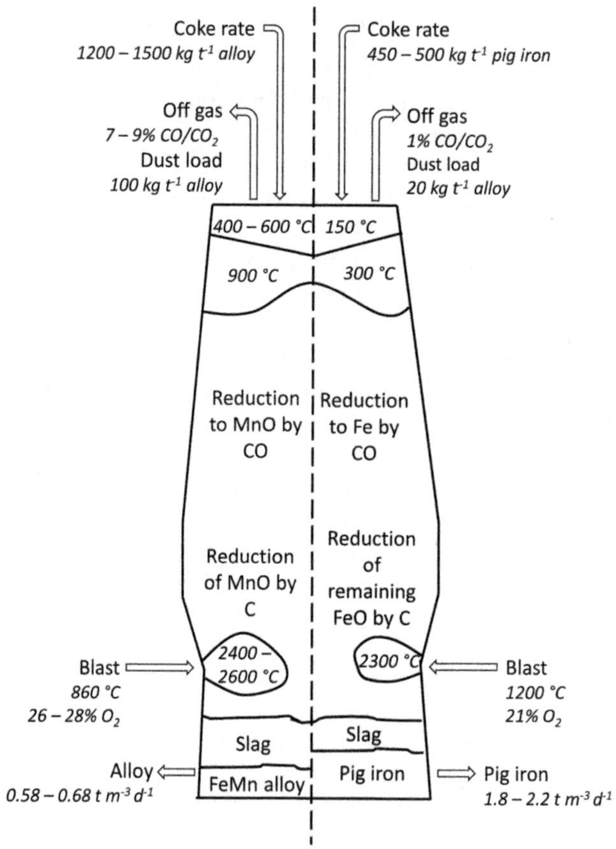

Figure 2.5 Comparison of process conditions when a blast furnace (BF) is used to produce HCFeMn (to the left) and pig iron (to the right). Adapted from ref. 52 with permission from JFE Steel Corporation, Copyright 1987.

the wet coke-bed. The wet coke-bed consists of a mixture of slag and carbonaceous reductant above the layer of alloy collected in the hearth of the furnace (Figure 2.6). In the SAF, the electrical energy is dissipated from the electrodes to the process through a combination of resistive heating of a wet coke-bed and arcing.[54] Although referred to as a coke-bed, a blend of coke, anthracite, and bituminous coal is fed into the SAF to sustain the reduction processes.[55] Gas permeability of the burden is imperative, and therefore raw materials have a lower screening size limitation, typically of 6 mm.

HCFeMn is produced either by using a discard slag practice or the duplex process.[46] In the discard slag practice, the slag produced contains 15–28 wt% MnO and is discarded as waste. In the duplex process, the slag produced contains 35–50 wt% MnO and is used as the main source of Mn in the subsequent SiMn production process. The choice of slag practice applied often depends on the type of Mn raw materials utilized. Basic Mn ores, such as the ores found in South Africa's Kalahari Manganese Field (KMF), and agglomerates

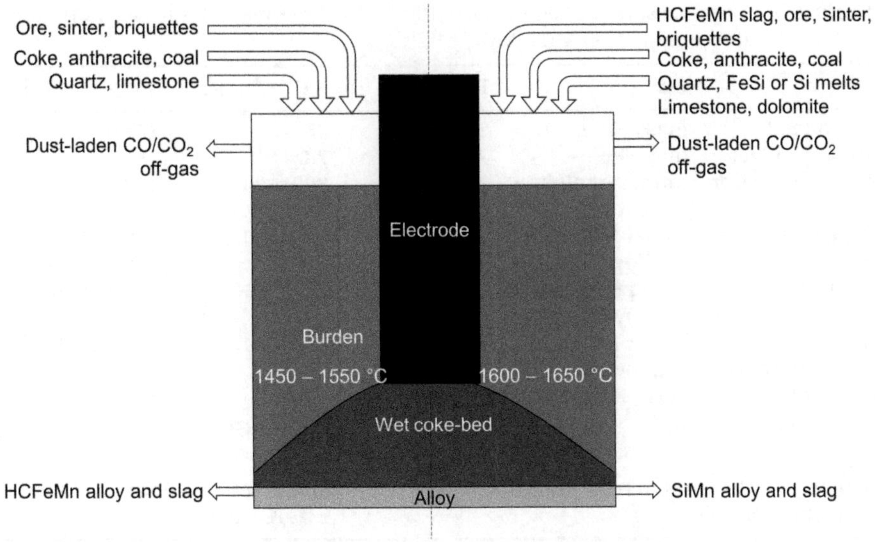

Figure 2.6 Simple schematic of the process flow in a submerged arc furnace (SAF) utilized in the production of HCFeMn (to the left) and SiMn (to the right).[46,50,56]

derived from them (sinter, briquettes) are rich in basic carbonates, which require fluxing with quartz in order to produce a liquid slag.[57,58] The recovery of Mn when slag is discarded is around 80% while the duplex process results in an overall recovery of 85–90% even though the recovery for the HCFeMn is much lower.[50] Table 2.5 compares the BF production of HCFeMn to that of SAFs following either the duplex or the discard route.

SAFs are typically circular in design with three Söderberg or self-baking electrodes in equilateral positions providing energy to the process.[62] For HCFeMn production, furnaces are rated at between 20 and 90 MVA.[63] The furnace designs can be closed or open. In closed furnaces, the roof is attached to the sidewalls and CO-rich off-gas is extracted and cleaned before being utilized elsewhere or is flared. In open furnaces, the CO-rich off-gas is combusted at the top of the furnace burden before being extracted, cleaned, and vented to the atmosphere.[63]

Furnaces can be equipped with single- or bi-level tap-holes.[64] For a single-level tap-hole, the slag and alloy are tapped simultaneously and separation occurs outside of the furnace. For bi-level tap-hole designs, the slag tap-hole is positioned 1 m or more above the alloy tap-hole and separation occurs inside the furnace. Post tap-hole, the alloy is cast in layers on the ground in casting pits or as pigs in a pig caster. Slag is typically cooled in slag pots. For the discard process, the cooled slag is often cleaned from entrained metal droplets in the metal recovery plant before being discarded on slag dumps. For the duplex process, the cooled slag is prepared as feed for the SiMn furnace by crushing and sieving.

Table 2.5 Comparison of blast furnace (BF) and submerged arc furnace (SAF) production of HCFeMn alloys.[46,50–54,59–61]

	Unit	BF	SAF – duplex	SAF – discard
Mn in feed	mass %	28–30	35–48	35–48
Mn in product	mass %	70–75	78	78
Alloy production	tonnes of alloy per day	270–550	300–400	400
Coke rate	kg per tonne of alloy	1500–2000	310–380	550–580
Electricity consumption	kWh per tonne of alloy	200	2200–3500	3300–3900
Ore consumption	kg per tonne of alloy	2500–3000	2750–2970	1950–2060
Mn yield	mass %	70–85	60–75	76–82
Flexibility on demand		Rigid (long campaigns)	Good	
Other potential products produced using technology		Pig iron, FeTi, FeP	SiMn, FeSi, HCFeCr, Si, CaSi, CaC_2, pig iron	

MCFeMn alloys are produced either by blowing HCFeMn with diluted oxygen (O_2) in a convertor similar to the BOF for steelmaking or by silicothermic reduction of a synthetic slag rich in MnO in a shaking or mixing ladle.[46,65–69] These processes can also be combined.[67] For the converter process, HCFeMn can be provided in the liquid state, either from a BF or a SAF.[67,68] Various converter designs exist, with O_2 either being blown from the top or the bottom of the reactor.[67] Pure O_2 can be used or O_2 diluted by steam (H_2O), carbon dioxide (CO_2), Ar, or N_2, or a combination of these, depending on the grade or quality requirements of the final product. The process temperature has to exceed 1750 °C, due to the nature of the refining process, which results in high refractory wear and high Mn losses owing to vaporization and oxidation, amongst other problems.[46,68] Separation of the slag from the alloy can also be a problem. The slag composition is thus managed by the addition of limestone and other fluxes to manage slag fluidity and refractory wear. FeSi, SiMn, metallurgical grade Si, or wrought Al is added to reduce the MnO in the slag to Mn and the corresponding oxidation products also contribute to changes in the slag composition.[46,67] To further control process temperatures and improve recovery of Mn, MCFeMn alloy fines can be added.[66–68]

For the silicothermic reduction process, a synthetic slag is produced by melting limestone and Mn ore in an open arc furnace.[66] The slag is reacted with SiMn (or sometimes FeSi) in a series of mixing ladles[46,52] or in a shaking ladle[69] to produce MCFeMn. The reaction temperature is lower than for the converter process, at around 1450–1600 °C.[46] Additional limestone, dolomite, or alumina can be added to flux the slag to improve Mn recovery.[46,66] The exothermic reactions provide sufficient energy to sustain the process. The alloy and slag produced in MCFeMn production are treated in a similar fashion as for HCFeMn production, namely the alloy is layer cast and the slag is collected in a slag pot.[67]

Table 2.6 summarizes the calculated averages and standard deviations for the chemical compositions of slags (wt%) and the amount of slag produced per tonne of alloy for HCFeMn produced in a BF or SAF and MCFeMn produced in a converter or shaking ladle. The results for BF slag[52,53,59,61,70–72] were calculated from 11 data sets, those for SAF duplex slag[46,73–75] from five data sets, and those for SAF discard slag[46,76] from four data sets. For the converter[65] and mixing or shaking ladle[69] slag, only one data set was included for each.

2.3.3 Production of Silicomanganese Alloys

The typical range of chemical compositions for SiMn alloys is presented in Table 2.7.

SiMn is produced by the carbothermic reduction of Mn-bearing ores in AC SAFs similar in design to the ones utilized in HCFeMn production.[46] The SAFs are rated at 15–63 MVA and produce 80 to 340 tonnes of alloy per day.

SiMn is also produced in two processes: ore-based or duplex. In the ore-based process, no slag produced during HCFeMn is included in the feed materials.[56] In the duplex process, the main source of Mn is a combination of

Table 2.6 Calculated average and standard deviation (SD) for the chemical compositions of slags (wt%) and the amount of slag produced per tonne of alloy for HCFeMn produced in a blast furnace (BF) or submerged arc furnace (SAF), using either the duplex or discard method, and MCFeMn produced in a converter or shaking ladle.[52,53,59,61,70–72]

Technology		Al_2O_3	CaO	FeO	MgO	MnO	SiO_2	Slag/alloy ratio
BF	Average	7.7	41.9	0.6	7.2	11.5	31.5	1153
	SD	1.1	3.3	0.5	2.3	4.6	2.1	547
SAF – duplex	Average	16.0	12.6	NR[a]	3.4	40.6	22.5	663
	SD	7.1	5.6	NR	2.5	3.8	2.6	32
SAF – discard	Average	8.2	29.9	NR	6.3	23.6	28.3	765
	SD	7.6	2.0	NR	0.6	6.3	3.8	70
Converter		1.0	32.0	NR	NR	30.0	27.0	122
Mixing or shaking ladle		1.4	NR	NR	2.1	26.4	32.7	1008

[a]NR: Not reported.

Table 2.7 Typical chemical composition of SiMn in wt% where the remainder not accounted for in the table is Fe.[77,78] Reproduced from ref. 78 with permission from JSAIMM, Copyright 2018.

	Mn	Si	C	P	S
ASTM Grade A	65.0–68.0	18.5–21.0	1.5	≤0.20	≤0.04
ASTM Grade B	65.0–68.0	16.0–18.5	2.0	≤0.20	≤0.04
ASTM Grade C	65.0–68.0	12.5–16.0	3.0	≤0.20	≤0.04

HCFeMn slag and ore.[46] The process temperatures required for SiMn production are higher than for HCFeMn production and the electricity consumption varies between 3500 and 4500 kWh per tonne of alloy.[50] The reductant requirement is addressed by a combination of coke, anthracite, and bituminous coal (Figure 2.6). Quartz, and sometimes off-grade FeSi and Si or their melts, is also added to achieve the required Si content of the alloy. Limestone and dolomite are added as fluxes in processes not based on basic carbonate ores.[50,56]

The slag and alloy are often tapped from the SAF from single-level tapholes with separation taking place through cascade tapping, where slag overflows from the ladle containing the alloy into a slag pot or the use of skimmers in the launders.[56,78] The alloy is cast in layers while the slag is cooled in slag pots. Cooled slag is taken to alloy recovery plants for the removal of entrained metal. The cleaned slag is discarded in waste dumps or used as fertilizer.[50,56,73,78] Table 2.8 summarizes the chemical compositions of slags (in wt%) and amounts of slag produced per tonne of alloy for SiMn produced in a SAF. The results were calculated from three data sets for the duplex process[50,56,79] and two for the ore-based process.[56,78]

2.3.4 Production of Ferrosilicon Alloys

FeSi produced commercially can be divided into a number of grades based on the Si content of the alloy produced. The typical chemical compositions of these alloys are summarized in Table 2.9 in four broad categories.[80] The most common grades utilized in steelmaking are the alloys containing 50 wt% Si and 75 wt% Si.

Table 2.8 Calculated averages and standard deviations (SD) of the chemical compositions of slags (in wt%) and the amount of slag produced per tonne of alloy for SiMn produced in a submerged arc furnace (SAF), using either the duplex[50,56,79] or ore-based[56,78] process.

Technology		Al_2O_3	CaO	FeO	MgO	MnO	SiO_2	Slag/alloy ratio
SAF – duplex	Average	15.8	22.0	NR[a]	7.5	10.2	39.7	1500
	SD	3.3	14.6	NR	NR	1.3	1.2	100
SAF – ore	Average	4.7	25.9	0.4	5.4	14.4	46.3	1025
	SD	0.1	1.0	0.1	0.5	2.3	0.1	35

[a]NR: Not reported.

Table 2.9 Categories of FeSi and their typical chemical compositions in wt% (balance Fe).[49,80] Reproduced from ref. 80 with permission from Elsevier, Copyright 2013.

Product	Si	Al	P	S	C	Mn	Cr	Ti
FeSi10–FeSi25	8–30	<1.5	<0.15	<0.06	<2	<3	<0.8	<0.3
FeSi45–FeSi65	41–68	<2	<0.05	<0.05	<0.2	<1	<0.05	<0.03
FeSi75Al1–FeSi75Al3	72–80	<3	<0.05	<0.04	<0.2	<0.5	<0.5	<0.3
FeSi90Al1–FeSi90Al2	87–95	<3	<0.04	<0.04	<0.15	<0.5	<0.2	<0.3

FeSi with up to 20 wt% Si can be produced in BFs by carbothermic reduction of quartz- and Fe-bearing raw materials.[81] This was the manner in which FeSi was produced for steelmaking until the application of EAFs in the production of high-Si FeSi was started in the United States in 1907. By 1920, FeSi was mainly produced in AC SAFs.[80,81] SAFs utilized at present are rated at 22 to 93 MVA and the process energy requirement ranges between 2100 and 8800 kWh per tonne of alloy.[80,81]

In the SAF, quartz is the source of Si, as indicated in Figure 2.7. Coal, charcoal, petroleum coke, and char are added as sources of C along with woodchips to ensure gas permeability. As sources of Fe, metallic Fe or Fe ores are added.[80,81] In ferroalloy production terms, FeSi production is nearly slag-free with 30 to 70 kg of slag tapped per tonne of alloy produced.[80]

The interior of a SAF producing FeSi appears very different from the interior of the so-called wet processes applied in HCFeMn, SiMn, and HCFeCr production. The dissipation of electrical energy in SAFs producing HCFeMn, SiMn, and HCFeCr is mainly through resistive heating, while the electrical energy in SAFs producing FeSi is mainly dissipated through arcing.[54] The reason for this is the formation of a large cavity, filled mainly with SiO gas and CO gas, below the electrode tip, as shown in Figure 2.7.

The cavity results from the sticking together of charge materials by Si, SiO_2, and SiC-rich condensate, resulting in a void as the raw materials are consumed during the production process.[82] The mechanical strength of the top of the cavity will decrease as the SiO_2, present in the condensate and the raw materials, starts to melt, resulting in temperatures in the region of 1650–1800 °C.[80–82] As the gas species are ascending into the charge at lower temperatures, the SiO gas formed lower down in the furnace will react either on the surface of the raw materials, to produce the condensate, or with the C

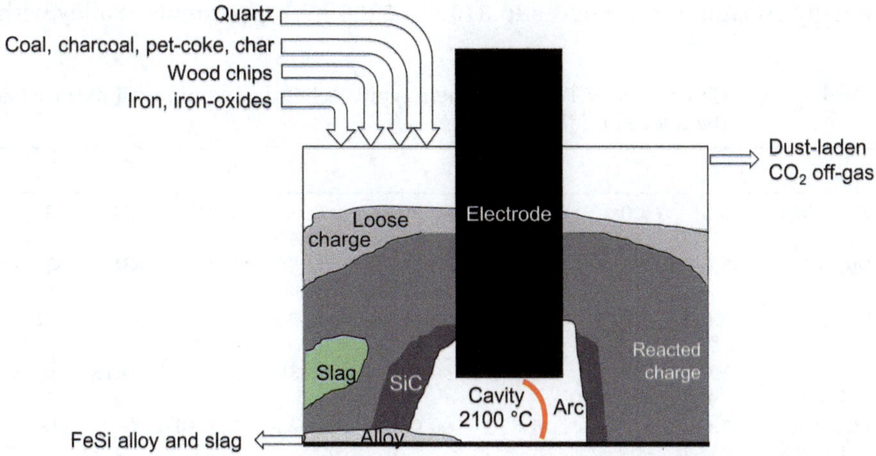

Figure 2.7 Simple schematic of the process flow in a submerged arc furnace (SAF) utilized in the production of FeSi containing 75 wt% Si (pet-coke: petroleum coke).[80–82]

added to produce SiC. The production of FeSi in the lower part of the furnace is a result of the intermediate formation of SiC and SiO.[82]

The SAFs are semi-open and the off-gas is combusted at the top of the furnace. Two products can be recovered from the off-gas: SiO_2 fume, utilized in the manufacturing of self-flowing refractory materials, and energy through energy recovery plants.[80,81] The alloy tapped at the bottom of the furnace is refined in the ladle and then cast in layers in beds made of FeSi fines or into iron molds, or granulated with water.[81]

2.3.5 Production of Ferrochromium Alloys

FeCr alloys produced commercially can be divided into: high-carbon ferrochromium (HCFeCr), medium-carbon ferrochromium (MCFeCr), low-carbon ferrochromium (LCFeCr), and LCFeCr with high Cr or N contents. Their typical chemical compositions are indicated in Table 2.10.[83,84] This section focusses on the production routes for HCFeCr, MCFeCr, and LCFeCr only.

HCFeCr is produced by the carbothermic reduction of Cr-bearing ores in SAFs, similar in design to those utilized in HCFeMn and SiMn production, or in direct current (DC) EAFs.[85] DC EAF technology is applied in the treatment of fine, friable raw materials (passing screening size of 6 mm) and is operated with an open bath and open arc, as shown in Figure 2.8. Commercial DC EAFs for FeCr production utilize a single graphite electrode as the cathode and a conductive hearth as the anode.[85]

To reduce the electrical energy required in SAF smelting, pellets are preheated in shaft kilns or pre-reduced, using solid carbonaceous reductants, in rotary kilns or rotary hearth furnaces.[85,89,90] Whereas electrical energy of 4500 kWh per tonne of alloy is required in a conventional SAF, the specific energy consumption reduces to 3100 to 3200 kWh per tonne of alloy with

Table 2.10 Categories of FeCr and their typical chemical compositions in wt% (balance Fe).[83,84]

	Cr	C	Si	P	S	N	Ni	Co
HCFeMn	45.0–75.0	4.0–10.0	1.5–10.0	0.030–0.050	0.05–0.10	NR[a]	NR	NR
MCFeCr	45.0–75.0	0.5–4.0	≤1.5	0.030–0.050	≤0.050	NR	NR	NR
LCFeCr	45.0–75.0	0.015–0.50	≤1.5	0.030–0.050	≤0.030	≤0.15	NR	NR
LCFeCr, high Cr	75.0–95.0	0.015–0.050	≤1.5	≤0.020	≤0.030	≤0.20	≤0.15	≤0.02
LCFeCr, high N	45.0–75.0	≤0.10	≥1.5	≤0.030	≤0.025	2.0–10.0	NR	NR
FeSiCr (typical)	28–45	0.1–6	10–45	0.03	0.02	NR	NR	NR

[a]NR: Not reported.

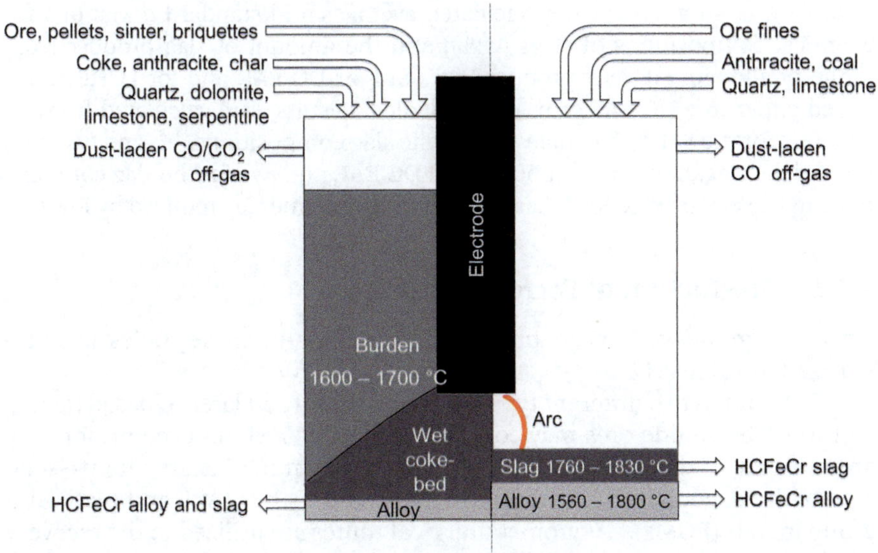

Figure 2.8 Schematic of the process flow in a submerged arc furnace (SAF) (to the left) and a direct current (DC) electric arc furnace (EAF) (to the right) utilized in the production of HCFeCr.[85–88]

preheating, and to 2000 to 2500 kWh per tonne of alloy with pre-reduction. SAFs are rated at 5–135 MVA. Chromium recovery is between 70 and 92%, with higher recoveries achieved when using DC EAF technology due to better flexibility in controlling slag chemistry and temperature.

Both single-level and bi-level tap-hole designs are found in SAFs, depending on the size of the furnace, and only bi-level tapping in DC EAFs.[85,89] Tapped alloy can be layer cast, sand cast in ingots, and granulated with water.[85,89] Tapped slag can be cooled in slag pots, layer cast, or granulated with water.[85,89,90] Cooled slag is taken to alloy recovery plants for the removal of entrained metal. The cleaned slag is discarded in waste dumps or is utilized as a building material, depending on the applicable local regulations.[85]

MCFeCr is mainly produced using top-blown oxygen converter technology to refine liquid HCFeCr.[84] Oxygen converter technology is also applied in LCFeCr production, but a number of other processes are in operation as well. In these processes, chromite ($FeCr_2O_4$) is reduced using HCFeCr, FeSiCr, FeSi, or Al alloys. As the reaction products for MCFeCr and LCFeCr production are SiO_2 or Al_2O_3, the slag basicity is adjusted by adding significant amounts of limestone. When liquid HCFeCr, liquid FeSiCr, or Al are utilized, the reactions can take place in mixing ladles, but most often the reactions take place in open-bath AC EAFs with a low load (*e.g.*, 5 MVA). FeSiCr is produced by reducing quartz with HCFeCr and coke, or chromite and quartz by coke only, in an open-bath AC EAF. FeSiCr is utilized as a reagent in LCFeCr production and as an alloy in steelmaking.[84]

Table 2.11 summarizes the calculated averages and standard deviations for chemical compositions of slags (wt%) and the amount of slag produced per tonne of alloy for HCFeCr produced in a SAF or DC EAF, and for LCFeCr produced either in a DC EAF or in mixing ladles. Results were calculated from 14 data sets for the SAF,[90–96] 11 data sets for the slag compositions,[86,87] and two data sets for the slag/alloy ratios[97] obtained in DC EAF, and two for the slag compositions in LCFeCr production[98] based on a chromite-lime slag reduced by FeSiCr.

2.3.6 Production of Ferronickel Alloys

Only a few grades of FeNi are produced commercially. These grades are summarized in Table 2.12.[99]

Ni is found in two different types of ores: sulfide- and laterite-based (oxide-silicate). The sulfide ores may contain up to 4 wt% Ni but account for only around 14% of the known Ni reserves. Ni sulfide minerals are often associated with Fe, copper (Cu), cobalt (Co), gold (Au), silver (Ag), and platinum-group metals (PGMs).[100] Pyrometallurgical routes are utilized in the recovery of Ni from sulfide ores. Two different smelting technologies are utilized: flash smelting and electric furnace smelting. The electric furnaces are operated in

Table 2.11 Calculated chemical compositions of slags (wt%) and the amount of slag produced per tonne of alloy for HCFeCr produced in either a submerged arc furnace (SAF)[90–96] or direct current (DC) electric arc furnace (EAF)[86,87,97] process and for FeCr refined in an EAF or in mixing ladles to produce LCFeCr.

Technology		Al_2O_3	CaO	Cr_2O_3	FeO	MgO	SiO_2	Slag/alloy ratio
SAF	Average	23.3	3.4	11.4	7.5	26.8	31.5	1138
	SD	2.3	0.6	4.5	8.8	5.0	4.5	48
DC EAF	Average	31.7	13.2	4.5	1.9	23.4	26.1	760
	SD	6.3	6.5	1.1	0.7	7.8	3.1	42
EAF or mixing ladles	Average	5.0	43.8	6.3	0.7	14.5	28.5	NR[a]
	SD	0.0	4.6	0.4	0.0	0.7	2.8	NR

[a]NR: Not reported.

Table 2.12 Categories of FeNi alloys and their typical chemical compositions in wt% (balance Fe).[99] Reproduced from ref. 99 with permission from Elsevier, Copyright 2013.

Product	Ni	Co	Si	C	Cr	S	P
FeNi10	5–15	0.3–0.6	<0.3	<0.1	<0.3	<0.08	<0.03
FeNi15	12–18	0.5–0.8	<0.4	1–2.5	<2	<0.1	<0.03
FeNi20	15–25	0.750–0.625	<0.4	1–2.5	<2	<0.4	<0.03

immersed electrode mode with the dissipation of electrical energy occurring through resistive heating, as shown in Figure 2.9.[101] In some instances, coke is added as a reductant.[100] Electrical energy consumption in the furnace is 440–1044 kWh per tonne of feed.[100] In some cases, a drying step or a fluidized bed roasting step is included prior to smelting to remove a significant amount of S as SO_2.[100] The matte tapped from the furnace is treated in an oxygen converter step to produce a low-Fe matte. Many operations include a slag-cleaning step where the matte is allowed additional settling time.[100] Slag is often granulated upon tapping.[102] Section 2.4.5 provides further details on the smelting and converting of Ni sulfide ores.

Laterite ores generally contain 1–2 wt% Ni and, depending on the compositions of the gangue minerals, are classified according to the type of gangue into magnesia, silicate, or alumina type. Limonite and saprolite are classified as laterite ores, with limonite ores containing 45–50 wt% Fe and 1.2–1.3 wt% Ni. Saprolite ores contain 15–20 wt% Fe and slightly higher Ni content of 2.0–2.3 wt%.[99] Pyrometallurgical processes are commonly utilized to recover Ni from saprolite ores.[103]

Traditionally, Ni was recovered from laterites utilizing BF technology; the first time this technology was applied was in 1879 in New Caledonia.[103] In the 1950s, the rotary kiln–electric furnace (RKEF) process was commercialized.[104,105] In the RKEF process, Ni laterite ores are dried and then calcined and partially reduced in rotary kilns using bituminous coal or anthracite.[103] Final reduction and separation of the FeNi from the slag occurs in a smelting step in large EAFs. Finally, the crude alloy is refined in ladles or refining

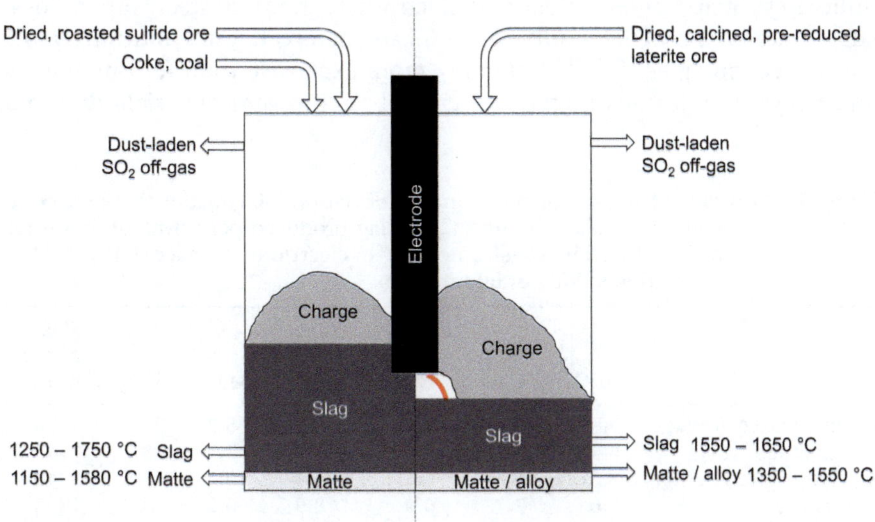

Figure 2.9 Simple schematic of the process flow in an electric furnace utilized in the production of FeNi, showing the immersed electrode smelting of sulfide ore to the left and the shielded arc smelting of laterite ore to the right.[100,101,103]

electric furnaces.[103] Initially, furnaces were rated at 10 MW, but today the largest furnace slightly exceeds 100 MW and these AC furnaces can be circular or rectangular.[105] From the 1970s onwards, the shielded arc mode of electrode operation was introduced, in which an arc is drawn between the electrode tip and the slag, and the arc is covered with feed calcine, which results in increased energy efficiency.[105] The electrical energy requirement for the smelting stage is 380–620 kWh per tonne of calcine.[103] During ore preparation, fine materials are produced that cannot readily be accommodated by the AC furnaces. DC EAF technology, similar to the technology described in FeCr production (Figure 2.8), is able to treat ore sized at less than 1 mm in diameter, which is fed together with coal or anthracite.[106–108]

Table 2.13 summarizes the calculated averages and standard deviations for chemical compositions of slags and the amount of slag produced per tonne of matte for FeNi produced in a flash smelter or EAF using either sulfide or laterite ores. The results for flash smelter slag were calculated from seven data sets,[100] those for the electric furnace smelting sulfide ores from 18 data sets[100,102] and slag/matte ratio from five,[102] and those for the electric furnace smelting laterite ore were calculated using 12 data sets.[103]

2.4 Production of Non-ferrous Metals

2.4.1 Overview

Non-ferrous metals refer to all metals that do not contain Fe in appreciable amounts.[18] These metals have irreplaceable applications in a number of industries, including electrical and electronics, stainless steel, superalloys, shape memory, magnetic alloys, chemicals, auto catalysts, and renewable energy technologies.[18,109,110] Generally more expensive than ferrous metals and alloys, non-ferrous metals are critical engineering materials due to a

Table 2.13 Calculated average and standard deviation (SD) for the chemical compositions of slags (in wt%) and slag produced per tonne of matte for FeNi produced in a flash smelter[100] or electric arc furnace (EAF),[100,102,103] using either sulfide or laterite ores.

Technology		Al_2O_3	CaO	Cr_2O_3	FeO	MgO	SiO_2	Slag/matte ratio
Flash smelter – sulfide	Average	NR[a]	NR	NR	50.0	5.2	32.6	3630
	SD	NR	NR	NR	5.4	3.5	3.2	1546
Electric furnace – sulfide	Average	3.3	8.8	1.5	21.4	15.2	43.5	5740
	SD	1.6	2.7	0.9	8.9	6.3	4.9	1905
Electric furnace – laterite	Average	NR	NR	NR	20.2	26.1	48.2	NR
	SD	NR	NR	NR	14.2	10.3	6.8	NR

[a]NR: Not reported.

number of desirable properties, such as low weight and malleability (*e.g.*, aluminum (Al) alloys), high strength-to-weight ratio (*e.g.*, Ti alloys), high conductivity (*e.g.*, Cu and its alloys), nonmagnetic properties, and resistance to corrosion (*e.g.*, zinc (Zn) and its alloys).[18,110,111] Non-ferrous metals are subdivided according to their geological, mineralogical, and technological properties into: base metals (*e.g.*, Cu, Co, Ni, Zn, and lead (Pb)), light metals (*e.g.*, Al and Ti), precious metals (Au and Ag), PGMs (platinum (Pt), rhodium (Rh), palladium (Pd), osmium (Os), iridium (Ir), and ruthenium (Ru)), refractory metals (*e.g.*, tantalum (Ta), niobium (Nb), Mo, vanadium (V), and zirconium (Zr)), and rare earth metals (*e.g.*, scandium (Sc), yttrium (Y), lanthanum (La), and cerium (Ce)).[111]

Non-ferrous metals predominantly occur in the Earth's crust in the form of sulfides, oxides, silicates, and carbonates.[111] The primary production of non-ferrous metals from natural ores and/or secondary materials involves a series of physical and chemical beneficiation steps followed by the recovery of pure metals or alloys, either by pyrometallurgical or hydrometallurgical methods, or a combination of both.[111] Hydrometallurgical processes capitalize on differences between constituent solubilities and/or electrochemical properties while in aqueous solution.[1,4] Pyrometallurgical technologies, discussed in more detail in the following subsections, use thermal energy to separate the desired metals from the gangue, based on differences between constituent oxidation potentials, melting points, vapor pressures, densities, and/or immiscibilities in the molten state.[1,4]

The smelting of non-ferrous metals from ores and/or secondary materials produces large volumes of slags.[4,112] Non-ferrous metal slags belong to the $FeO–SiO_2–Al_2O_3–CaO$ slag system and are formed from the reaction between the gangue in ore, ash in reductants, and fluxes in the primary smelting units, or from the reactions between fluxes and impurity oxides in the refining units.[112] This section provides an overview of primary production processes for selected non-ferrous metals, namely, Al, Pb, Zn, Cu, Ni, and PGMs, using conventional pyrometallurgical methods.

2.4.2 Production of Aluminum

The low density, high strength-to-weight ratio, ductility, and corrosion resistance properties of Al make it one of the leading non-ferrous metals in use, with unique applications in industries such as transportation, construction, packaging, and chemical.[113,114] Aluminum exists in nature in minerals such as gibbsite (α-$Al_2O_3·3H_2O$), böhmite (α-$Al_2O_3·H_2O$), diaspore (β-$Al_2O_3·H_2O$), all minerals found in bauxite, and cryolite (Na_3AlF_6).[114] Cryolite has been reported to be rare to such a point it could be the only mineral that has been mined to commercial extinction, leaving bauxite as the only primary source of Al.[113]

Aluminum is commercially produced *via* two main processes, namely, the Bayer thermochemical digestion process for refining bauxite containing approximately 30–60 wt% Al_2O_3 to high-grade alumina (Al_2O_3) and the Hall–Héroult process for electrolytic smelting of alumina dissolved in synthetic

Na_3AlF_6 to produce liquid Al metal.[113,114] The Bayer and Hall–Héroult processes account for over 95% of primary Al production, albeit being associated with significant environmental consequences.

The high energy intensity (13 000–15 000 kWh per tonne of Al) of the Hall–Héroult process accounts for up to 40% of the cost of primary Al production.[114] Additionally, the Bayer process produces large volumes of waste in the form of bauxite residue or red mud, a highly alkaline waste stream with complex mineralogy and composition (*e.g.*, 5–60 wt% Fe_2O_3, 5–30 wt% Al_2O_3, up to 15 wt% TiO_2, 2–14 wt% CaO, 3–50 wt% SiO_2, and 1–10 wt% Na_2O).[114,115] Approximately 4 tonnes of bauxite ore produces 2 tonnes of Al_2O_3 and 2 tonnes of red mud.[114] In addition to being an intensive emitter of carbon dioxide (CO_2), the Hall–Héroult process also generates significant amounts of solid waste in the form of drosses (*i.e.*, mixtures of Al metal, Al_2O_3, AlN, AlC, Na_3AlF_6, and other oxides, ranging up to 80 wt% Al) and spent pot lining (SPL) contaminated with Al and fluorides.[114] The drosses are recycled to recover the Al while the SPLs, consisting of 50 wt% refractory and 50 wt% carbon, are usually consumed as fuel in cement and ferrosilicon furnaces.

Aluminum is one of the most recyclable non-ferrous metals.[114] The recycling of end-of-life and new Al scrap, drosses, and salt cake (containing 3–10 wt% Al, 20–60 wt% Al_2O_3, and 20–80 wt% NaCl–KCl salt) requires 95% less energy than the primary requirements for bauxite ore processing.[6,114–117] Aluminum scrap is recycled by melting in induction, rotary, or reverberatory furnaces under a bath of supernatant molten salt (*i.e.*, a eutectic or near-eutectic mixture of NaCl and KCl containing some cryolite).[113–117] The recycling of Al scrap produces large amounts of salt slag containing oxides, nitrides, chlorides, and some residual metal or alloys such as Fe, Zn, Cu, Si, Mn, and Mg.[6,116,117] The discard salt slags are treated in a series of steps, including the separation of entrained metal, leaching, and crystallization of soluble salts to regenerate the salt fluxes, and the recycling of the leach residues in the process or in other industries.[6,116,117]

2.4.3 Production of Lead and Zinc

Lead and Zn are probably the two most commonly used non-ferrous metals after Al and Cu.[118] Over 60% of the primary Pb produced globally is consumed in the manufacture of lead–acid batteries. A significant amount of Pb is also used in the production of industrial alloys and ammunitions, and in the production of architectural ornaments and rolled extrusions for various applications.[119] Galena (PbS) is the principal mineral exploited in the primary production of Pb. Other primary sources include secondary minerals, such as cerussite ($PbCO_3$) and anglesite ($PbSO_4$).[120]

Zinc is predominantly used in the production of galvanized steel.[121] Large quantities of Zn are also used to produce die-castings for automobiles, electrical and hardware applications, and in the production of alloys such as brass, Ni–Ag, and Al solder.[120,121] In its oxide form, Zn is widely used in the manufacture of chemical products such as paints, rubber, cosmetics, pharmaceuticals, plastics, batteries, and electrical equipment.[122] Zinc is primarily produced from sphalerite (ZnS) as well as from secondary minerals, such as franklinite ($Zn(Fe,Mn)_2O_4$), willemite (Zn_2SiO_4), smithsonite ($ZnCO_3$), and hemimorphite ($Zn_4Si_2O_7(OH)_2{\cdot}H_2O$).[120,123]

Sulfide minerals hosting Pb and Zn naturally occur with each other and as polymetallic mineralizations containing Cu, Au, Ag, arsenic (As), and bismuth (Bi).[123–125] Lead- and Zn-bearing sulfides occur combined with Fe and S in the form of complex polymetallic sulfide systems associated with Fe-bearing minerals, such as bornite (Cu_5FeS_4), chalcopyrite ($CuFeS_2$), pyrrhotite (Fe_7S_8), and/or pyrite (FeS_2).[112,124–127] This means that slags from the pyrometallurgical processing of Pb and Zn ores would invariably contain significant amounts of iron oxides.[112]

2.4.3.1 *Primary Production of Lead*

Lead is produced predominantly by pyrometallurgical processing of primary sulfide ores and recycled scrap.[118,124,128,129] Lead production from ores involves crushing, milling, and concentration by flotation followed by sintering and BF smelting. The conventional sinter–BF route accounts for over 70% of global production, with the rest coming from direct smelting and/or hydrometallurgical processes.[124,128,129]

Figure 2.10 shows a simplified flowsheet for the primary production of Pb from sulfide ores. The first step involves the solid-state oxidation of

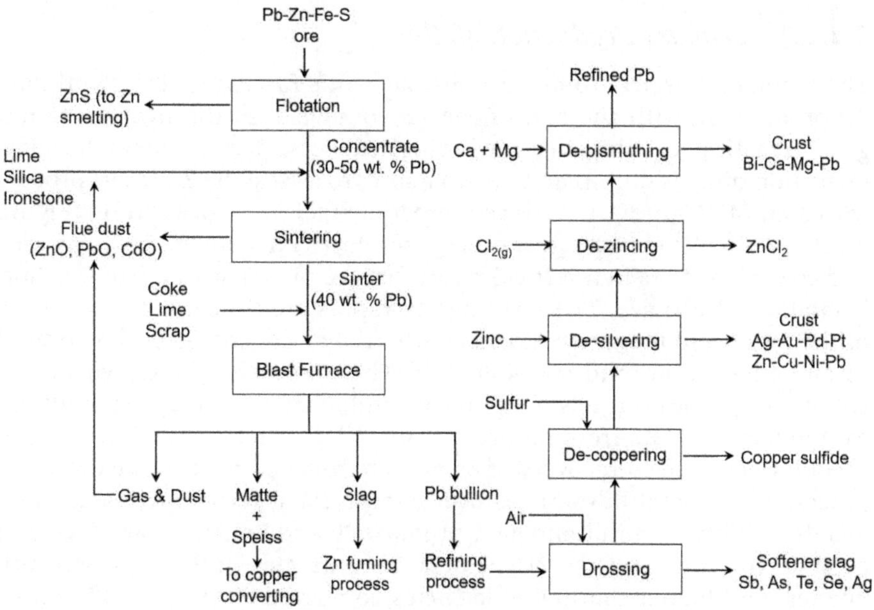

Figure 2.10 Simplified flowsheet for primary production of Pb from sulfide ore. Reproduced from ref. 129 with permission from The Authors, Copyright 2008.

fluxed sulfide concentrates in the sinter plant to produce PbO-rich sinter and SO_2-rich off-gas.[129,130] Fluxes are added to control the amounts and types of sinter phases and, ultimately, the chemistry and volume of the final BF slag.[129,130] Lead sinters often consist of complex mixtures of melilite ((Ca,Na, $Pb)_2(Al,Mg,Fe^{2+},Zn)(Al,Si)_2O_7$), spinel ($(Zn,Fe^{2+},Mg)(Fe^{3+},Al)_2O_4$), lead oxide (PbO), larsenite ($PbZnSiO_4$), and dicalcium silicate $(Ca,Pb)_2SiO_4$ phases.[130] The second step involves the carbothermic smelting of Pb-bearing phases in sinter in the BF to produce crude Pb and final slag. The final slag is then tapped from the furnace and is either recycled or sent for Zn fuming to recover the contained Zn. The crude Pb tapped from the furnace goes through a series of pyrometallurgical steps for removal of impurities to produce a final product containing 99.9 wt% Pb.[129,130]

In recent years, the contribution of direct smelting technologies, such as the Ausmelt/Isasmelt, Queneau–Schuhmann–Lurgi (QSL), and flash-cyclone-oxygen-electric smelting (KIVCET) processes, has rapidly increased due to the need to overcome the challenges of large capital investments, high operating costs, and environmental footprints associated with the conventional sinter–BF processes.[127,128,131] Direct smelting technologies integrate the oxidation and reduction stages in single units and exploit the exothermicity of sulfide oxidation to achieve sufficiently high temperatures. These processes are based on the fact that the smelting of concentrates under controlled oxygen potential can result in over 95% of the Pb and Zn being volatilized in metallic form and recovered.[112,126,127]

2.4.3.2 Primary Production of Zinc

The pyrometallurgical route currently accounts for about 20% of primary Zn production, with the rest of the Zn coming from the hydrometallurgical (roasting-leaching-electrowinning) process. The pyrometallurgical extraction of Zn is dominated by two main processes: the Zn retort process with a Zn BF (Figure 2.11A) or the Imperial Smelting Process (ISP) (Figure 2.11B).[18,124,127,132] The Zn retort process produces Zn metal by reducing ZnO and coke briquettes in a vertical retort furnace or, to a lesser extent, a horizontal retort furnace. The resulting Zn vapors are collected from the top of the retort and fed to a condenser, where they are cooled and recovered as molten Zn at around 420–450 °C.[132] The coexistence of Zn with Pb in sulfide ores, however, has a significant influence on the pyrometallurgical production of Zn from natural Zn ores.[124,133] The ISP was developed to accommodate the co-recovery of Zn and Pb from complex Zn–Pb ores. The process has several advantages over conventional retort processes, *inter alia*, the ability to simultaneously produce Pb and Zn from low-grade and complex mixed charges of Zn–Pb ores, Pb–Zn sinter, and hot-pressed briquettes, and higher thermal efficiencies compared to either of the retort processes.[18,127]

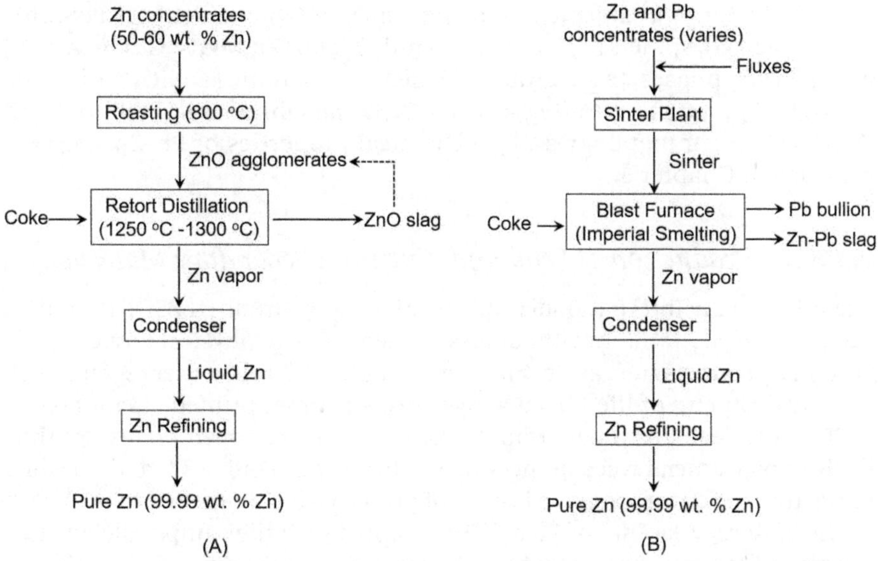

Figure 2.11 Comparison of processes for the production of Zn using (A) the conventional Zn retort process and (B) the zinc blast furnace or Imperial Smelting Process (ISP). (A) Adapted from ref. 18 with permission from Elsevier, Copyright 2016. (B) Adapted from ref. 124 with permission from Elsevier, Copyright 2014.

The production of Zn from complex Pb–Zn ores using the ISP involves the sintering of pre-fluxed Zn–Pb–S concentrates to produce sinters containing varying amounts of zincite [(Zn,Fe)O], spinel, larsenite, dicalcium silicate, and complex Pb–Zn–S phases.[130] The ZnO–PbO sinter is then reduced by solid preheated coke in the furnace shaft to form metallic Pb and Zn.[124] Molten slag and Pb are continually tapped from the bottom of the furnace into the fore hearth, where separation occurs. The Zn vapor in the off-gas (5–7 wt% Zn) is recovered in the Pb splash condenser and is further refined by stepwise distillation.[18,124]

The properties of Zn and Pb–Zn slags depend on the gangue content, the ash content of the coke, and the fluxing regime used to modify the slag behavior in the smelting units. Compared to Pb BF slags (2–4 wt% Pb and 2–12 wt% Zn), ISP slags contain up to 2 wt% Pb and up to 4 wt% Zn.[6,18,127,130,133] The Pb–Zn slags tapped from the furnaces are first cleaned to recover the Pb and Zn in fuming EAFs or Waelz kilns. The slag-cleaning process involves the carbothermic reduction of PbO and ZnO in slags to form metallic vapors, which are then collected as PbO and ZnO in the flue dust and returned to the hydrometallurgical processing plant.[124,128,133]

Depending on the process, the discard Zn and Pb–Zn slags solidify as heterogeneous and complex mixtures of entrained metallic elements (*e.g.*, Pb, Zn, Cu, Fe), oxides (*e.g.*, PbO, (Fe,Zn)O, Fe_3O_4), sulfides (*e.g.*, PbS, ZnS, $CuFeS_2$, $Fe_{1-x}S$), spinels (*e.g.*, $ZnFe_2O_4$, $ZnAl_2O_4$ and $(Zn,Mg,Fe,Cu)(Fe,Al)_2O_4$) and silicate phases (*e.g.*, fayalite ($FeSiO_4$), willemite ($ZnSiO_4$), diopside ($CaMgSi_2O_4$), augite ([Ca(Mg,Al,Fe)Si_2O_6)]), larsenite and melilite.[134–136] Details on the mineralogy and geochemical properties of Pb–Zn slags are provided in Chapter 3.

2.4.3.3 *Production of Lead and Zinc from Secondary Materials*

According to the International Lead and Zinc Study Group (ILZSG), Pb and Zn can be recycled infinitely without loss of chemical and physical properties.[118] Secondary sources such as Pb–Zn smelter slags, lead–acid battery scrap, steel-making dust, end-of-life Pb–Zn alloys, postconsumer printed circuit boards (PCBs), and leach residues from Zn roast–leach–electrowin plants are thus likely to play crucial roles in meeting the future demand of Pb and Zn while mitigating the anthropogenic impact of primary Pb–Zn processes.[118,127,128,137]

The high recyclability of Pb and Zn, coupled with their unparalleled environmental toxicity, has prompted the urgent development of a number of smelting and refining technologies to process Pb–Zn-containing technogenic materials. A significant portion of Pb–Zn secondary raw materials are treated *via* co-smelting in direct smelting furnaces, BFs, EAFs, reverberatory furnaces, and plasma arc furnaces.[127,128,131,137,138] Most notably, direct smelting processes (*e.g.*, Ausmelt/Isasmelt, QSL, and KIVCET) and direct current plasma arc technologies (*e.g.*, EnviroPlas™ developed by Mintek) are not only capable of treating Pb–Zn sulfide concentrates but are also compatible with a plethora of secondary resources.[127,128,131,139] The commercial recycling processes, applied either individually or in combination with hydrometallurgical technologies, have demonstrated high flexibility to economically process highly variable feed charges.

2.4.4 Production of Copper

Primary Cu-bearing minerals can be divided into three generic groups: primary sulfides (*e.g.*, $CuFeS_2$, Cu_5FeS_4, and Cu_3AsS_4), oxides (*e.g.*, Cu_2O), silicates ($CuSiO_3{\cdot}2H_2O$), and secondary sulfides (*e.g.*, Cu_2S and CuS).[4,140] Copper predominantly occurs in nature as sulfides disseminated in igneous rocks and, in most cases, occurs in association with Fe sulfides. The run-of-mine (ROM) sulfide ore containing 0.3–2 wt% Cu is processed by crushing, milling, and flotation to produce concentrates (20–30 wt% Cu, 25 wt% S, 25 wt% Fe, and varying amounts of gangue).[4,140,141] Copper is produced from the concentrates using two complementary approaches, namely: hydrometallurgical and pyrometallurgical. Hydrometallurgical approaches include unit steps such as leaching, solvent extraction, and electrowinning and are widely applied in the processing of oxidized ores and/or roasted sulfide concentrates.[4,140]

Primary sulfide minerals are more abundant in nature and stoichiometrically contain more Cu than either the oxide or disseminated sulfide deposits.[141] Over 80% of global primary Cu production comes from the pyrometallurgical processing of sulfide deposits.[4,140–143] Pyrometallurgical approaches include high-temperature smelting of roasted concentrates in reverberatory or electric furnaces, or the smelting of sulfide concentrates and/or secondary materials, both of which produce Cu matte (45–75 wt% Cu) and fayalite-rich slag (Figure 2.12).[4,140–143] Alternative smelting approaches include the high-intensity bath smelting of sulfide concentrates and secondary materials directly to crude Cu and fayalite-rich slag.[140,144] Slag from the matte smelting furnaces is reprocessed in slag-cleaning furnaces to recover the entrained matte before disposal in slag dumps.

Molten matte tapped from the smelting furnaces is further refined in matte converting units. Matte converting involves blowing O_2-enriched air through the matte in cylindrical Teniente or Peirce–Smith converters to oxidize the Fe and S and produce a supernatant FeO-rich slag layer and crude Cu (97–99 wt% Cu).[140,143] Quartz is added as a flux in the converter to control

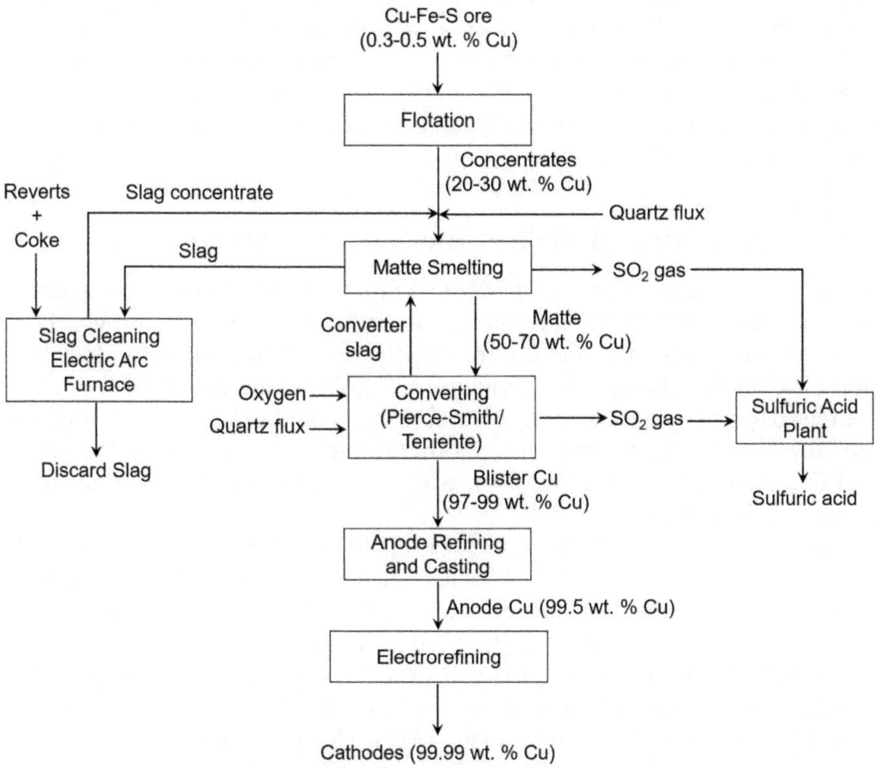

Figure 2.12 Flowsheet for producing Cu from the pyrometallurgical processing of Cu sulfide ores. Reproduced from ref. 127 with permission from Elsevier, Copyright 2014.

the basicity, liquidus temperature, and viscosity of converter slags, and ultimately the slag-matte immiscibility and loss metal values by mechanical entrainment.[127,140] Converter slag is mostly recycled to the smelting furnace or, in some cases, is reprocessed in slag-cleaning electric furnaces to recover entrained and/or dissolved values before disposal.[140,145]

Controlling the SiO_2/Fe ratio is important as it minimizes metal losses to slag by controlling the precipitation of magnetite (Fe_3O_4) and spinel inclusions.[140,146] In fact, it has been demonstrated industrially that an SiO_2/Fe ratio of 0.5 can limit the formation of Fe_3O_4 to acceptable levels of 12–18 wt%.[140] More recently, the use of basic CaO fluxes has also increased, mostly due to the ability to control the precipitation of Fe_3O_4 by forming calcium ferrite phases. Basic slags also possess high affinity for acidic oxide impurities, such as As_2O_3 and Sb_2O_3.[140]

The crude Cu tapped from the converter furnace requires further refining to produce Cu of a quality suitable for casting into anodes.[140] In this case, the molten Cu is mixed with fluxes and directly charged into rotary or hearth furnaces maintained at 1200 °C.[140] The first step involves injecting air through the molten metal to lower the level of residual S and other impurities. Inevitably, the oxidation stage results in the formation of CuO-rich slag.[140,143] The second step then involves the reduction of CuO in the slag by injecting natural gas (*e.g.*, propane) into the Cu melt. The refined Cu (approximately ~0.003 wt% S; ~0.16 wt% O) is then cast into anodes and sent for electrorefining to produce cathode Cu (approximately 99.99 wt% Cu).[140,143]

2.4.5 Production of Nickel from Ni Sulfide Ores

Nickel is produced from sulfide deposits (0.15–8 wt% Ni) and Ni laterites (1.0–1.6 wt% Ni).[147] Nickel laterites account for over 70% of land-based Ni deposits but only contribute approximately 40% of Ni production (see Section 2.2.6 for details on FeNi production).[147] Sulfides are the most important sources of primary Ni, currently contributing about 60% of the Ni produced globally.[148] Pentlandite ($(Ni,Fe)_9S_8$) is the dominant Ni-bearing sulfide mineral and exists in close association with Fe, Cu, Co, PGMs, Au, and Ag in mafic and ultramafic rocks.[147,148]

Nickel sulfide ore undergoes a series of beneficiation steps to produce Ni concentrates (10–20 wt% Ni).[147,148] The concentrates are dried and processed *via* two broad approaches: partial roasting/electric furnace smelting/Peirce–Smith converting or flash smelting/Peirce–Smith converting.[142,147,148] The first approach entails partially oxidizing the Ni concentrates in rotary, multi-hearth, or fluidized bed roasters to produce NiO-rich calcine. The roasted calcine is mixed with fluxes and smelted in electric furnaces, as shown in Figure 2.9, or in some cases in natural gas or coal-fired reverberatory furnaces, to produce Ni matte and discard slag.[142,147,148] Alternatively, the dried concentrates are flash-smelted at 1300–1350 °C in oxygen-enriched air in the presence of SiO_2 fluxes to produce Ni matte and slag (Figure 2.13A).[127,142,147,148] The flash smelting chemistry is similar to that of Cu sulfide processing.[127] The Ni matte tapped from furnaces is converted to a high-Ni (50–70 wt% Ni, 1 wt% Co) and low-Fe (0.5–2 wt% Fe) converter matte by blowing oxygen-enriched air in the presence of fluxes in a Peirce–Smith converter (Figure 2.13B).[127]

Converter matte is refined by either hydrometallurgical or pyrometallurgical means, or a combination of both, to produce pure Ni, Co, PGMs, and by-products Co, Au, and Ag.[148] The high oxygen potential in the converter process often results in the loss of valuable metals to slag. In fact, studies by Toscano and Utigard[142] and Nagamori[149] explicitly demonstrated an increase in Ni solubility in converter slags with increasing oxygen partial pressures (pO_2). Invariably, the converter slag is recycled back to the smelting furnace and/or is treated in separate slag-cleaning electric furnaces to recover the values.[142]

2.4.6 Production of Platinum-group Metals from Sulfide Ores

Platinum-group metals (PGMs) refers to a group of chemically similar elements: Pt, Pd, Ir, Os, Rh, and Ru.[102,150–153] PGMs have unique physical and chemical properties that make them indispensable to modern technologies, and they are extensively used in a number of commercial and industrial applications, including as catalysts in the chemical, fuel cell and automobile industries, and in the manufacture of jewelry, medical devices, and high-temperature alloys.[150–154] The primary production of PGMs is dominated by South Africa, Russia, Zimbabwe, the United States, and Canada.[155]

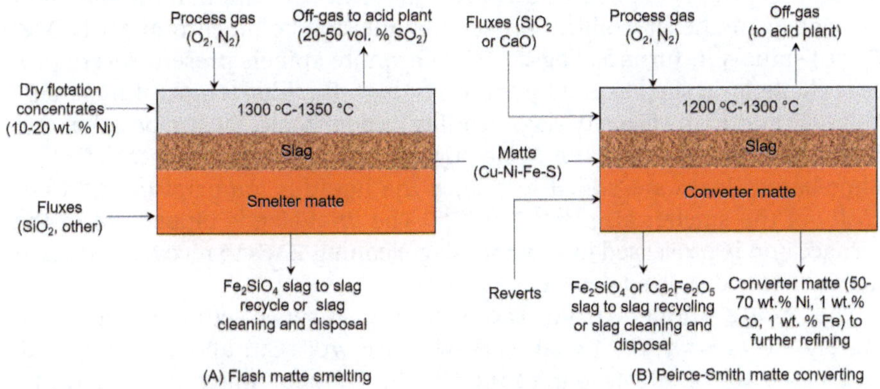

Figure 2.13 Schematic representation of (A) matte smelting and (B) matte converting processes to recover Ni from primary sulfide concentrates. Adapted from ref. 127 with permission from Elsevier, Copyright 2014.

The six PGMs exhibit strong siderophilic and chalcophilic tendencies.[151] This means that PGMs tend to exist preferentially bonded with Fe, Ni, Cu, and S rather than with oxygen.[145,150,151] In nature, the PGMs are chiefly held either in base metal sulfide minerals (*e.g.*, Fe_7S_8, $(Ni,Fe)_9S_8$ and $CuFeS_2$) in Ni–Cu–S magmatic rocks or in PGE-bearing accessory minerals (*e.g.*, braggite ((Pt,Pd,Ni)S), cooperite (PtS), and Fe–Pt alloys).[151] The Ni–Cu-dominant PGM deposits, such as the Sudbury (Canada), Jinchuan (China), and Norilsk (Russia) Igneous Complexes, produce PGMs as by-products from the matte smelting of Ni–Cu–Fe–S concentrates. In contrast, the processing of PGM-dominant deposits produces PGMs as the main economic products with Ni and Cu as by-products.[150,151] Examples of PGM-dominant deposits include Stillwater Igneous Complex (United States), the Main Sulphide Zone hosted in the pyroxenite mineralization in the Great Dyke (Zimbabwe), Merensky reef types, chromitite reef or UG2 type, contact type (*e.g.*, Platreef), and dunite pipes. The last four types of deposits are typically well developed in the Bushveld Igneous Complex in South Africa.[154,156,157]

Similar to Cu and Ni, the primary production of PGMs is dominated by matte smelting of flotation concentrates.[102,151–158] The ROM ore (3–5 g PGMs per tonne) undergoes a series of comminution, gravity, and flotation steps to produce PGE-enriched concentrates (100–400 g PGMs per tonne).[102,154] The dried concentrates are pneumatically transferred to six in-line SAFs and smelted at approximately 1350 °C to produce PGM matte (600–1500 g PGMs per tonne) and immiscible silicate slag.[151,156]

The smelting of UG2 chromitite reef types is, however, slightly different from the predominantly sulfide ores. The UG2 deposit has a particularly high chromite content and relatively low quantities of base metal sulfides.[102,156–158] The high levels of chromitite (1.6–1.7 wt% Cr_2O_3) and pyroxenite or ultramafic (MgO-rich) gangue in UG2 ores leads to Cr saturation in the slags and ultimately in the crystallization of high-melting-point chromite ($(Fe,Mg)Cr_2O_4$) spinels in furnace slags.[157] The chromite spinels present severe process challenges, such as the formation of high-liquidus-temperature (1450–1600 °C) and high-viscosity slags, leading to poor matte–slag separation and increased reversion of matte to slag by mechanical entrainment.[102,156,157,159] Lime-based fluxes are added to reduce the liquidus temperatures and viscosity of the smelter slag.[102,151–153,156–159] The final slag is tapped out of the furnace and is processed in separate slag-cleaning EAFs to recover entrained matte and/or dissolved base metals.

The matte tapped from the smelting furnace contains approximately 30–41 wt% Fe, 12–20 wt% Ni, 7–16 wt% Cu, and 25–30 wt% S, and 650–2400 g PGMs per tonne.[152] The tapped matte is then transferred to a Peirce–Smith converter furnace to oxidize Fe and S impurities and further enrich the values in matte to about 2100–6000 g per tonne of PGMs.[102,152,154] The converter matte is then refined sequentially by hydrometallurgical means to produce individual PGMs with >99.9% purity (Figure 2.14).[152,154]

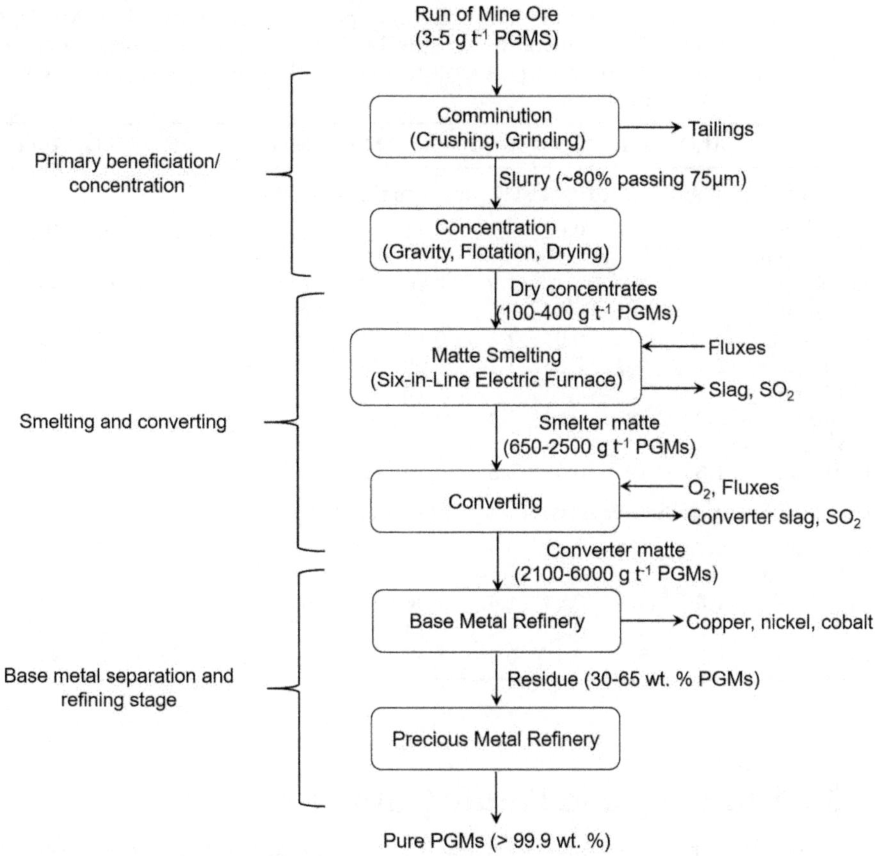

Figure 2.14 Process flowsheet for the primary production of platinum-group metals (PGMs) from sulfide ores. Reproduced from ref. 154, DOI: 10.5772/intechopen.73214, under the terms of the CC BY 3.0 license, https://creativecommons.org/licenses/by/3.0/.

Turbulent conditions in Peirce–Smith converters increase the mechanical entrainment of matte in the converter slag.[157,159] The proliferation of chromite spinels in the converter slag also impedes matte settling and further exacerbates PGM losses to the slag as matte inclusions entrained in silicate slag phases.[157,159] The high pO_2 in the converter also results in significant losses of Co, Ni, and Co as oxides dissolved in the slag. Invariably, the converter slag is recycled back into the primary smelting furnace or is reprocessed in slag-cleaning electric furnaces to recover the entrained and/or dissolved values before disposal in slag dumps.[102,149,153,157,158] However, the direct recycling of converter slags from the processing of UG2 blends is problematic due to the presence of elevated amounts of dissolved Cr_2O_3 (Table 2.14) as this would naturally result in the recycling and build-up of Cr in the process.[102,156,157]

Table 2.14 Chemical composition (in wt%) of platinum group metal (PGM) slags from various operations in the Bushveld Igneous Complex. Adapted from ref. 102 with permission from South African Journal of Science, Copyright 1999.

	Al_2O_3	CaO	Co	Cr_2O_3	Cu	MgO	FeO	Ni	S	SiO_2	Total
Composition of smelter furnace slag (unnormalized, wt%)											
Amplats: Waterval	3.3	6.4	0.05	0.8	0.11	31	15	0.19	0.50	46	103
Amplats (Union)	3	5.8	—	2.8	0.10	20	13	0.16	0.33	41	86
Impala	6	8	0.03	1.2	0.11	21	18	0.11	0.25	47	101
Lonmin: Merensky	2	9.8	0.05	1.2	0.09	28	19	0.15	—	44	104
Lonmin: UG2	3.9	13	0.02	2.4	0.13	9.2	22	0.11	—	47	98
Northam	1.5	10	0.03	0.8	0.10	21	20	0.2	—	44	98
Composition of converter furnace slag (unnormalized, wt%)											
Amplats: Waterval	0.7	0.4	0.45	0.4	1.17	1.1	63	2.25	2.4	27	99
Impala	1.8	0.3	0.43	1.4	1.06	0.71	64	1.90	1.0	27	100
Lonmin	0.7	0.5	0.39	1.4	0.94	0.78	65	1.43	1.7	28	100
Northam	1.3	0.7	0.4	0.36	1.37	0.82	64	2.18	—	27	98

2.5 Summary and Future Outlook

This chapter provides an overview of the metallurgical production of selected metals, mattes, and alloys from primary ores and/or their secondary resources. Metallurgical slags constitute a high-volume by-product and, depending on the process design and conditions, tend to contain appreciable amounts of entrained and/or dissolved metals, mattes, and alloys. If not properly disposed of, metallurgical slags can result in long-term environmental challenges. Increasing the valorization potential of metallurgical slags is thus beneficial in reducing the volumes of landfilled materials and, ultimately, mitigating the storage and toxicity concerns associated with their disposal.

The depletion of high-grade reserves inevitably means that exploring the sustainable production of metals from secondary resources and/or anthropogenic stocks has never been more important for a sustainability-driven economy.[1,160] In line with the circular economy and, more precisely, the principles of "ubuntu", the future of metal extraction processes is strongly dependent on the efficient recovery of metals from both primary and secondary sources.[161–163] These circular economy and ubuntu discourses are mutually inclusive – they emphasize the mutual coexistence between humanity and nature, decoupling the economy from the linear consumption of natural resources, and maintaining the value of resources and products in the economy for as long as is possible.[1,161–163]

Anthropogenic stocks such as electronic waste, auto catalysts, end-of-life batteries, and steel scrap possess both resources and pollution potential.[164,165] For example, approximately 44.7 million tonnes of e-waste, containing various amounts of precious and critical metals, was generated in 2016.[165] This means that the recovery of metals from e-waste may partially be able to fulfill the global demand for precious and critical metals. Ultimately, the future best available technologies will likely be able to process combined feeds consisting of both primary and secondary streams, such as postconsumer scrap, process residues, and reprocessed slags. Obviously, new systems thinking approaches to materials, processes, and equipment designs are required to shape innovations in primary metal production and leverage the circularity and recyclability of metals and alloys.[166,167]

Acknowledgements

This chapter was published with permission and support from Mintek. The authors also acknowledge Carlo Mapelli of the Polytechnic University of Milan for reviewing the chapter.

References

1. E. Matinde, G. Simate and S. Ndlovu, *J. South. Afr. Inst. Min. Metall.*, 2018, **118**, 825.
2. D. R. Gaskell, in *Methods for Phase Diagram Determination*, ed. J.-C. Zhao. Elsevier, Amsterdam, 2007, p. 442.
3. N. M. Piatak, M. B. Parsons and R. R. Seal II, *J. Appl. Geochem.*, 2015, **57**, 236.
4. S. Ndlovu, G. S. Simate and E. Matinde, *Waste Production and Utilization in the Metal Extraction Industry*, CRC Press, Boca Raton, 2017.
5. L. Holappa and Y. Xiao, *J. South. Afr. Inst. Min. Metall.*, 2004, **104**, 429.
6. M. Reuter, Y. Xiao and U. Boin, presented at VII International Conference on Molten Slags Fluxes and Salts, Johannesburg, South Africa, 2004, p. 349.
7. Y. Yang, K. Raipala and L. Holappa, in *Treatise on Process Metallurgy*, ed. S. Seetharaman, Elsevier, Amsterdam, 2014, p. 2.
8. A. K. Biswas, *Principles of Blast Furnace Ironmaking: Theory and Practice*, Cootha Publishing House, Brisbane, Australia, 1981.
9. M. Dıez, R. Alvarez and C. Barriocanal, *Int. J. Coal Geol.*, 2002, **50**, 389.
10. N. Andriopoulos, C. E. Loo, R. Dukino and S. J. McGuire, *ISIJ Int.*, 2003, **43**, 1528.
11. S. Kuang, Z. Li and A. Yu, *Steel Res. Int.*, 2018, **89**, 1700071.
12. S.-y. Kitamura, in *Treatise on Process Metallurgy*, ed. S. Seetharaman, Elsevier, Amsterdam, 2014, p. 177.
13. F. N. H. Schrama, E. M. Beunder, B. Van den Berg, Y. Yang and R. Boom, *Ironmak. Steelmak.*, 2017, **44**, 333.

14. F. N. Schrama, F. Ji, A. Hunt, E. M. Beunder, R. Woolf, A. Tuling, P. Warren, J. Sietsma, R. Boom and Y. Yang, *Ironmak. Steelmak.*, 2020, 1.
15. R. J. Fruehan, in *The Making, Shaping and Treating of Steel, 11th Edition, Steelmaking and Refining Volume*, ed. R. J. Fruehan, AISE Steel Foundation, Pittsburgh, 1998, p. 1.
16. J. Zhang, H. Matsuura and F. Tsukihashi, in *Treatise on Process Metallurgy*, ed. S. Seetharaman, Elsevier, Amsterdam, 2014, p. 1507.
17. M. Geerdes, R. Chaigneau and I. Kurunov, *Modern Blast Furnace Ironmaking: An Introduction*, IOS Press, Amsterdam, Netherlands, 2015.
18. G. C. Wang, *The Utilization of Slag in Civil Infrastructure Construction*, Woodhead Publishing, Duxford, CB22 4QH, UK, 2016.
19. P. Scott, S. Critchley and F. Wilkinson, *Mineral. Mag.*, 1986, **50**, 141.
20. H. Jalkanen and L. Holappa, presented at VII International Conference on Molten Slags Fluxes and Salts, Cape Town, 2004, p. 71.
21. H. Jalkanen and L. Holappa, in *Treatise on Process Metallurgy*, ed. S. Seetharaman, Elsevier, Amsterdam, 2014, p. 223.
22. L. Holappa, in *Treatise on Process Metallurgy*, ed. S. Seetharaman, Elsevier, Amsterdam, 2014, p. 301.
23. L. De Vos, I. Bellemans, C. Vercruyssen and K. Verbeken, *Metall. Mater. Trans. B*, 2019, **50**, 2647.
24. E. Turkdogan and R. Fruehan, in *The Making, Shaping and Treating of Steel, 11th Edition, Steelmaking and Refining Volume*, ed. R. J. Fruehan, AISE Steel Foundation, Pittsburgh, 1998, p. 121.
25. I. Z. Yildirim and M. Prezzi, *Adv. Civ. Eng. Mater.*, 2011, **2011**, 463638.
26. R. Hüsken, R. Fechner and J. Cappel, *Iron Steel Technol.*, 2011, **8**, 46.
27. C. Shi, *J. Mater. Civ. Eng.*, 2004, **16**, 230.
28. M. Kirschen, V. Risonarta and H. Pfeifer, *Energy*, 2009, **34**, 1065.
29. M. Kirschen, K. Badr and H. Pfeifer, *Energy*, 2011, **36**, 6146.
30. T. Engh, *Principles of Metal Refining*, Oxford University Press, Walton St, Oxford OX 2 6 DP, UK, 1992, vol. 473, p. 280.
31. J.-H. Wei and D.-P. Zhu, *Metall. Mater. Trans. B*, 2002, **33**, 111.
32. D. Swinbourne, T. Kho, B. Blanpain, S. Arnout and D. Langberg, *Miner. Process. Extr. Metall.*, 2012, **121**, 23.
33. R. Fruehan, presented at VII International Conference on Molten Slags Fluxes and Salts, The South African Institute of Mining and Metallurgy, Cape Town, South Africa, 2004, p. 263.
34. H. Matsuura and R. J. Fruehan, *ISIJ Int.*, 2009, **49**, 1530.
35. H. Shen, E. Forssberg and U. Nordström, *Resour., Conserv. Recycl.*, 2004, **40**, 245.
36. S. R. Rao, *Resource Recovery and Recycling from Metallurgical Wastes*, Elsevier, Amsterdam, 2006, p. 269.
37. F. A. Lopez, A. Lopez-Delgado, and N. Balcazar, Afinidad LIII, Barcelona, Spain, 1996, p. 39.
38. S. Fathy, G. Liping, M. Rui, G. Chuping and S. Wei, presented at Proceedings of Academics World International Conference, Cairo, Egypt, 2016, p. 8.
39. A. Rađenović, J. Malina and T. Sofilić, *Adv. Mater. Sci. Eng.*, 2013, 1.

40. J. Waligora, D. Bulteel, P. Degrugilliers, D. Damidot, J. Potdevin and M. Measson, *Mater. Charact.*, 2010, **61**, 39.
41. M. I. Gasik, in *Handbook of Ferroalloys – Theory and Technology*, Butterworth-Heinemann, Oxford, United Kingdom, 2013, p. 3.
42. L. Holappa and S. Louhenkilpi, presented at XIII International Ferro-alloys Congress, Almaty, Kazakhstan, 2013, p. 1083.
43. M. I. Gasik, *Handbook of Ferroalloys: Theory and Technology*, Butterworth-Heinemann, Oxford, United Kingdom, 2013.
44. R. F. Schulte and C. A. Tuck, *Minerals Yearbook – Ferroalloys*, United States Geological Survey, 2016.
45. H. Bessemer, in *Sir Henry Bessemer, F.R.S. An Autobiography*, Offices of Engineering, London, UK, 1905, p. 256.
46. S. E. Olsen, M. Tangstad and T. Lindstad, *Production of Manganese Ferroalloys*, Tapir Academic Press, Trondheim, Norway, 2007.
47. J. H. Downing, *Encyclopedia Britannica*, 2012, vol. 2021.
48. L. Holappa, in *Handbook of Ferroalloys – Theory and Technology*, Butterworth-Heinemann, Oxford, United Kingdom, 2013, p. 9.
49. ASTM A99–03: ASTM International, West Conshohocken, PA, USA, 2009.
50. M. Tangstad, in *Handbook of Ferroalloys – Theory and Technology*, Butterworth-Heinemann, Oxford, United Kingdom, 2013, p. 221.
51. J. Madias, *The Iron & Steel Technology Conference and Exposition*, Indianapolis, USA, 2011, p. 401.
52. T. Morimoto, M. Yoshimoto, H. Takahashi, T. Hamada, N. Fukushima and S. Sakurai, *Construction of Shaft Type Ferromanganese Smelting Furnace*, Kawasaki Steel Technical Report 17, JFE Steel Corporation, 1987.
53. Y. N. Kozhemyachenko, S. I. Pervushin, V. I. Varava, A. I. Strelets, G. G. Vasyura, V. P. Osadchii, K. A. Nikolaev, Y. V. Anishchenko and N. S. Shubravyi, *Metallurgist*, 1987, **31**, 354.
54. J. D. Steenkamp, C. J. Hockaday, J. P. Gous and T. W. Nzima, *JOM*, 2017, **69**, 1712.
55. M. Tangstad, J. P. Beukes, J. D. Steenkamp and E. Ringdalen, in *New Trends in Coal Conversion – Combustion, Gasification, Emissions, and Coking*, ed. I. Suarez-Ruiz, M. A. Diez and F. Rubiera, Woodhead Publishing, Elsevier, Duxford, United Kingdom, 2019, p. 405.
56. J. D. Steenkamp, P. Maphutha, O. Makwarela, W. K. Banda, I. Thobadi, M. Sitefane, J. P. Gous and J. J. Sutherland, *J. South. Afr. Inst. Min. Metall.*, 2018, **188**, 309.
57. D. Chetty and J. Gutzmer, presented at XV International Ferro-alloys Congress, Cape Town, South Africa, 2018, p. 1.
58. J. D. Steenkamp, presented at 11th International Symposium on High-temperature Metallurgical Processing, San Diego, California, 2020, p. 303.
59. I. L. Kolesnik, A. L. Berin, V. F. Beloshapka, P. K. Lebed and V. I. Timoshenko, *Metallurgist*, 1971, **15**, 498.
60. J. Basson, T. R. Curr and W. A. Gericke, presented at XI International Ferro-alloys Congress, New Delhi, India, 2007, p. 3.

61. I. M. Mishchenko, V. I. Varava, N. M. Bondarenko, V. I. Malkin, A. Z. Krizhevskii and V. P. Osadchii, *Metallurgist*, 2000, **44**, 522.
62. M. O. Sem, *J. Electrochem. Soc.*, 1954, **101**, 487.
63. J. D. Steenkamp and J. Basson, *J. South. Afr. Inst. Min. Metall.*, 2013, **113**, 667.
64. J. D. Steenkamp, W. G. Bam and S. P. Tsebe, *JOM*, 2018, 2422.
65. J. Nell and I. Nolet, presented at XII International Ferro-Alloys Congress, Helsinki, Finland, 2010, p. 579.
66. N. A. Barcza and D. P. O'Shaughnessy, *Can. Metall. Q.*, 1981, **20**, 285.
67. A. Kitera, S. Sakurai, K. Saito and S. Suzuki, presented at V International Ferro-Alloys Congress, New Orleans, USA, 1989, p. 177.
68. R. Burger and M. Masukawa, presented at X International Ferro-Alloys Congress, Cape Town, South Africa, 2004, p. 254.
69. S. Onouye and K. Saitoh, presented at III International Ferro-Alloys Congress, Tokyo, Japan, 1983, p. 194.
70. G. N. Mul'ko, A. A. Bondar, V. A. Zaitsev, E. A. Nitskii and E. G. Cherkasov, *Metallurgist*, 2000, **44**, 51.
71. A. N. Red'ko, *Metallurgist*, 1958, **2**, 454.
72. X. Xiaoxing and L. Daolin, presented at VIII International Ferro-Alloys Congress, Beijing, China, 1998, p. 226.
73. T. Tomioka, M. Misawa and T. Hanano, presented at I International Ferro-Alloys Congress, Johannesburg, 1974, p. 129.
74. G. Tranell, S. Gaal, D. Lu, M. Tangstad and J. Safarian, presented at XI International Ferro-Alloys Congress, New Delhi, India, 2007, p. 231.
75. M. Yastreboff and S. Ganguly, presented at VIII International Ferro-Alloys Congress, Beijing, China, 1998, p. 263.
76. D. M. Kazadi, D. R. Groot, J. P. R. De Villiers, J. D. Steenkamp and H. Pöllmann, *Hydrometallurgy*, 2016, **166**, 214.
77. ASTM A483, *Standard Specification for Silicomanganese*, ASTM International, West Conshohocken, PA, 2003, 2010.
78. N. A. Sithole, N. Rambuda, J. D. Steenkamp, D. A. Hayman and C. J. Hockaday, *J. South. Afr. Inst. Min. Metall.*, 2018, **118**, 1205.
79. E. Ringdalen and M. Ksiazek, *Furnace Tapping 2018*, Kruger National Park, South Africa, 2018, p. 167.
80. M. Tangstad, in *Handbook of Ferroalloys – Theory and Technology*, Butterworth-Heinemann, Oxford, United Kingdom, 2013, p. 179.
81. A. Schei, J. K. Tuset and H. Tveit, *Production of High Silicon Alloys*, 1st edn, Tapir Forlag, Trondheim, Norway, 1998.
82. M. Ksiazek, M. Tangstad and E. Ringdalen, *Silicon for the Chemical and Solar Industry XIII*, Kristiansand, Norway, 2016, p. 33.
83. ISO 5448: International Organization for Standardization, Geneva, Switzerland, 1981.
84. M. I. Gasik, in *Handbook of Ferroalloys – Theory and Technology*, Butterworth-Heinemann, Oxford, United Kingdom, 2013, p. 268.
85. J. Basson and J. Daavittila, in *Handbook of Ferroalloys - Theory and Technology*, Butterworth-Heinemann, Oxford, United Kingdom, 2013, p. 317.

86. M. Ford and J. F. Oosthuizen, *VII International Ferro-Alloys Congress*, Trondheim, Norway, 1995, p. 263.
87. I. J. Geldenhuys, *XIII International Ferro-Alloys Congress*, Almaty, Kazakhstan, 2013, p. 31.
88. P. C. Hayes, *X International Ferro-Alloys Congress*, Cape Town, South Africa, 2004, p. 1.
89. I. Mc Dougall, in *Handbook of Ferroalloys – Theory and Technology*, Butterworth-Heinemann, Oxford, United Kingdom, 2013, p. 83.
90. P. Niemela and M. Kauppi, *XI International Ferro-Alloys Congress*, New Delhi, India, 2007, p. 171.
91. J. D. Steenkamp, *J. South. Afr. Inst. Min. Metall.*, 2019, **119**, 537.
92. R. Kopong, *Dissertation for MSc in Engineering*, University of the Witwatersrand, Johannesburg, 2015.
93. E. Ringdalen, Doctoral thesis, Norwegian University of Science and Technology, Trondheim, 1999.
94. E. Ringdalen, M. Rocha, R. Neto and T. Malvik, presented at XIII International Ferro-Alloys Congress, Almaty, Kazakhstan, 2013, p. 115.
95. C. N. Harman and N. S. S. Rama Rao, presented at XIII International Ferro-Alloys Congress, Almaty, Kazakhstan, 2013, p. 103.
96. M. A. Nkohla, *Masters Dissertation*, Cape Peninsula University of Technology, 2006.
97. H. Oterdoom and J. H. Zietsman, presented at XI International Heavy Minerals Conference, Cape Town, South Africa, 2019, p. 1.
98. R. N. Statnykh, Y. I. Ostrovskii, A. A. Tsernikel and V. I. Zhuchkov, *Steel Transl.*, 2013, **43**, 309.
99. O. Polyakov, in *Handbook of Ferroalloys – Theory and Technology*, Butterworth-Heinemann, Oxford, United Kingdom, 2013, p. 367.
100. A. E. M. Warner, C. M. Díaz, A. D. Dalvi, P. J. Mackey, A. V. Tarasov and R. T. Jones, *JOM*, 2007, **59**, 58.
101. A. G. Matyas, R. C. Francki, K. M. Donaldson and B. Wasmund, *CIM Bull.*, 1993, **86**, 92.
102. R. Jones, *S. Afr. J. Sci.*, 1999, **95**, 525.
103. A. E. M. Warner, C. M. Díaz, A. D. Dalvi, P. J. Mackey and A. V. Tarasov, *JOM*, 2006, **58**, 11.
104. P. J. Mackey and E. Jak, presented at Proceedings of the First Global Conference on Extractive Metallurgy, Ottawa, Ontario, Canada, 2018, p. 3.
105. R. J. Hundermark and L. R. Nelson, *JOM*, 2017, **69**, 335.
106. R. T. Jones, presented at In Celebrating the Megascale: Proceedings of the Extraction and Processing Division Symposium on Pyrometallurgy in Honor of David G. C. Robertson, San Diego, California, 2014, p. 129.
107. I. J. Reinecke and H. Lagendijk, presented at XI International Ferro-Alloys Congress, New Delhi, India, 2007, p. 781.
108. C. P. Naudé and M. D. Shapiro, presented at XII International Ferro-Alloys Congress, Helsinki, Finland, 2010, p. 839.
109. IGMNiR, *Non-ferrous Metals Industry*, http://www.igmnir.pl/en/non-ferrous-metals-industry, last accessed 12 February 2021.

110. A. Boulamanti and J. A. Moya, *Resour. Policy*, 2016, **49**, 112.
111. J. Rubinstein and L. Barsky, *Non-Ferrous Metal Ores: Deposits, Minerals and Plants*, Taylor & Francis Group, New York, 2002, p. 1.
112. I.-K. Suh, Y. Waseda and A. Yazawa, *High Temp. Mater. Processes*, 1988, **8**, 65.
113. RSC, *Periodic Table-Aluminium*, https://www.rsc.org/periodic-table/element/13/aluminium, last accessed 12 February 2021.
114. A. T. Tabereaux and R. D. Peterson, in *Treatise on Process Metallurgy*, ed. S. Seetharaman, Elsevier, Amsterdam, Netherlands, 2014, p. 839.
115. W. B. Frank, W. E. Haupin, H. Vogt, M. Bruno, J. Thonstad, R. K. Dawless, H. Kvande and O. A. Taiwo, *Ullmann's Encyclopedia of Industrial Chemistry*, 2000.
116. Y. Xiao, M. A. Reuter and U. Boin, *J. Environ. Sci. Health, Part A: Toxic/Hazard. Subst. Environ. Eng.*, 2005, **40**, 1861.
117. U. Boin and M. Bertram, *JOM*, 2005, **57**, 26.
118. ILZSG, *Lead and Zinc Statistics*, https://www.ilzsg.org/static/statistics.aspx?from=1, last accessed 12 February 2021.
119. RSC, Periodic Table-Lead, https://www.rsc.org/periodic-table/element/82/lead, Last accessed 12 February 2021.
120. Government of South Australia, *Lead–Zinc*, https://energymining.sa.gov.au/minerals/mineral_commodities/lead_-_zinc, last accessed 12 February 2021.
121. RSC, Periodic Table-Zinc, https://www.rsc.org/periodic-table/element/30/zinc, Last accessed 12 February 2021.
122. D. Gregurek, Z. Peng and C. Wenzl, *JOM*, 2015, **67**, 1986.
123. S. Moradi and A. Monhemius, *Miner. Eng.*, 2011, **24**, 1062.
124. H. Sohn and M. Olivas-Martinez, in *Treatise on Process Metallurgy*, ed. S. Seetharaman, Elsevier, Amsterdam, Netherlands, 2014, p. 671.
125. G. M. Mudd, S. M. Jowitt and T. T. Werner, *Ore Geol. Rev.*, 2017, **80**, 1160.
126. A. Yazawa, *Metall. Trans. B*, 1979, **10**, 307.
127. H. Sohn, in *Treatise on Process Metallurgy*, ed. S. Seetharaman, Elsevier, Amsterdam, Netherlands, 2014, p. 533.
128. J. Hoang, M. Reuter, R. Matusewicz, S. Hughes and N. Piret, *Miner. Eng.*, 2009, **22**, 742.
129. K. Sahu and A. Agrawal, *Proceedings of Extraction of Nonferrous Metals and Their Recycling-A Training Programme*, Jamshedpur, India, 2008.
130. B. Zhao, in *Sintering Applications*, ed. B. Ertug, InTech, Rijeka, 2013, p. 165.
131. T. W. Ellis and A. H. Mirza, *J. Power Sources*, 2010, **195**, 4525.
132. F. Habashi, *Metall-Forschung*, 2012, **66**, 149.
133. A. Richards, *Can. Metall. Q.*, 1981, **20**, 145.
134. V. Ettler, Z. Johan, B. Kříbek, O. Šebek and M. Mihaljevič, *Appl. Geochem.*, 2009, **24**, 1.
135. V. Ettler, Z. Johan, B. Kříbek and H. Nolte, *Communs Geol. Surv. Namibia*, 2009, **14**, 3.
136. A. T. Awoniran, A. Ketheson, S. Piazolo and D. B. Gore, *Powder Diffr.*, 2017, **32**, S70.

137. P. B. Queneau, R. Leiby and R. Robinson, *World Metall.--Erzmet.*, 2015, **68**, 149.
138. B. Blanpain, M. Reuter and A. Malfliet, *Socrates Policy Brief*, 2019, p. 1.
139. G. Denton, N. Barcza, P. Scott and T. Fulton, presented at John, Floyd International Symposium on Sustainable Developments in Metals Processing, Melbourne, Australia, 2005, p. 273.
140. M. E. Schlesinger, K. C. Sole and W. G. Davenport, *Extractive Metallurgy of Copper*, Elsevier, Amsterdam, Netherlands, 2011.
141. P. Lusty and S. Hannis, *Commodity Profile: Copper*, British Geological Survey, Nottingham, UK, 2009.
142. P. Toscano and T. Utigard, *Metall. Mater. Trans. B*, 2003, **34**, 121.
143. R. Moskalyk and A. Alfantazi, *Miner. Eng.*, 2003, **16**, 893.
144. P. Voigt, A. Burrows, M. Somerville and C. Chen, presented at 8th International Symposium on High-Temperature Metallurgical Processing, 2017, p. 261.
145. A. Warner, C. Diaz, A. Dalvi, P. Mackey, A. Tarasov and R. Jones, *JOM*, 2007, **59**, 58.
146. E. Jak, B. Zhao and P. Hayes, *Miner. Proc. Extract. Metall. (Trans. Inst. Min. Metall. C)*, 2008, **117**, 147.
147. M. Henckens and E. Worrell, *J. Cleaner Prod.*, 2020, 121460.
148. M. S. Moats and W. G. Davenport, in *Treatise on Process Metallurgy*, ed. S. Seetharaman, Elsevier, Amsterdam, Netherlands, 2014, p. 625.
149. M. Nagamori, *Metall. Trans.*, 1974, **5**, 539.
150. A. Gunn and A. Benham, *Platinum*, British Geological Survey, Nottingham, UK, 2009.
151. G. Gunn, in *Critical Metals Handbook*, ed. G. Gunn, British Geological Survey, Nottingham, UK, 2014, p. 284.
152. K. Nose and T. H. Okabe, in *Treatise on Process Metallurgy*, ed. S. Seetharaman, Elsevier, Amsterdam, Netherlands, 2014, p. 1071.
153. R. Jones, in *Nickel and Cobalt 2005: Challenges in Extraction and Production*, ed. H. Schonewille, and J. Donald, Canadian Institute of Mining, Metallurgy and Petroleum, Calgary, Alberta, Canada, 2005, *44th Annual Conference of Metallurgists*, p. 147.
154. B. M. Thethwayo, in *Noble and Precious Metals: Properties, Nanoscale Effects and Applications*, ed. M. Seehra and A. Bristow, IntechOpen, London, UK, 2018, p. 393.
155. IPA, *PGM Facts*, https://ipa-news.de/index/platinum-group-metals/, last accessed 12 February 2021.
156. L. A. Cramer, *JOM*, 2001, **53**, 14.
157. J. Nell, *J. South. Afr. Inst. Min. Metall.*, 2004, **104**, 423.
158. R. T. Jones, *JOM*, 2004, **56**, 59.
159. L. Andrews and P. Den Hoed, *J. South. Afr. Inst. Min. Metall.*, 2011, **111**, 675.
160. F. Ongondo, I. Williams and G. Whitlock, *Waste Manage.*, 2015, **45**, 4.
161. M. Geissdoerfer, P. Savaget, N. M. Bocken and E. J. Hultink, *J. Cleaner Prod.*, 2017, **143**, 757.

162. V. Elia, M. G. Gnoni and F. Tornese, *J. Cleaner Prod.*, 2017, **142**, 2741.
163. M. F. Murove, *Mank. Q.*, 2004, **45**, 195.
164. Y. Sugimura and S. Murakami, *Resour., Conserv. Recycl.*, 2016, **112**, 93.
165. A. Borthakur, M. Govind and P. Singh, in *Electronic Waste Management and Treatment Technology*, ed. M. N. V. Prasad and M. Vithanage, Elsevier, 2019, p. 35.
166. B. K. Reck and T. E. Graedel, *Science*, 2012, **337**, 690.
167. M. A. Reuter, A. van Schaik, J. Gutzmer, N. Bartie and A. Abadías-Llamas, *Annu. Rev. Mater. Res.*, 2019, **49**, 253.

CHAPTER 3

Geochemistry and Mineralogy of Slags

NADINE M. PIATAK*[a], VOJTĚCH ETTLER[b] AND DARRYL HOPPE[a]

[a]US Geological Survey, 954 National Center, Reston, Virginia, 20192, USA;
[b]Institute of Geochemistry, Mineralogy and Mineral Resources, Faculty of Science, Charles University, Albertov 6, 128 00 Prague 2, Czech Republic
*E-mail: npiatak@usgs.gov

3.1 Introduction

Slags are a by-product in extractive pyrometallurgical operations that have been considered as an environmental contaminant, an inert material, and a valuable resource. Historically, slags were considered inert and were used in the construction of buildings and roads without consideration of their potential deleterious effects. Ferrous slags remain a viable raw material for use in construction and cement and have more recently found use in some environmental applications. However, the leaching of potentially toxic elements from legacy stockpiles of slags, most from base metal production but also from steel and ferroalloy production, have brought slags into focus as a potential ecological risk. The high metal content of some slags has also been examined for secondary metal recovery; metals of interest include Fe and commonly extracted base metals, as well as by-products and critical elements used in new and green technologies. The environmental toxicity or resource potential is directly related to the physical, chemical, and mineralogical characteristics of the slag. These characteristics depend on the natural or

Chemistry in the Environment Series No. 2
Metallurgical Slags: Environmental Geochemistry and Resource Potential
Edited by Nadine M. Piatak and Vojtěch Ettler

Published by the Royal Society of Chemistry, www.rsc.org

anthropogenic materials being pyrometallurgically processed, the targeted metal or compound for extraction, the processing steps, fluxes added to the furnace, furnace type and conditions, and cooling rates. The combination of these variables results in diversity among and within the different slag types that are examined in this chapter.

Slags, in general, can be divided into three main types: ferrous slags, ferroalloy slags, and non-ferrous slags. Analytical techniques to determine mineralogy and physical and chemical characteristics are utilized irrelevant of slag type and were generally developed to analyze natural rocks. Therefore, Section 3.2 details sampling strategies, bulk analytical techniques for both chemical and mineralogical characterization, and microanalytical techniques used to determine the chemistry, structure, and elemental speciation of phases present in slags. Chemistry and mineralogy are generally unique to each slag type and therefore Section 3.3 on elemental composition and Section 3.4 on mineralogy are subdivided into ferrous, ferroalloy, and non-ferrous categories. In this chapter, ferrous slags include blast furnace (BF) slags and steel slags. Ferroalloy slags are generated in the production of bulk ferroalloys, that is, when elements such as Cr, Mn, Ni, and Si are combined with Fe. Non-ferrous slags discussed include those generated from the production of Cu, Ni, Pb, and Zn, with less of a focus on platinum-group metals (PGMs), Al, and Sn. A brief discussion of slags resulting from recycling is also included. Content in this chapter has been adapted from previous review papers on slag by Piatak *et al.*[1] and Piatak,[2] and augmented with additional information from the literature from more recent reviews and case studies.

As previously stated, the characteristics of slags result from the smelting conditions and cooling rates. The metallurgical processes that generate slags were discussed in Chapter 2, providing the background formation conditions that result in the primary mineralogy and chemistry presented in this chapter. The following chapter, Chapter 4, on weathering details the processes that affect slags with time in the environment, including a discussion of secondary phases. To note, although crystalline phases found in slag are anthropogenic and not technically minerals, mineralogical terminology is used in this manuscript for the sake of clarity. This chapter provides the basis for subsequent chapters on the environmental stability and impacts of slags and their numerous valuable uses. In addition, the slag characterization provided herein may provide insight into smelting conditions at legacy sites. If a particular type of slag can be thoroughly characterized, unique characteristics identified, and elements of enrichment determined, this information may also be used to predict which legacy sites pose higher environmental risks or are of value to warrant reprocessing. For example, Riley *et al.*[3] integrated chemical knowledge of iron and steel slag in the United Kingdom with spatial information on stockpiles of slag to evaluate its resource potential. Environmental impacts from slag deposits can be predicted in the same way. Predictions can only be made if we understand the chemical and mineralogical characteristics of slags and their diversity.

3.2 Methods Used to Characterize Slag

Slags are chemically and mineralogically analogous to rocks and, therefore, the methods used to characterize both slags and rocks are generally the same. The sampling methods and bulk analytical and microanalytical techniques discussed in this section were mostly developed to investigate Earth science materials of various matrices or to characterize synthetic compounds. These methods are well suited to characterizing slag as well as other types of mine waste or processing by-products that are, like rocks, made up of an inorganic solid mass consisting of one or more mineralogical or glass phase(s).

3.2.1 Sampling

Modern smelters can generate hundreds of tonnes of slag per day. At modern and legacy disposal sites, slag piles can range from meters to kilometers in length and from meters to hundreds of meters in depth. Of the hundreds of slag studies that reported chemistry and mineralogy, the vast majority are based on a limited number of samples, from as little as a few to, usually at the most, dozens of samples. In most cases, the samples are typically collected from the surface of the slag pile.[1] For modern smelter studies, slag samples commonly represent a "snapshot" in time, during which a particular ore was processed with a certain combination of fluxes optimized for that feed chemistry. How can we accurately summarize and discuss the mineralogy and chemistry of the various types of slag knowing that individual samples in case studies likely do not accurately represent the overall composition of a slag pile? The analytical error associated with the methods used to analyze a slag sample is routinely reported; yet, the error associated with sampling is difficult to quantify and rarely discussed. One value of review papers, such as those by Piatak *et al.*,[1] Potysz *et al.*,[4] Naidu *et al.*,[5] Pan *et al.*,[6] and Yildirim and Prezzi,[7] is that chemical compositions and mineralogy from multiple studies are compiled, more fully illustrating slag variability that may be missed from examining a few samples from one site.

The goals of the study should first be identified when defining a successful sampling plan. Moreover, the type of samples to collect, the analytical methods to be used to characterize the slag, the minimum amount of sample needed for each analysis, and how many samples to analyze should all be considered. A sampling plan may involve more than one sampling strategy. For example, a study aimed at understanding the mobility of trace elements from slag dumps might focus on collecting grab samples in weathered areas where secondary phases occur, whereas, a study to determine the potential for a slag dump to be used or reprocessed may require systematic composite sampling that will accurately represent the pile. A composite sampling plan should consider if the target population is homogeneous or heterogeneous;[8] for studies involving slag dumps, the latter is nearly always the case. Studies on slags have employed all of the following sampling strategies: unbiased random sampling (which is not the same as haphazard sampling or sampling without a plan), selective or judgmental sampling using knowledge of a slag dump to help guide selection of samples, systematic sampling or sampling at regular intervals, convenience sampling when sampling error and variance are not taken into consideration, or a combination of any of these together.[8] The types of samples collected are most frequently grab samples providing a "snapshot" of the target population and composite samples where information on the individual samples is lost but the average composition is of interest. Both types of sample collection methods have been used in studies involving slags but grab sampling is most often employed. Owing to the generally stable nature of slags over the timeframe of sample collection to analysis, sample storage and preservation methods usually only involve placing slag samples in air-tight containers to avoid cross-contamination among samples and avoid fluctuations in humidity that may alter or enhance weathering and weathering products.

3.2.2 Bulk Analytical Techniques

The bulk chemical and mineralogical compositions are typically determined as the first step in slag characterization and are imperative for understanding the metallurgical origin of slags, their behavior in the environment, and their potential for reuse. Bulk analytical techniques inherently use the entire sample split and therefore are not beam-size or position-sensitive like microanalytical techniques (Section 3.2.3). For most bulk analytical techniques, the slag samples or composite samples are divided into subsamples for individual analyses. The hard and dense (and commonly massive, if air-cooled) nature of slags requires crushing, pulverizing, and/or grinding to reduce the particle size before analysis. Once the slags are reduced to a uniform size, the sample is mixed thoroughly to ensure homogeneity and then split into representative subsamples, preferably using a riffle splitter or the coning and quartering method. The subsample may then be additionally crushed, pulverized, fluxed, and/or digested (total or partial), depending on the analytical methods employed.

3.2.2.1 Chemistry

For bulk chemical analyses, elemental concentrations are typically reported. Sample preparation depends on the method being used and the matrix composition of the sample. The most regularly used techniques to determine the chemical composition of slag are described in more detail in this section with information sourced from Tessadri,[9] Harvey,[8] and Helaluddin *et al.*,[10] who are referenced, where applicable, at the end of each paragraph.

Some bulk chemistry techniques analyze a solid form; others require the solid sample to be converted to a liquid before analysis. There are various digestion methods depending on the sample matrix. A fine-grained representative split of a sample may be combined with digestion solutions, commonly composed of acids, and reacted at room or elevated temperatures.

To aid in decomposition, microwave (or infrared) energy is sometimes employed to enhance the thermodynamics or kinetics of digestion. Some elements are volatile upon dissolution, especially at high temperatures, and may be lost during digestion. *Aqua regia*, a mixture of concentrated nitric acid and hydrochloric acid in a 1 : 3 volumetric ratio, is a solution frequently used for digestion of organic matter, oxidation of sulfides, and decomposition of some oxides. However, these acids are not effective at breaking down silicates and some oxides, both of which are abundant in slags. Total or near-total digestion using hydrofluoric and perchloric acids, typically also combined with nitric and hydrochloric acids, known as the four-acid digestion, breaks down many minerals, including silicates. However, some phases, such as spinel-group phases (*e.g.*, chromite ($FeCr_2O_4$) and gahnite ($ZnAl_2O_4$)), which are commonly found in slags, may be only partially dissolved in acid mixtures. Residuals remaining after digestion indicate partial dissolution and should be examined to determine refractory phases and report analytes as partial digestions. One method commonly utilized to overcome partial dissolution by acids is to create a fused material by melting the fine-grained homogenous sample with a flux such as sodium peroxide or lithium metaborate, followed by rapid cooling. Fusion can be performed at low to moderate temperatures (~500 °C), avoiding losses of volatile elements. Fusions are also generally effective at solubilizing refractory minerals. The fused cake is then dissolved by dilute acids, such as nitric acid, before being introduced into an analyzer.

Inductively coupled plasma mass spectrometry (ICP-MS) and inductively coupled plasma atomic (optical) emission spectrometry (ICP-AES or ICP-OES) involve decomposed samples in a liquid state that are vaporized using an argon plasma source, during which atoms are converted to ions in the excited state. For ICP-MS, the ions then flow into a mass spectrometer, where they are isolated by atomic mass-to-charge ratios using a quadrupole or magnetic sector analyzer. A detector then measures the metal ions from the mass spectrometer and calculates abundance based on signal intensity. In contrast, ICP-AES is based on measuring light emitted from excited atoms and ions instead of a mass measurement. During AES, the excited ions emit electromagnetic radiation in the ultraviolet to infrared region when they return to the ground state. A spectrometer is used to grate and detect the light emitted, the wavelength of which is characteristic of an electronic configuration or element and the intensity of which is proportional to its abundance. ICP-MS and ICP-AES are capable of measuring 70 elements simultaneously with atomic masses ranging from Li to U, and sometimes heavier (excluding a few due to interference); ICP-MS has lower levels of detection than ICP-AES.[9,10]

X-ray fluorescence (XRF) is a method in which a solid sample does not need to be decomposed for analysis, overcoming the issue of incomplete dissolution of refractory minerals, and it also belongs among non-destructive techniques. Solid samples can be run as is or, to create a homogeneous sample, they can be powdered and prepared either as a pressed pellet or fused with a flux such as lithium metaborate/lithium tetraborate to create a glass disk. The sample is then irradiated with X-rays, which dislodge inner orbital electrons in the atoms. As electrons from an outer shell replace these displaced electrons, energy in the form of secondary X-rays or fluorescence is emitted, which is characteristic of the configuration of a specific atom with an intensity relative to its abundance. Two main types of detectors are used with XRF analysis: wavelength detection systems (WDXRF) and energy dispersive systems (EDXRF). For WDXRF, crystal monochromators are used to separate the characteristic wavelengths emitted from a sample before hitting the detector. In contrast, fluorescent photons are detected without being separated with solid-state detectors or semiconductor crystals in EDXRF systems. WDXRF is more sensitive, provides better energy resolution, and can determine a wider range of elements, and is therefore the more frequently used method of the two for determining bulk chemical compositions.[9,10]

Loss on ignition (LOI) is typically reported along with major element oxide percentages determined using WDXRF. The sum of major elements (in oxide form if appropriate) from WDXRF and LOI is expected to be 100 wt% for most geologic samples. For LOI calculations, the mass loss is measured upon heating the sample to temperatures that evaporate water (105 °C) and break down organics (550 °C) and carbonates (1000 °C); sulfides and ferrous iron also oxidize during heating. Most slags form at temperatures near or above the maximum temperature attained during the maximum-temperature LOI analysis. However, carbonate or hydrated phases may form during cooling, especially if enhanced techniques are used, or during weathering. For example, ferrous slags may contain significant amounts of carbonates, which form while the slag cools or during weathering. Secondary hydrated and volatile phases form on the surface and in cracks and pores where interaction with water and air is possible. During smelting, the organic matter used as a fuel source may not be completely combusted; the residual fragments may volatilize during LOI determination. Occasionally, a component of the sample combines with oxygen during heating so a gain in mass is observed after heating. This is referred to as gain on ignition.

For atomic absorption spectroscopy (AAS) analysis, a solid sample, a digested solid sample, or an aerosol generated from the sample by a nebulizer is converted to free atoms by different methods depending on the specific AAS technique. Atomizers include a flame (FAAS) or a graphite-coated furnace (GFAAS), both of which dissociate a sample into its atomic components to generate a cloud of atoms. In contrast to the multi-element analysis of ICP-MS and ICP-AES, AAS determines the concentrations of a particular element. However, both FAAS and GFAAS can be used to determine the concentrations of over 50 different elements. Some AAS methods of vaporization are element-specific, particularly for highly volatile elements. For example, Hg is usually determined using a cold vapor flameless method (CVAAS) whereas As, Se, Sb, and Te analyses use the hydride generation method (HGAAS). During AAS analysis, a beam of electromagnetic radiation irradiates the free atoms and when the energy is equivalent to the energy needed to excite an atom from its ground state to an excited state, the energy is absorbed. The radiation then passes through a monochromator to isolate the adsorption line from background light and is then measured by a detector. Elemental quantities can be determined based on measuring the absorbance at a wavelength that is specific to the ground state of the atom being measured. Basically, AAS uses the absorption of light to measure the concentration of gas-phase atoms.[8,9]

During instrumental neutron activation analysis (INAA), a sample is irradiated using neutrons in a nuclear reactor, which generates radioactive isotopes for targeted elements. Gamma rays are emitted as the radioactive isotopes decay with energy that is characteristic of a specific element with intensities proportional to the element abundance. The method is non-destructive; however, the sample may remain radioactive for months to years after analysis. Analyses performed at a nuclear reactor have the highest flux, however, accessibility may be limited. The method can detect over 70 elements.

Compared to most of the methods described so far in this section, where an instrument either simultaneously or consecutively quantifies the abundance of numerous elements, some methods are element-specific. Methods that are commonly applied to slag samples include determining total sulfur or total carbon using an infrared detector. The sample is combusted in an oxygen-rich environment to form sulfur dioxide and carbon dioxide. The resulting gases are purified and then quantified with a solid-state infrared detector. The gas evolved is normalized to the starting mass of the sample. Similarly, elemental analyzers are used to measure N and/or H, also after combustion.

The techniques discussed in the preceding paragraphs are those most frequently applied to studies on slags,[1] and determine the bulk elemental compositions. Only a few studies have applied synchrotron-based X-ray absorption spectroscopy (XAS) to investigate the speciation of elements in slags. XAS can be used to determine the location of atoms within matter, as well as their oxidation states and bond distances, and is generally divided into two energy regions, X-ray absorption near-edge structure (XANES) and extended X-ray absorption fine structure (EXAFS), each of which provide different information about an element and its local atomic environment. The reader is referred to Calvin[11] for the theory and a more detailed explanation of synchrotron-based techniques. Analyses can be conducted on bulk samples, as discussed here, or on a micrometer (μm)-scale, as discussed in Section 3.2.3. Some examples of the application of XAS to slag studies include bulk XANES analyses that provided insight into the speciation and potential toxicity of elements and the environmental and engineering stability of slags. Chaurand *et al.*[12] found less toxic, less mobile Cr^{3+} in basic oxygen furnace (BOF) slag that leached minimally compared to the significant leaching of V from the slag. The V was present as V^{4+} in the slag and oxidized to V^{5+}, the most toxic form, during leaching; these results provided a better understanding of the impact of the use of the material as aggregate in road construction.[12] Another study by Roy[13] utilized bulk XANES of S to identify the presence of mainly sulfide with minor amounts of sulfate in ground granulated BF slag compared to mostly sulfate present in air-cooled BF slag. Analysis of activated granulated BF slag containing mostly sulfate provided insight into the stability of this material to be used in cementitious materials.[13] Additionally, Ettler[14] discussed several studies where XAS was used to determine the solid speciation of contaminants in smelter-contaminated soils, some of which contained slag particles.

Another method that provides speciation information is Mössbauer absorption spectroscopy, which is based on the absorption and re-emission of gamma rays to provide information about an atom's local environment in a material. The technique is typically used to determine the oxidation state of Fe in crystalline phases. The presence of both Fe^{2+} and Fe^{3+} hosted in silicate/aluminosilicate structures and in the magnetite structure in Zn–Pb slag from Poland was confirmed using Mössbauer absorption spectroscopy; this information along with the presence of sulfides and intermetallic compounds provided insight into the smelting conditions.[15]

3.2.2.2 *Mineralogy*

Powder X-ray diffraction (XRD), a bulk technique, is by the far the most commonly used method in slag studies to determine bulk mineralogy and gain insight into the phase compositions. This non-destructive method is used for identification and quantification of mineral phases, as well as unit cell and crystal structure information. For powder XRD, samples should be fine-grained particles in random orientations so diffraction occurs from all lattice plane families simultaneously.[16] The detection limit is typically in the order of one or more weight percent. Incident beams of monochromatic X-rays, typically from a Cu, Co, or Mo radiation source, are scattered when they interact with a crystalline structure. Some of the scattered X-rays undergo constructive interference or diffraction and allow for the identification of the structure of a mineral based on Bragg's Law.[17] Abundance estimates for minerals in a mixture are determined based on the intensity of the diffraction at diagnostic lattice spacings (d-spacings).[17] Crystalline mineral phases as well as amorphous phases, including glass, can be quantified using techniques such as Rietveld analysis and diffraction profile fitting.

Glass is generally ubiquitous in slags, especially in those that are quickly cooled. Figure 3.1A illustrates the differences between XRD signatures for a slowly cooled, mostly crystalline slag and a quickly cooled glassy slag. Determining the presence of glass in slag and challenges associated with its quantification are discussed in work by Piatak *et al.*[1] and Piatak and Seal.[18] For example, the composition of the glass affects the diffractograms, as shown in Figure 3.1B, complicating quantification. In general, the accuracy of abundance estimates for crystalline and glass phases in slag based on XRD or other methods should be assessed using synthetic mixtures that are similar in composition to the slag phases.[1]

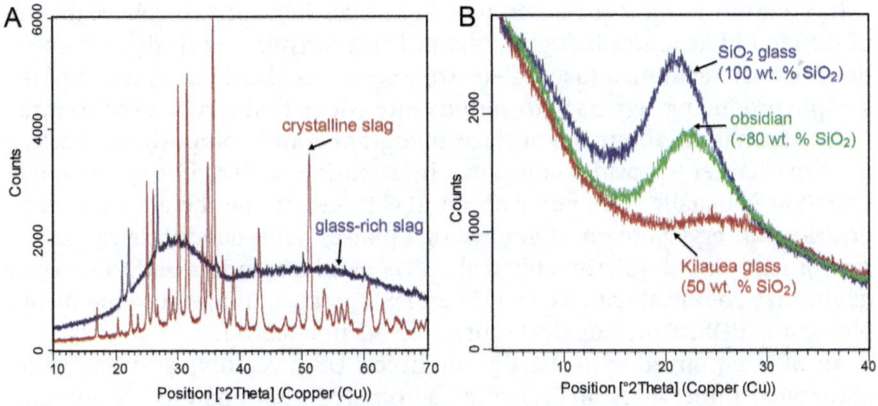

Figure 3.1 X-ray diffraction patterns of (A) slow cooled crystalline Pb slag (red trace: fayalite, spinel, galena, quartz, litharge, glass) and quenched glass-rich Pb slag (blue trace: glass, quartz, and cristobalite) (samples discussed in Ettler *et al.*[19]) and (B) glass of varying SiO_2 compositions. Reproduced from ref. 1 with permission from Elsevier, Copyright 2014.

3.2.3 Microanalytical Techniques

Microanalytical techniques are used to determine the major, minor, and trace element chemistry of phases within a crystalline slag or of the amorphous phase within a partially crystalline or glassy slag. In addition, these techniques are used to examine the distribution or speciation of elements within individual phases or glass. Microanalytical techniques are position-sensitive in that the beam is focused on a particular area in the sample targeting a particular location in a mineral. Some methods, such as electron probe microanalysis (EPMA), require a polished surface; the sample is commonly embedded in epoxy or prepared as thin or thick sections before polishing a flat surface for analysis. Typically, transmitted and/or reflected-light microscopy of the polished surface is used to identify phases, determine textural relationships among phases, and locate areas of interest for microanalytical techniques such as EPMA. Other microanalytical methods, such as scanning electron microscopy (SEM), have more flexibility with required sample preparation and can accommodate millimeter-sized pieces of slag or micrometer-sized grains. Ideally, for energy dispersive X-ray spectrometer (EDS) analysis with an SEM, a polished surface is utilized. Due to the excitation volume or nature of the method, some techniques require a sample thickness greater than the standard 30 μm thickness of a thin section. For example, depending on the instrument conditions, laser ablation inductively coupled plasma mass spectrometry (LA-ICP-MS) may ablate through this standard thickness.

In addition to optical microscopy, SEM is widely used to collect images of mineral/phase morphologies, textural relationships, and their compositions. Under vacuum, a focused electron beam scans over the surface of the sample producing signals with various intensities that can be used to interpret information about the surface topography and composition. Section 6.2.1 in Chapter 6 provides additional information on SEM image collection. The SEM is usually equipped with an EDS to determine the chemical compositions of crystalline and amorphous phases semi-quantitatively on the micrometer or sub-micrometer scale; EDS can be standardized to produce qualitative chemical data as well. The EDS operates under the same principles as the EDXRF previously discussed in Section 3.2.2.1.

An SEM equipped with usually multiple EDS detectors can be used for automated mineralogy analysis (*i.e.*, automated SEM analysis). Automated mineralogy provides rapid and, generally, automated mineral identification, quantitative or semi-quantitative mineral chemistry, mineral modal abundances, grain-size distribution, and mineral association information. This technique is useful for determining the mineralogical and morphological characteristics of slags. Tuhý *et al.*[20] used automated mineralogy to determine the distribution of trace elements including As, Cu, Pb, and Zn among phases in slag-containing soils near the Tsumeb Cu smelter in Namibia; these authors verified automated mineralogy results with bulk chemical analysis. In addition, Pietranik *et al.*[21] were able to rapidly classify phases into detrital, primary, or secondary categories based on automated mineralogy for weathered slag-derived material in a Zn–Pb slag heap in Poland. Lastly, the degree of access for infiltrating fluids to the surfaces of Pb-containing phases and the effect of grain size on fluid accessibility in Pb–Zn slag from Australia quantified by automated mineralogy allowed Morrison *et al.*[22] to evaluate the susceptibility of slags to leach Pb over time.

EPMA is the most widely used analytical technique for determining the elemental composition of crystalline and amorphous phases found in slags. The electron microprobe is fundamentally similar to an SEM but is equipped with wavelength dispersive spectrometers. The concept of wavelength dispersive systems was previously discussed in Section 3.2.2.1. There are several limitations of EPMA. These limitations include the inability or limited ability to measure light elements such as H, Li, Be, and B, the inability or limited ability to determine the oxidation state of an element, a detection limit of 100 mg kg^{-1} or higher for most elements,[9] beam interaction volume, and peak overlap. Light elements do not generally partition into the slag during smelting so this limitation is less important. Moreover, trace elements present at concentrations of hundreds of mg kg^{-1} or less may not be as important as more abundant elements from both an environmental contaminant perspective and for resource recovery potential. For example, Zn present at tens of weight percent in a sulfide phase and a glass phase will be targeted for extraction over Zn present in an oxide at hundreds of mg kg^{-1} for a given slag dump mineralogy. As for oxidation state, the inability to differentiate between Fe^{2+} and Fe^{3+} with EPMA becomes an issue for determining mineral formulae for spinels (AB_2O_4, where A = Fe^{2+}, Mg, Mn, Zn, and B = Al, Cr, or Fe^{3+}) and pyroxenes ($XY(Si,Al)_2O_6$, where X = Ca, Na, Fe^{2+}, or Mg and Y = Cr, Al, Fe^{3+}, Mg, Mn, or Ti, among others), which may contain Fe in both states. In this case, the Fe oxidation state in these phases must then be estimated according to charge balance when calculation of the structural formulae is done. Fortunately, the peak location difference of S^{2-} and S^{6+} is generally measurable using EPMA; therefore, insights into S present as a sulfide or sulfate may be achievable. This may be applicable to interpreting either reduced or oxidized conditions for the formation of S-containing glass. For example, Piatak and Seal[23] reported S to be in the reduced state in glass and melilite ($(Ca,Na)_2(Al,Mg,Fe^{2+})(Al,Si)_2O_7$) in historical BF slag from the USA, consistent with the reducing conditions in the furnace. Another issue with EPMA is the unavoidable peak overlaps that occur despite optimizing instrument and analysis settings. For example, Piatak *et al.*[24] corrected for phantom Co in Fe-rich sulfides in base metal slags. Lastly, the interaction volume of the electron beam depends on the spot size, which is several micrometers in width and depth at its minimum. The fine grain size of crystalline phases found in some slags can cause challenges in acquiring EPMA data for a single phase. Furthermore, micrometer to nanometer sulfide and metallic inclusions in the glass and crystalline phases of slag are common. Therefore, EPMA data reported in the literature may be a mixture of more than one phase or contaminated by these inclusions. For example, Piatak *et al.*[24] suggested that the high levels of S reported in EPMA analyses of interstitial glass were due to micrometer- to nanometer-sized sulfide inclusions for Cu slags. For higher resolution analyses, a focused ion beam transmission electron microscope (FIB-TEM) may be used. The FIB uses a focused beam of ions to make electron-transparent foils, which are then analyzed with a beam of electrons that pass through the sample (TEM). Using FIB-TEM, Ettler *et al.*[25] determined that K-rich pyroxenes reported in Cu slag from Zambia[26] were actually composed of nanometer-size leucite ($KAlSi_2O_6$) inclusions entrapped within the pyroxene ($(Na,Ca)(Mg,Fe,Al)(Al,Si)_2O_6$). The FIB technique can also be combined with SEM to obtain three-dimensional (3-D) imaging, 3-D element distribution, and 3-D textural analysis.[27]

LA-ICP-MS, like EPMA, provides quantitative elemental compositions of minerals and glasses. The technique involves a laser beam focused to as narrow as a few tens of micrometers that surface ablates particles, which are then transported to an ICP-MS instrument by a carrier gas. The ICP-MS was explained in Section 3.2.2.1. An advantage of LA-ICP-MS is that the detection limits are lower when compared to EPMA; however, the spatial resolution may be less narrow (tens of micrometers compared to a few micrometers). For example, Tyszka *et al.*[28] determined the distribution of Cd in various phases present in Pb–Zn slags from Poland using a combination of EPMA (for major element phase chemistry and to quantify trace elements present at a few hundred mg kg^{-1} or more) and LA-ICP-MS measurements (for trace elements present at lower than a few hundred mg kg^{-1}). In addition, Parbhakar *et al.*[29] used LA-ICP-MS instead of EPMA, the routinely used technique, to determine the chemistry of solid phases in base metal slags in Tasmania; mineralogical hosts of potentially valuable elements (*i.e.*, Zn, Pb, Ag, and Cu) and potentially toxic elements (*i.e.*, As and Cd) were determined.

Section 3.2.2.1 discussed synchrotron-based methods for bulk analyses. Experiments can also be conducted on the micrometer scale to gain insight into the speciation of elements and atomic structural information (μ-XAS) or the crystalline structure of phases (μ-XRD) in slags. For example, μ-XANES was utilized to determine the oxidation state and coordination of V in primary Ca silicates and weathering products in BOF slags to better understand the processes that control the mobilization of this potentially toxic element.[30] In a study on weathering of Sb-rich mining waste, μ-XRD, in conjunction with μ-Raman spectroscopic analyses, aided in the identification of Sb host phases in slags.[31] Some Sb phases were present as micrometer-sized inclusions, which may have been missed by more standard techniques, such as EPMA and SEM.

Another molecular-based technique, Raman spectroscopy, gives insight into the chemical structure of matter, its crystallinity, and molecular interactions. Similar to synchrotron-based techniques, Raman spectroscopy has been less frequently utilized for investigating slags when compared to more standardized microanalytical techniques (*e.g.*, EPMA, SEM, LA-ICP-MS). This technique detects vibrational, rotational, and other states in a molecular system and is based on the scattering of light with a change in frequency. Raman microspectrometry was used by Vanaecker *et al.*[32] to investigate base metal slags and determine mineralogical hosts of Zn and their weathering evolution to gain insight into the release of potentially toxic elements. The Raman microspectrometric analyses revealed fine intergrowths of more than one phase (*i.e.*, goslarite ($ZnSO_4 \cdot 7H_2O$) and smithsonite ($ZnCO_3$)) and that some seemingly anhydrous phases (*i.e.*, willemite (Zn_2SiO_4)) were hydrous (*i.e.*, hemimorphite ($Zn_4Si_2O_7(OH)_2 \cdot H_2O$)).[32] These types of observations required specialized micrometer-scale techniques. Similarly, Sobanska *et al.*[33] investigated weathering of slags and used Raman mapping to determine the distribution of partially degraded Fe–S phases and ferric hydroxide and oxide weathering products in Pb–Zn slags from France. In another application, μ-Raman spectroscopy was used to identify graphitic C, Fe sulfide, Fe oxides, and a Ca silicate glass in a BF slag; however, the fluorescent background in some analyses was problematic.[34] Lastly, Riboldi *et al.*[35] used μ-Raman spectroscopy to identify crystalline phases in electric arc furnace (EAF) slags and postulated using a portable real-time Raman instrument to use during steel production.

3.3 Elemental Composition

The bulk chemistry reported in over 110 studies is compiled in Tables 3.1 and 3.2, illustrating the diverse compositions of slags. A previous version of these compilations that included 70 studies was published by Piatak *et al.*[1] and then reproduced in Piatak.[2] The Appendix provides the list of references used in Tables 3.1 and 3.2. These tables include minimum, maximum,

Table 3.1 Summary of major, minor, and trace element chemistry of ferrous slags by slag type. See the Appendix for sources.

	BF						Pre-1900 BF						BOF					
	Min[a]	Max	Average	SD	*n*1[b]	*n*2[c]	Min	Max	Average	SD	*n*1	*n*2	Min	Max	Average	SD	*n*1	*n*2
wt%																		
Al_2O_3	4.01	41.20	12.5	6.4	56	18	0.72	22.28	7.7	5.3	63	5	0.02	20.56	3.6	4.6	30	19
CaO	0.15	46.80	36.9	8.9	56	18	0.88	40.12	12.1	9.7	63	5	19	60.4	44.2	8.5	32	19
FeO_{total}	0.02	22.1	2.6	5.4	51	17	0.08	89.05	33.9	32.2	63	5	10	35	24.6	6.2	32	19
K_2O	0.04	2.12	0.7	0.4	38	12	0.05	3.67	1.1	1.0	58	4	0.027	1.23	0.2	0.4	9	7
MgO	1.66	19.6	8.2	4.0	56	18	0.14	24.7	6.2	7.5	63	5	0.82	28	7.5	5.1	30	19
MnO	0.01	3.12	0.8	0.8	39	12	0.01	11.56	0.7	1.6	63	5	0.28	8.48	3.9	1.8	28	17
Na_2O	0.04	1.73	0.3	0.3	38	12	0.01	0.76	0.3	0.2	58	4	0.036	0.25	0.1	0.1	8	6
S	0.04	3.15	1.0	0.5	52	15	0.02	0.96	0.2	0.3	30	2	0.02	0.5	0.1	0.1	20	12
SiO_2	8.21	46.1	34.2	7.0	56	18	2.15	61.28	33.2	19.3	63	5	0.23	24.41	12.0	4.2	32	19
TiO_2	0.02	3.7	0.7	0.7	41	13	0.01	1.1	0.5	0.4	56	4	0.3	2	0.8	0.4	16	13
LOI	0.10	4.52	1.8	1.3	8	6	0.03	2.26	0.7	0.8	9	1	0.3	7.54	2.5	2.7	6	6
mg kg^{-1}																		
As	0.5	52	9.7	13.9	14	5	<1	12	2.1	2.5	20	1	—	—	—	—	—	—
Ba	52.8	1110	417	322	14	5	90	2030	579	526	51	3	24	260	103	91.1	5	3
Cd	<0.2	<0.4	—	—	6	2	0.3	0.3	0.3	—	1	1	0.8	15	6.1	—	3	1
Co	<0.5	33.0	7.5	9.8	14	5	1	210	58.6	79.1	30	2	2	12	5.9	—	3	1
Cr	8	1760	328	517	17	6	2	39	9.1	7.7	20	1	240	23 000	3172	5954	13	10
Cu	1.7	30	14.4	8.5	17	6	2.7	777	139	165	51	3	8	120	42.5	52.5	4	2
Ni	1.4	30	10.0	9.4	11	5	1.1	17.1	7.2	4.6	14	1	2.1	25	10.5	10.2	4	2
P	40	2300	460	589	23	10	0.0	15 300	1450	2234	46	4	470	20 000	6445	5027	27	16
Pb	<0.4	150	19.7	36.3	16	6	0.6	250	73.9	102	29	2	2.4	330	134	157	5	3
Sb	<0.1	10	4.4	4.8	10	3	0.06	0.39	0.2	0.1	9	1	3	8.8	5.0	—	3	1
Sn	1.6	45	9.4	12.0	14	5	0.1	0.6	0.3	0.2	8	1	2.7	26	11.7	—	3	1
V	8	923	205	251	17	6	3	900	310	364	30	2	68	25 100	5094	8292	10	8
Zn	3	183	67.7	58.6	17	6	1	40	15.4	13.0	29	2	8.5	200	105	82.9	7	5

(*continued*)

Table 3.1 (*continued*)

	EAF						Steel and legacy sites						Refining/ladle					
	Min[a]	Max	Average	SD	*n*1[b]	*n*2[c]	Min	Max	Average	SD	*n*1	*n*2	Min	Max	Average	SD	*n*1	*n*2
wt%																		
Al_2O_3	2	22.59	6.9	4.0	37	18	0.66	18.8	5.3	4.4	35	12	4.3	35.3	18.3	11.5	11	6
CaO	2.28	60	31.0	12.1	37	18	1.63	70.1	32.9	13.5	37	12	23.8	60.0	46.6	11.3	11	6
FeO_{total}	1	50.9	27.8	12.0	36	18	1.07	40	19.6	10.5	37	12	0.68	16.3	4.3	5.7	11	6
K_2O	<0.01	1.52	0.3	0.5	14	11	<0.01	1.28	0.3	0.7	32	9	0.01	0.1	0.1	0.1	7	3
MgO	2.91	15.2	7.6	3.6	35	18	0.43	20	6.3	4.1	37	12	1	14	8.6	3.9	11	6
MnO	0.4	15.58	4.4	2.9	34	17	0.30	21.5	4.9	4.1	35	10	0.2	11	2.0	3.3	10	6
Na_2O	<0.01	0.84	0.1	0.2	14	11	<0.01	0.84	0.2	0.2	32	9	0.03	0.1	0.1	0.0	6	2
S	0.04	0.91	0.2	0.2	17	10	0.06	1.06	0.4	0.3	31	8	0.03	1.5	0.6	0.5	9	4
SiO_2	4.6	32.7	15.9	6.7	34	17	7.61	61.1	21.1	12.2	37	12	2	35	16.4	9.1	11	6
TiO_2	0.33	1.32	0.6	0.3	13	10	0.06	1.96	0.6	0.4	37	11	0.14	0.89	0.4	0.3	4	2
LOI	<0.1	6.2	2.7	2.2	5	3	0.2	11.2	3.4	3.7	13	4	1.2	5.5	2.9	—	3	1

mg kg⁻¹																		
As	0.5	7.12	4.3	2.3	7	3	1	244	20.8	52.5	22	5	—	—	—	—	—	—
Ba	160	3600	1195	1169	10	5	34	2120	415	479	29	6	41.3	101	75.8	—	3	1
Cd	0.1	19	4.0	7.2	7	3	0.1	128	11.5	29.5	23	4	—	—	—	—	—	—
Co	2.32	11	5.4	2.8	10	4	0.8	36	7.9	10.5	26	4	—	—	—	—	—	—
Cr	320	200 000	16 873	38 920	25	12	4	9580	1988	2482	32	7	70	13 100	2997	4148	9	4
Cu	62	540	197	116	12	5	3	400	53.9	75.1	31	6	20	20	20.0	—	1	1
Ni	0.07	3180	298	833	14	6	<0.5	176	32.9	44.6	30	5	70	70	70.0	—	1	1
P	100	5400	2161	1575	21	12	<40	8260	2526	2163	36	11	<4	4000	1095	1438	7	4
Pb	2.9	3000	533	1065	10	5	2	1040	124	281	31	6	6.8	19.4	14.2	—	3	1
Sb	1.09	18	5.2	7.2	5	2	0.07	26	2.1	5.4	25	4	—	—	—	—	—	—
Sn	3.2	34	17.5	10.4	6	3	0.2	48	7.0	9.4	27	5	—	—	—	—	—	—
V	170	1710	873	555	18	8	14	1600	472	437	30	6	205	5060	1473	2392	4	2
Zn	31	6800	648	1718	15	8	1	11 000	650	2236	32	7	112	476	302	157	4	2

[a] Minimum and maximum values are generally reported to the significant digits reported in the original reference. For the calculated values of average and standard deviation, numbers were reported to one decimal place for values less than 100, whereas values equal to or greater than 100 were rounded to the nearest whole number. Standard deviations (SD) were not calculated if three or less values were available and averages and standard deviations were not calculated if all values were less than the detection limit. Values below the detection limit were included as the detection limit in calculations.

[b] Number of samples.

[c] Number of studies.

Table 3.2 Summary of major, minor, and trace element chemistry of non-ferrous slags by slag type. See the Appendix for sources.

	Cu[b]						Ni (laterite ore)						Ni (sulfide ore)					
	Min[a]	Max	Average	SD	*n*1[c]	*n*2[d]	Min	Max	Average	SD	*n*1	*n*2	Min	Max	Average	SD	*n*1	*n*2
wt%																		
Al_2O_3	0.01	18.9	8.1	5.3	137	22	2.79	17.96	10.9	5.4	11	4	1.7	7.00	4.7	2.1	13	4
CaO	0.15	27.4	9.8	7.4	140	24	0.308	57.58	10.6	14.6	21	5	0.98	3.96	2.2	1.0	14	5
FeO_{total}	0.67	64.40	29.0	18.6	160	26	5.40	39.4	16.2	9.7	21	5	18	53.1	46.0	8.8	14	5
K_2O	0.01	6.7	2.3	2.0	134	20	<0.05	2.31	1.0	0.8	9	2	0.18	0.18	0.2	—	1	1
MgO	0.09	6.48	2.1	1.5	136	21	2.04	31.2	16.5	12.1	21	5	0.55	9.70	2.6	2.2	14	5
MnO	0.026	6.55	0.5	0.7	133	19	0.205	1.72	0.5	0.4	21	5	0.028	0.08	0.1	0.0	12	3
Na_2O	0.01	4.31	0.4	0.6	134	20	<0.07	2.24	0.6	0.7	10	3	0.3	0.3	0.3	—	1	1
S	0.006	6.51	1.1	1.3	140	20	0.025	0.2	0.1	0.1	3	3	0.868	1.78	1.2	0.3	12	3
SiO_2	9.82	70.72	38.2	11.3	136	23	18.17	50.5	39.7	8.7	11	4	29	39.3	37.0	2.9	13	4
TiO_2	0.1	1.66	0.5	0.3	126	17	0.070	0.73	0.4	0.2	10	3	0.230	0.239	0.2	0.0	6	1
LOI	−2	11.8	1.3	2.9	31	6	—	—	—	—	—	—	—	—	—	—	—	—
mg kg^{-1}																		
As	<0.5	75 865	2927	8777	97	16	26.2	26.2	26.2	—	1	1	—	—	—	—	—	—
Ba	28	29 000	1749	3889	97	11	112	112	112	—	1	1	—	—	—	—	—	—
Cd	0.12	14 000	570	2068	76	11	4.58	4.58	4.6	—	1	1	—	—	—	—	—	—
Co	13.5	24 104	1872	4479	99	14	5	700	199	254	7	2	750	2480	1699	623	13	4
Cr	13	7510	356	851	103	10	330	21 210	6751	6544	11	4	15	15	15.0	—	1	1
Cu	1400	353 580	20 296	46 331	151	24	<5	2200	965	1059	6	2	110	13 450	4744	5500	13	4
Ni	2	935	80.4	134	84	13	74	6907	1989	1916	20	4	1100	11 850	5748	4013	13	4
P	100	9540	1254	1375	109	15	90	4096	1388	1441	10	3	—	—	—	—	—	—
Pb	<1.5	183 800	11 907	25 831	118	18	8	501	246	—	3	1	50	50	50.0	—	1	1
Sb	0.1	8200	785	1646	70	11	79.2	79.2	79.2	—	1	1	—	—	—	—	—	—
Sn	0.7	1400	164	355	59	10	—	—	—	—	—	—	—	—	—	—	—	—
V	11	1228	229	309	25	4	60	84.8	70.4	—	2	2	—	—	—	—	—	—
Zn	44	280 000	30 922	52 071	120	20	13	2500	919	986	8	3	180	1410	698	622	12	3

	Pb (primary)						Pb (secondary)						Zn and Pb[e]						Zn					
	Min	Max	Average	SD	*n*1	*n*2	Min	Max	Average	SD	*n*1	*n*2	Min	Max	Average	SD	*n*1	*n*2	Min	Max	Average	SD	*n*1	*n*2
wt%																								
Al_2O_3	1.22	11.11	4.2	1.7	39	16	1.42	11.52	5.5	—	3	3	0.9	21.9	8.3	5.8	63	7	10.03	10.4	10.2	—	2	2
CaO	0.45	23.11	9.5	7.2	42	18	1.8	21.8	9.4	9.1	4	4	0.18	32.23	17.3	7.2	65	8	18.1	18.55	18.3	—	2	2
FeO_{total}	3.2	41	26.3	7.8	51	18	31.9	52.5	43.4	8.1	5	5	0.88	59.6	16.7	10.1	66	8	31	33.7	32.6	—	2	2
K_2O	0.05	2.87	1.0	0.7	36	14	0.1	0.48	0.2	—	3	3	0.02	3.91	0.6	0.7	64	7	—	—	—	—	—	—
MgO	0.36	5.44	1.5	1.2	38	15	0.26	2.88	1.5	—	3	3	0.61	15.9	5.4	4.1	63	7	1.81	1.81	1.8	—	1	1
MnO	0.09	9.4	3.1	2.8	38	15	0.32	0.88	0.6	—	3	3	0.01	3	0.5	0.5	68	9	1.04	1.04	1.0	—	1	1
Na_2O	0.02	3.43	0.4	0.6	42	13	0.32	4.12	1.4	1.8	4	4	<0.01	3.93	0.3	0.6	62	7	—	—	—	—	—	—
S	0.12	22.02	2.4	3.8	39	16	0.36	11.2	5.6	4.2	6	6	0.02	6	1.2	1.4	50	7	1.58	1.58	1.6	—	1	1
SiO_2	1.99	54.6	29.4	14.9	41	17	4.71	30.06	14.9	10.5	6	6	2.04	57.1	28.6	11.8	50	7	21.9	24.38	23.1	—	2	2
TiO_2	0.01	5	0.4	0.8	34	12	0.19	0.6	0.4	—	3	3	0.07	1.14	0.4	0.3	63	6	—	—	—	—	—	—
LOI	−4.38	7.1	0.8	5.6	4	4	−0.9	7.48	3.3	—	2	2	−4.59	29.6	7.0	7.5	44	5	—	—	—	—	—	—

(continued)

Table 3.2 *(continued)*

	Pb (primary)						Pb (secondary)						Zn and Pb[e]						Zn					
	Min	Max	Average	SD	*n*1	*n*2	Min	Max	Average	SD	*n*1	*n*2	Min	Max	Average	SD	*n*1	*n*2	Min	Max	Average	SD	*n*1	*n*2
$mg\ kg^{-1}$																								
As	87	13 300	1534	3232	45	16	6.04	1000	503	—	2	2	1	10 710	1181	1951	66	7	1500	1500	1500	—	1	1
Ba	169	190 000	16 168	33 649	44	12	5393	5393	5393	—	1	1	76	17 914	1126	2300	62	6	—	—	—	—	—	—
Cd	0.3	3500	698	1265	19	11	200	200	200	—	1	1	0.38	575	31.2	84.5	53	6	—	—	—	—	—	—
Co	6.1	450	79.7	108	38	9	—	—	—	—	—	—	8.5	242	35.7	45.2	42	4	—	—	—	—	—	—
Cr	19	4000	290	655	39	10	700	4432	2505	1985	4	4	4	700	155	114	32	5	—	—	—	—	—	—
Cu	88.45	165 200	18 059	44 534	36	13	27.4	10 000	3922	4368	4	4	16	7400	802	1347	60	7	—	—	—	—	—	—
Ni	5.6	430	135	115	32	11	27.8	1100	564	—	2	2	13	240	59.7	42.3	58	6	—	—	—	—	—	—
P	300	6284	1614	1349	30	9	1000	1000	1000	—	1	1	43.57	26 400	5138	8213	63	6	—	—	—	—	—	—
Pb	5000	319 190	88 637	87 321	42	18	7900	219 700	89 462	97 422	5	5	1.9	62 300	8052	12 268	69	10	6900	15 400	11 155	—	2	2
Sb	5.7	5490	1538	1498	41	13	147	2700	1424	—	2	2	0.16	245	42.6	52.7	57	6	—	—	—	—	—	—
Sn	23	3900	1066	1457	6	6	230	3100	1777	—	3	3	0.1	500	23.4	104	23	3	—	—	—	—	—	—
V	20	805	161	249	13	6	—	—	—	—	—	—	6	9980	2294	2832	44	4	—	—	—	—	—	—
Zn	701	174 000	49 864	45 794	51	18	964	11 000	4763	4492	4	4	212	379 700	39 697	66 299	70	9	69 200	70 300	69 749	—	2	2

[a]Minimum and maximum values are generally reported to the significant digits reported in the original reference. For the calculated values of average and standard deviation, numbers were reported to one decimal place for values less than 100, whereas values equal to or greater than 100 were rounded to the nearest whole number. Standard deviations (SD) were not calculated if three or less values were available. Values below the detection limit were included as the detection limit in calculations.
[b]Cu slags include those generated from the recovery of Cu from ores that may also contain recoverable amounts of other base metals and precious metals.
[c]Number of samples.
[d]Number of studies.
[e]Pb and Zn slags include those generated from the recovery of Zn or Zn and Pb from ores that may contain other base metals and precious metals that were simultaneously recovered.

average, and standard deviation values for the major, minor, and trace elements reported in slags, as well as the number of samples (*n*1) and the number of studies (*n*2). A few notes regarding the tables are: (1) values below the detection limit were reported as the detection limit, and (2) analyses were included from studies that reported data for individual samples and for compiled datasets such that, for example, minimum, maximum, and average values representing 45 EAF slags from the United States and Canada were included as only three values (not individual analyses), as reported by Proctor *et al.*[36] In addition, most major element concentrations are reported as oxides in Tables 3.1 and 3.2, a common convention, regardless of how they were reported in the cited article. However, some elements listed as an oxide do not exclusively occur in the oxide form. For example, the total Fe concentration is reported as FeO but Fe has been shown to be present in slags as FeO, Fe_2O_3, Fe metal, and FeS.[1] The major element compositions of slags, compiled from the literature in Tables 3.1 and 3.2, are illustrated in the SiO_2–FeO–CaO ternary diagram in Figure 3.2. On average, the ternary represents approximately three-fourths, or 75%, of the total mass of the samples. For each sample represented in the ternary, the sum of the three ternary components (*e.g.*, SiO_2, FeO, and CaO) in weight percent was calculated and then the sum values for all samples averaged; this average is given in parentheses after each slag type. These ternary diagrams, with indicated phase boundaries and isotherms, have also been frequently used for estimating the melting temperatures in the furnace, especially in the reconstruction of historical conditions of smelting.[1,15,19,24,26,37–41] Phase diagrams as well as geothermometry, thermodynamic modeling with MELTS software, and temperature XRD analyses have all been used to reconstruct temperatures in smelting furnaces to understand pyrometallurgical technologies at historical sites; a review of these methods and their application to K-rich Cu slags can be found in the article by Kądziołka *et al.*[41]

In addition to Figure 3.2, the concentrations of major elements and trace elements in slags are illustrated in scatter plots in Figures 3.3 and 3.4, respectively. In order to evaluate element enrichment in slags, either from an environmental contaminant or resource potential perspective, concentrations can be compared to regional soil screening levels for site remediation for residential use and for industrial use; criteria are from the USEPA[43] except for Cr. The USEPA has not established a guideline for total Cr concentrations, only separate criteria for Cr^{3+} and Cr^{6+}; therefore, the Canadian guideline is used for total Cr.[44] These guidelines are shown for comparison purposes only in Figure 3.4.

3.3.1 Ferrous Slags

The major, minor, and trace element compositions of each type of ferrous slag depend on the furnace feed, fluxes added, fuel source, and furnace conditions. As discussed in Chapter 2, ferrous slags are generated in the

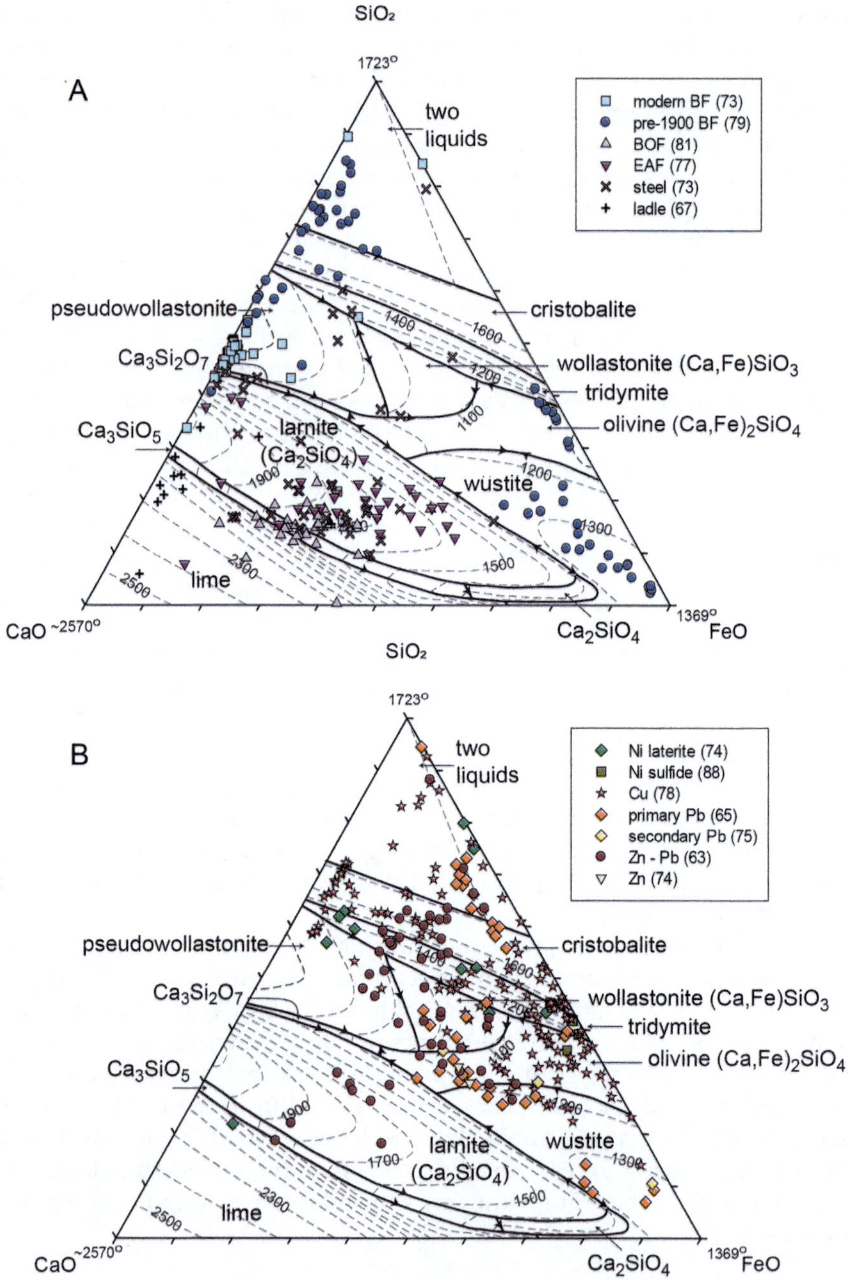

Figure 3.2 Bulk chemical compositions of slags in weight percent on FeO–SiO_2–CaO ternary diagrams for (A) ferrous slags and (B) non-ferrous slags. The various symbols represent different slag types. The average total weight percentages represented by the symbols are given in parentheses after the slag type. Phase boundaries and isotherms, in Celsius, from work by Levin *et al.*[42] See the Appendix for sources of data.

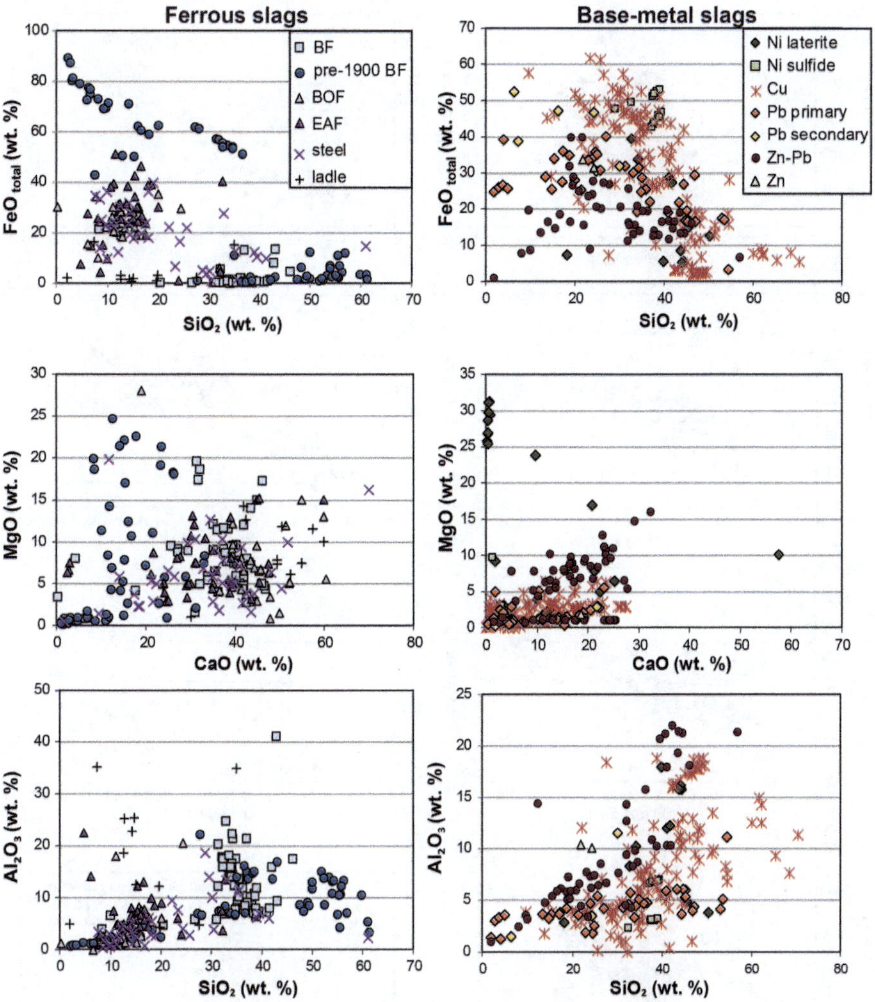

Figure 3.3 Concentrations of major elements in various slag types. See the Appendix for sources of data. Concentrations reported at the detection limit were included as the detection limit.

manufacture of iron and steel products. Ferrous slags are produced at various steps of the ironmaking and steelmaking process and include the following: BF slags generated during smelting of iron ores to produce pig iron; BOF slags from refining of pig iron and lesser amounts of steel scrap and/or direct reduction iron (DRI) to make crude steel; and EAF slags produced as a by-product during the smelting of steel scrap and DRI to produce steel. Additionally, a ladle furnace or argon oxygen decarburization (AOD) vessel is used to further refine the crude steel generated in a BOF or EAF, generating ladle and AOD slags as by-products.

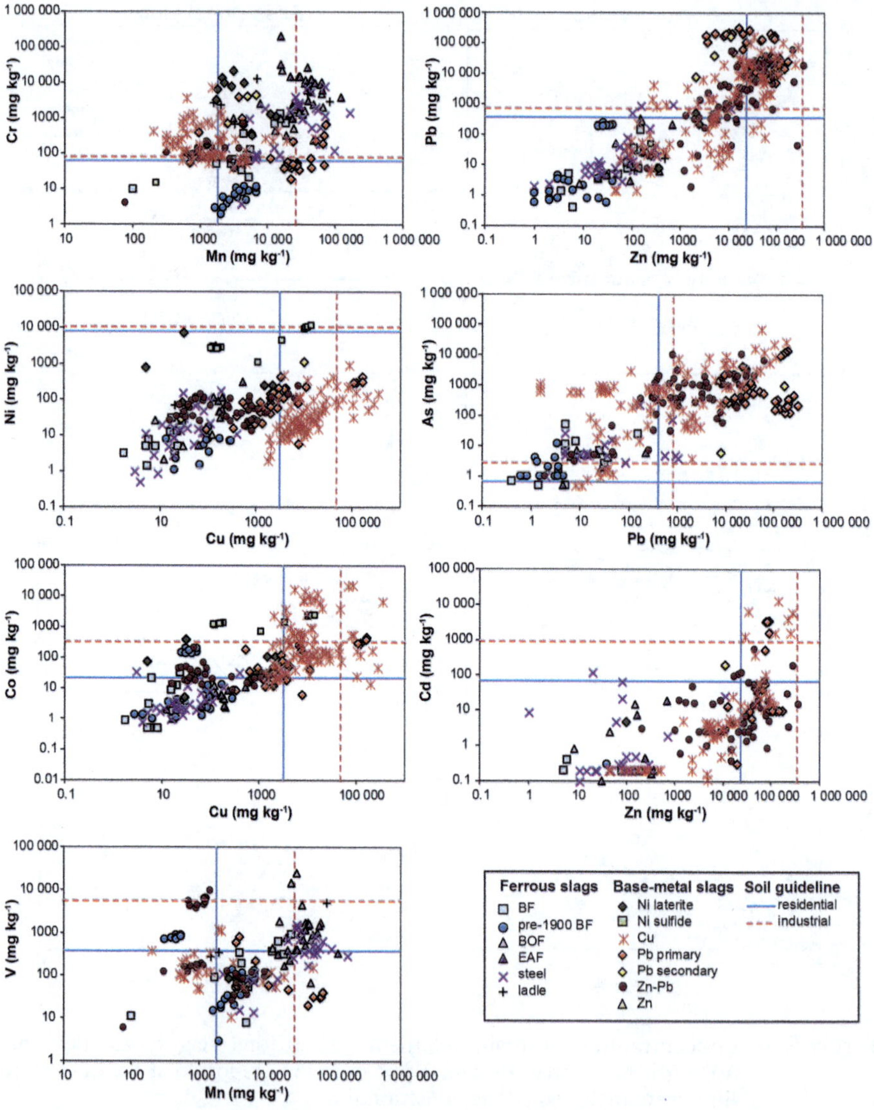

Figure 3.4 Concentrations of minor and trace elements in various slag types. See the Appendix for sources of data. Residential use (solid blue lines) and industrial use (red dashed lines) soil screening levels are shown. Guidelines are from ref. 43 for all elements except Cr, which are Canadian guidelines.[44] Concentrations reported at the detection limit were included as the detection limit.

3.3.1.1 Blast Furnace Slag

The concentrations of the major, minor, and trace elements in BF slags can vary widely but, in general, the most abundant components are CaO and SiO_2 for both modern and historical BF slags, with some historical slags also containing significant amounts of FeO (Figures 3.2 and 3.3). Based on the compositions compiled in Table 3.1, BF slags generated during ironmaking in the 20th or 21st century (*i.e.*, modern BF slags) contained the following elemental/oxide averages (and standard deviations) in decreasing abundance: 36.9 (±8.9) wt% CaO, 34.2 (±7.0) wt% SiO_2, 12.5 (±6.4) wt% Al_2O_3, 8.2 (±4.0) wt% MgO, 2.6 (±5.4) wt% FeO, and 1.0 (±0.5) wt% S. Other major elements reported as oxides with averages (given in parentheses) of less than 1 wt% include K_2O (0.7 wt%), MnO (0.8 wt%), TiO_2 (0.7 wt%), and Na_2O (0.3 wt%). The historical BF slags, those generated before 1900 C.E., in Table 3.1 represent by-products generated during iron production and include a study of slag generated in the late 1700s to late 1800s during colonial times in the USA,[23] slags generated before 1900 from several locations around the world,[45] slags from the 3rd/4th century C.E. Ethiopia,[46] slags from several centuries from Europe (3rd/4th, 16th, 19th centuries),[47] and 1st to 7th century C.E. slags from Turkey.[48] These pre-1900 slags from ironmaking contain the following major elements in decreasing order of abundance (on average (standard deviation)): FeO (33.9 (±32.2) wt%), SiO_2 (33.2 (±19.3) wt%), CaO (12.1 (±9.7) wt%), Al_2O_3 (7.7 (±5.3) wt%), MgO (6.2 (±7.5) wt%), and K_2O (1.1 (±1.0) wt%) (Table 3.1). However, as shown in Figure 3.2, pre-1900 BF slags can be divided into two chemically distinct groups. The compositions of Fe slags that formed before 700 C.E. reported by Severin *et al.*[46] and Eekelers *et al.*[48] are significantly different to the other pre-1900 Fe slags, most of which formed after 700 C.E.; the pre-700 C.E. slags contain higher amounts of Fe (up to 89.1 wt% FeO) and lower amounts of Ca (up to 14.8 wt% CaO) and Mg (up to 1.7 wt% MgO) (Figures 3.2 and 3.3). The remaining elements are present in historical BF slags, on average, at less than 1 wt% in Table 3.1.

In BF slags, Ca, Mg, and Si are generally sourced from the fluxes added, such as lime (CaO), limestone, dolomite ($CaMg(CO_3)_2$), olivine ($(Ca,Mg,Mn,Fe)_2SiO_4$)-bearing rocks, and quartzite or quartz (SiO_2) sand; Si, Al, and S may be from the gangue minerals in iron ore or the fuel source, such as coke. The major compositional difference between the modern BF slags and those generated pre-1900 is that the latter may contain appreciably more Fe, likely from less optimal slag formation conditions and non-ideal furnace feed and fluxes leading to insufficient Fe recovery (Figures 3.2 and 3.3). The reducing conditions in a modern BF result in slags containing low amounts of Fe, with the majority of BF slags containing 2 wt% FeO or less in Figure 3.3. However, a few modern BF slags contain appreciable amounts of Fe, up to 22.1 wt% FeO (Table 3.1 and Figure 3.3); the highest Fe contents were reported for the fine-fractions (generally less than 9.5 mm in diameter)[49] and the sorting by size may have led to fines enriched in Fe metal.

The concentration of Ca is, on average, less than half in pre-1900 BF slags (12.1 wt% CaO) compared to modern BF slags (36.9 wt% CaO), whereas the ranges and average concentrations of SiO_2 for both types are similar (averages of 34.2 wt% SiO_2 for modern and 33.2 wt% SiO_2 for pre-1900) (Figures 3.2 and 3.3). The conditions in the historical blast furnaces were generally less basic (*i.e.*, CaO to SiO_2 ratio of less than 1) than modern blast furnaces (*i.e.*, CaO to SiO_2 ratio of 1 or greater) (Table 3.1). Sulfur from coal or coke, and potentially trace amounts present as sulfides in iron ore, added to the furnace is an undesirable element in steel and blast furnaces are designed to entrap S in the slag. High basicity and reducing conditions in the furnace facilitate S entrapment in slag by the formation of CaS.[50] Therefore, modern blast furnaces operating under more basic conditions compared to historical blast furnaces likely results in the higher S content of modern BF slag. Sulfur reaches up to 3.2 wt% and averages 1.0 wt% S in modern BF slags, whereas the maximum and average for pre-1900 BF slags are 0.96 and 0.2 wt% S, respectively (Table 3.1).

Elements that are high in BF slags when compared to soil guidelines are Cr and Mn, and locally Co (Figure 3.4). Furthermore, Cr concentrations are generally lower in the historical Fe slags compared to modern BF slags due to steel scrap not being added to the furnace feed in the former (Figure 3.4). For example, Cr concentrations average 328 (±517) mg kg^{-1} in modern BF slags compared to 9 (±8) mg kg^{-1} in pre-1900 BF slag samples (Table 3.4). Among the historical BF slags, the concentrations of Co are anomalous and between 140 and 210 mg kg^{-1} for 3rd/4th C.E. Ethiopian slag,[46] which exceeds the residential-use soil guidelines of 23 mg kg^{-1} (Figure 3.4). The lateritic iron ore related to the Ethiopian smelting contained similar concentrations of Co, which may explain its enrichment in the slag.

3.3.1.2 Steel Slags

The maximum, minimum, and average concentrations of major, minor, and trace elements reported in steel slags are given in Table 3.1. Steel slags include BOF and EAF slags from primary steel production and refining/ladle slags from further refining of the steel. In Table 3.1 and Figures 3.2–3.4, BOF, EAF, and ladle slags were compiled and plotted separately based on the furnace type. The remaining samples were grouped in the general category of steel slags when the type was not specified or not known. At legacy steelmaking sites, slag piles may be composed of mixtures of different types of slag; these were also included in the steel slag category and may include BF slags. Each slag-generating process involves adding fluxes rich in Ca and Mg, such as limestone, lime, or dolomite, to the furnace to optimize slag formation. Ladle furnaces also require the addition of alloys, desulfurizing agents (such as Ca, Mg, CaSi, and CaC_2), and deoxidizing agents (such as Si and Al to form SiO_2 and Al_2O_3, respectively).[7] EAFs involve the addition of a solid fuel source such as coke. The furnace feed and fuel source influence the chemistry of the resulting slag.

The chemical compositions of the different steel slag types (BOF, EAF, and ladle), and the general category of steel slag are illustrated in Figures 3.2–3.4. Similar to modern BF slags, CaO and SiO_2 are the main chemical components of all steel slag types; however, steel slags also contain major amounts of FeO (Figures 3.2 and 3.3). BOF and EAF slags are similar in composition, yet the concentrations of elements in EAF slag can vary widely depending on the types of recycled steel scrap added to the furnace (Figures 3.3 and 3.4).[7] Furthermore, the composition of the ladle slags depends on the alloys added to the furnace to achieve a specific steel grade.[7] In general, for all categories of steel slags in Table 3.1, CaO, by mass, dominates the major element chemistry and is usually found in amounts equal to or higher than FeO, which is, on average, higher than the SiO_2 concentration (*i.e.*, CaO > FeO > SiO_2; Figure 3.2); the exception is ladle slags, which are generally lower in FeO compared to the other types. The average FeO in EAF slags is 27.8 (±12.0) wt% and in BOF slags is 24.6 (±6.2) wt%, compared to ladle slags with an average of 4.3 (±5.7) wt% FeO (Table 3.1). Depending on the furnace conditions, FeO concentrations in steel slags can reach as high as 50.9 wt%, as reported for an EAF slag from Sweden,[51] and as low as 0.7 wt%, as reported for a ladle furnace slag from Romania[52] (Table 3.1).

The next most abundant components after CaO, FeO, and SiO_2 are Al_2O_3 and MgO, which are frequently found in concentrations of a few weight percent or more and, usually, are higher in concentration than MnO. To note, Al_2O_3 is higher in ladle slag compared to the other types of steel slag (Table 3.1). As noted by Yildirim and Prezzi,[7] Al and Si are added to the ladle furnace during deoxidation forming Al_2O_3 and SiO_2, which report to the slag. The variation among the steel slag types for MgO and MnO content is not remarkable (Table 3.1). Other major elements such as K_2O, Na_2O, TiO_2, and S are present at less than 1 wt% on average (Table 3.1). The concentrations of Cr, Mn, and V on average are similar among the steel slag types and, as shown in Figure 3.4, may exceed soil guidelines. Vanadium reaches concentrations of 2.5 wt% in BOF slag (Table 3.1), whereas, Cr, Cu, Ni, and Zn are on average higher in EAF slags than BOF slags, possibly due to the higher contribution of steel scrap to EAF feed compared to BOF feed, which is mostly pig iron with some steel scrap. In contrast to Cr, Mn, and V, the concentrations of Cu, Ni, and Zn do not exceed soil guidelines, except on rare occasions (Figure 3.4).

The steel slags contain similar amounts of Ca but higher amounts of Cr, Fe, Mn, and Zn and lower amounts of Al and Si compared to modern BF slags. The high Fe in certain types of steel slag is the result of oxidizing furnace conditions oxidizing impurities to the slag compared to the reducing conditions of BFs. The higher Fe content of the feed into the steelmaking furnaces compared to the blast furnace may also impact the Fe content in steel slags. Lower amounts of Al_2O_3 and SiO_2 in most steel slags compared to modern BF slags may be due to the feed being predominantly pig iron, steel scrap, and DRI in steelmaking furnaces and not iron ore containing aluminosilicate gangue minerals as in the latter. The exception is ladle slags, which contain, on average, the highest Al_2O_3 for all ferrous slag types due to Al-bearing compounds being added to the furnace for de-oxidation. The use of fluxes with variable amounts of Al, Mg, and Si will influence their content in the slag. As for trace elements, Cr, Mn, and Zn may be sourced from recycled steel scrap and from the addition of alloy elements, leading to higher concentrations of these elements in steel slags compared to BF slags. Moreover, the concentrations of V are higher in steel slags than in BF slags with the following averages (and standard deviations) reported in Table 3.1: modern BF slags 205 (±251) mg kg^{-1}; BOF slags 5094 (±8292) mg kg^{-1}; EAF slags 873

(±555) mg kg^{-1}; ladle slags 1473 (±2392) mg kg^{-1}. In addition to the reprocessing of steel slags to recover residual Fe, other elements such as Cr, Mn, and V are potential resources for recovery and will be discussed in Chapter 9.[53,54]

3.3.2 Ferroalloy Slags

Ferroalloys are overwhelmingly utilized in steelmaking to impart specific material properties. For example, FeCr, FeMn, FeNi, and FeSi are usually produced in an EAF to make different grades of steel (*i.e.*, alloyed, carbon, stainless, and tool). Ferroalloy slags are generated in the production of these bulk ferroalloys. The major elemental compositions of the various ferroalloy slags discussed in this section are reported in Chapter 2 and include Table 2.6 (FeMn slags), Table 2.8 (SiMn slags), and Table 2.11 (FeCr slags); see Chapter 2 for the sources of data.

The compositions of FeMn slags compiled in Table 2.6 reveal that their major element chemistry is variable depending on the smelting conditions. Silica and CaO are consistently a significant component of these slags whereas Al_2O_3 and MgO can vary from a weight percent to over 15 wt%. Moreover, due to the highly reducing conditions in the furnaces, Fe should partition to the alloy phase resulting in slags containing little Fe; Fe is generally not a major component of these slags. As for other elements, the Mn content averages 11.5 wt% MnO in FeMn slag generated in a BF and 23.6 wt% MnO and higher for the other FeMn slag types (Table 2.6). Despite some steel slags containing up to 21.5 wt% MnO, the average amounts of Mn in FeMn slags are anomalously higher than in any other types of slag (Table 3.1 and Table 3.2). The FeMn slag to be discarded is typically processed to recover entrained metal droplets before final disposal, a practice that is common to other ferroalloy slags. In contrast to cleaning and disposal, some FeMn slags (*i.e.*, "duplex" slag in Table 2.6) are used for feed in another furnace. In particular, FeMn duplex slag may be the source of Mn in SiMn alloy production when Mn is not sourced from ore. Slags generated from the production of SiMn are also rich in Mn with an average composition of 10 wt% MnO or higher (Table 2.8).

Ferrochromium constitutes the bulk of ferroalloys due to its appreciable consumption in stainless steel manufacturing. The major element compositions of slags generated in the production of FeCr alloys are given in Table 2.11. Silica is consistently a major component of these slags. Additionally, MgO and Al_2O_3 are significant (averages of 23.3 wt% or higher) for some FeCr slag types (submerged arc furnace (SAF) and direct current (DC) arc furnace), whereas CaO dominates for others (EAF or ladle furnaces) (Table 2.11). The Fe content is variable among the types and Cr_2O_3 is, on average, 4.5 wt% or greater in all types (Table 2.11). The element being alloyed with Fe, in this instance Cr, ends up in significant concentrations in the slag; the average concentration of Cr in FeCr slag is generally higher than for any other slag types (Table 2.11, Table 3.1 and Table 3.2). A similar trend was noted for Mn concentrations in FeMn slag.

Nickel slags generated from the processing of either Ni sulfide or Ni laterite ores may be considered in the ferroalloy slag category owing to the use of Ni in FeNi or discussed with other base metal slags due to the use of the refined metal in catalysts and chemicals. In this chapter, Ni slags are discussed in Section 3.3.3.3.

3.3.3 Non-ferrous Slags

Base metal slags are generated from the smelting of natural ores and/or secondary materials to recover non-ferrous metals, such as Cu, Pb, Zn, Ni, Sn, Al, and Mn. As discussed in Chapter 2, non-ferrous metals, with some rare exceptions, occur in the Earth's crust as sulfide, silicates, oxides, and carbonates that require processing and refining to generate pure metals. The smelting of non-ferrous ores produces a silicate slag and a crude alloy or matte. The alloy or matte is further refined to high-purity metals. Many of the base metals, such as Co, Cu, Ni, Pb, and Zn, are found in the Earth's crust as sulfide minerals that are invariably associated with Fe-bearing sulfides. Oxidizing conditions during smelting result in the breakdown of sulfides and the reporting of Fe to the slag. The smelting processes, ore compositions, types of furnace, and fluxes added for each grouping of base metals are discussed in Chapter 2. The compositions of non-ferrous slags compiled from the literature are given in Table 3.2 and bulk elemental compositions are illustrated in Figures 3.2–3.4. The major element composition of base metal slags is represented by the SiO_2–FeO–CaO system illustrated in Figure 3.2. Considerable overlap in compositions among the non-ferrous slag types exists, illustrating their chemical similarity. The ternary diagram generally represents greater than two-thirds of the mass of the slag sample and the average of the sum of the three components in weight percent for the individual analyses is given in parentheses after each slag type in Figure 3.2. Other major elements in base metal slags are generally lower in abundance than Si, Fe, and Ca and include Al, K, Mg, Mn, and S (Figure 3.3 and Table 3.2). Various sources may contribute to the final amount of each element in the slag. For example, the SiO_2 content may reflect the amount of gangue silicate minerals in the ore and concentrate or the amount of SiO_2-containing flux added to the furnace or both. The Fe content likely depends on the amount of gangue iron sulfide minerals in the ore that are retained in the base metal concentrates and, less typically, on non-sulfide gangue minerals. Aluminum is likely from silicate minerals in the gangue, which may also host alkali elements. The amount of S depends on the amount of entrained matte. The content of trace elements, especially those being targeted, depends on whether the smelting conditions facilitate partitioning of trace elements to the matte phases instead of the partitioning to slag or off-gas. The abundance of major, minor, and trace elements depends on the type of slag, as discussed in the following sections.

3.3.3.1 Cu Slags

Copper can occur in ores as sulfides, carbonates, sulfates, and oxides. Copper slags result from the direct smelting of Cu-bearing sulfide flotation concentrates using a flash furnace (*i.e.*, combined roasting and smelting) or from the smelting of roasted sulfide concentrates in reverberatory and electric furnaces to form a Cu-rich matte. The matte is then further purified in a converter furnace to create blister Cu, which is then refined to produce pure Cu. Globally, the majority of Cu originates from pyrometallurgical processing of Cu sulfide ores. As for the oxidized ores, extraction of Cu is generally performed with hydrometallurgical methods.[55]

Copper sulfide minerals in ores may be associated with other base and precious metals, such as Ag, Au, Co, Ni, Pb, PGMs, and Zn, which may be targeted for recovery during processing. Slags generated during the recovery of Cu and associated metals are grouped together in Table 3.2. The chemical composition of Cu slags is highly variable, yet the main chemical components, like other base metal slags, are FeO (average 29.0 (±18.6) wt%) and SiO_2 (average 38.2 (±11.3) wt%) (Figures 3.2 and 3.3 and Table 3.2). Copper slags also contain CaO (average 9.8 (±7.4) wt%) and Al_2O_3 (average 8.1 (±5.3) wt%), but often at concentrations lower than FeO and SiO_2. K_2O, MgO, MnO, and S all reach maximum concentrations of approximately 6.5 wt%, whereas Na_2O only reaches up to 4.3 wt% (Table 3.2 and Figure 3.3).

Copper slags contain the highest concentrations of Cu when compared to all other types of slag (Figure 3.4). The concentrations of Cu reach 8.6 wt% and 35.4 wt% in two historical Cu–Co slags from the Copperbelt Province, Zambia, that are interpreted to be a mixture of slag and matte.[26] The other slags from Zambia in the study only contained between 0.2 and 2.0 wt% Cu, more similar to the amounts of Cu reported in most other Cu slag studies (Table 3.2 and Figure 3.4). Converter slag generated during the refining of matte may contain more than a few weight percent of Cu. These slags, however, are typically recycled into the smelting furnace to recover the Cu and, therefore, do not usually end up in the waste piles. For example, Khorasanipour and Esmaeilzadeh[56] reported 8.1 and 9.0 wt% Cu in converter slags generated at the Sarcheshmeh industrial complex in Iran. Other Cu slags that have over a few weight percent of Cu may have resulted from smelting processes that were not efficient at recovering Cu. For example, 23–29th century B.C.E. crucible, or refining, slags from Huelva Providence, Spain, contain 20.2 to 28.1 wt% Cu; Cu was not effectively recovered during this ancient smelting according to Sáez *et al.*[57] Furthermore, 10.6 and 12.2 wt% Cu were reported for two historical Cu and Cu–Pb slags from the Tsumeb smelter in Namibia[58] with the enrichment attributed possibly to low recovery efficiencies. Other elements besides Cu that may be found in significant concentrations in Cu slag include As, Cd, Cr, Co, Pb, and Zn, all of which have been reported in concentrations that may exceed soil remediation guidelines (Figure 3.4). The following are the average concentrations (and standard deviations) of these elements found in Cu slags: 2927 (±8777) mg kg^{-1} for As, 570 (±2068) mg kg^{-1}

for Cd, 1872 (±4479) mg kg^{-1} for Co, 356 (±851) mg kg^{-1} for Cr, 11 907 (±25 831) mg kg^{-1} for Pb, and 30 922 (±52 071) mg kg^{-1} for Zn. Arsenic and Pb are highest in several slag samples from the Tsumeb smelter, Namibia, with concentrations reaching up to 7.6 wt% and 18.4 wt%, respectively.[58] The two samples from this study with the highest As content, one of which also has the high Pb, are also those reported to have anomalous Cu concentrations of 10.6 and 12.2 wt%. Zinc also reaches 12.1 wt% in slags from the Tsumeb smelter.[58] Parsons *et al.*[59] also reported atypical trace element content in Cu slags from a historical Cu–Zn mine in California, USA. In Table 3.2, the maximum Ba (2.9 wt%), Cd (1.4 wt%), Sb (0.8 wt%), and Zn (up to 28.0 wt%) contents for Cu slags are from Parsons *et al.*;[59] these slags also contain significant levels of Cu (up to 6.4 wt%) and Pb (up to 11.0 wt%). Lastly, several trace elements are anomalously high in previously mentioned slag/matte mixtures generated from smelting Cu–Co ores in the Copperbelt Province of Zambia;[26] the maximum values reported in Table 3.2 are from one of these samples with 2.4 wt% Co, 7510 mg kg^{-1} of Cr, and 935 mg kg^{-1} of Ni. The sample also contained 8.6 wt% Cu and greater than 1000 mg kg^{-1} of Pb and Zn. Several slag samples not mixed with matte from this study contained over 1 wt% Co and are used for the pyrometallurgical recovery of Co.[26]

As metallurgical processing techniques have improved over time, an expectation is that modern slags should contain lower concentrations of targeted elements than historical slags that may have used more primitive techniques, as discussed by Sáez *et al.*[57] for Cu in slags from Huelva, Spain. However, as mentioned in a review of Cu slags by Potysz *et al.*,[4] weathering of slag dumps may leach trace elements leaving metal-depleted slag. Roman-era slags from São Domingos, Portugal, in the Iberian Pyrite Belt contain lower concentrations of base metals such as Cu and Zn compared to modern slag from the same location.[60] Álvarez-Valero *et al.*[60] suggested that the concentrations of these trace elements in both slag types may have been similar when each formed and that weathering has leached them from the Roman slags.

3.3.3.2 Pb and Zn Slags

This section discusses slags from both Pb and Zn ores due to the common co-occurrence of these elements in nature. Slags in this section are divided based on which element(s) are the priority for recovery, either Pb or Zn, or Pb and Zn. Other elements, such as Ag, As, Au, Bi, and Cu, are commonly geologically associated with Pb and Zn ores and may be recovered simultaneously so slags generated during processing of these types of ore are also included in this section. Differentiating between Pb, Zn, and Pb–Zn slags may be complicated by the disposal of more than one type of slag at legacy sites where reconstructing the exact smelting history is not possible, which is often the case. Therefore, Zn and Pb slags were categorized in Table 3.2 based on the element(s) recovered as identified by the authors of the source reference; if the commodity was not specified, categorization was made based on the site and sample descriptions. Most slags are from primary production but slags from recycling are also mentioned.

Primary Pb slag is from the smelting of Pb, Pb–Ag, or Pb–Zn roasted ore concentrates in furnaces, usually blast or reverberatory; less often, Pb slag is produced from direct smelting. These slags include those generated from Pb–Zn ores where only Pb is mentioned as being recovered. Lead slag may be fumed to volatilize and then condense the Zn for recovery. Slags remaining after the Zn fuming process are categorized as Zn–Pb slags in Table 3.2. Electric furnaces, Imperial Smelting Furnaces (ISF), Waelz furnaces, or direct smelting technologies can be used to recover both Pb and Zn. Slags from smelting Pb–Zn ores from which Zn or Zn and Pb were recovered are also included in the Zn–Pb slag category in Table 3.2. In other words, Zn–Pb slags include slags generated from the processing of Pb–Zn or Zn–Pb ores where Pb and Zn were recovered, or where only Zn was recovered. Lastly, for slag from primary sources, Zn slags are generated from the processing of Zn ores without a mention of the co-occurrence or recovery of Pb. Pyrometallurgical processing of Zn ores can occur in a Zn blast furnace, ISF, or vertical retort process. However, currently, most Zn ores are processed *via* hydrometallurgical processing. Only two studies are categorized as Zn slags. Barna *et al.*[61] characterized slags generated during Zn extraction in an ISF at a site in France while Puziewicz *et al.*[62] examined slags from Zn production at the Świętochłowice smelter, Poland, which used a series of horizontal furnaces to oxidize, reduce, vaporize, and then condense Zn. Both of these slags contain around 1 wt% Pb, suggesting that the Pb was geologically associated with the processed ores. Nonetheless, these two studies are included in a separate Zn-only slag group, even though they might be more appropriately grouped with some samples in the Zn–Pb slag group (Table 3.2 and Figures 3.2–3.4). Secondary Pb slags are included in Table 3.2 and are generated from the recycling of mainly Pb–acid batteries, but also steelmaking dust and Pb scrap. The recycled materials are processed by direct smelting reduction, electric furnaces, reverberatory furnaces, and plasma arc furnaces.

Lead and Zn slags from natural ores are mainly composed of FeO and SiO_2 with major oxide components shown in Figures 3.2 and 3.3 and listed in Table 3.2. In general, CaO is the next most abundant component. There is a wide range of concentrations of FeO, SiO_2, and CaO that generally overlap for Pb and Zn slags generated from the smelting of natural ores (Figure 3.2). Lead slags on average have a slightly higher amount of FeO (26.3 (±7.8) wt% FeO) and a lower amount of CaO (average 9.5 (±7.2) wt% CaO) than Zn–Pb slags (averages of 16.7 (±10.1) wt% FeO and 17.3 (±7.2) wt% CaO); the few Zn slag samples are similar in Fe and Ca composition to Zn–Pb slags (Table 3.2 and Figure 3.2). Aluminum, Mg, and Mn are the next most abundant major elements followed by S with concentrations varying based on the type of slag (Table 3.2 and Figures 3.3 and 3.4). Aluminum and Mg are generally higher, Mn lower, and S similar in Zn–Pb slag when compared to primary Pb slag and Zn slag (Table 3.2). Potassium (as K_2O), Na (as Na_2O), and Ti (as TiO_2) are present in trace to minor amounts in all Zn and Pb slags with averages commonly of 1 wt% or less.

The compositions of primary and secondary Pb slags are slightly different, as shown in Table 3.2 and Figures 3.2 and 3.3, and are similar to those reported by Pan *et al.*[6] Secondary Pb slags generally contain higher amounts of FeO, averaging 43.4 (±8.1) wt%, compared to primary Pb slags with an average of 26.3 (±7.8) wt% FeO (Table 3.2). Silica is usually lower in secondary Pb slags (average of 14.9 (±10.5) wt% SiO_2), compared to primary Pb slag with an average of 29.4 (±14.9) wt% SiO_2, whereas Ca concentrations are similar among both types (averages of 9.5 (±7.2) wt% CaO for primary and 9.4 (±9.1) wt% CaO for secondary) (Table 3.2). In addition, the Al_2O_3, MgO, and S concentrations are comparable (Table 3.2).

Lead and Zn slags are enriched in Pb and Zn with concentrations that nearly always exceed soil guidelines, as illustrated in Figure 3.4. The concentrations of Pb in primary and secondary Pb slags reach 31.9 wt% and 22.0 wt%, respectively, and both average 8.9 wt%. The primary Pb slags most enriched in Pb are from the 14th century C.E.: Pb–Ag slag from the Czech Republic (2.8 to 31.9 wt% Pb)[19] and a slag generated during the refining of Pb bullion from Pb–Zn ores in Poland (13.9 to 20.2 wt% Pb).[63] Primary Pb slags for the other studies compiled in Table 3.2 contain 6.3 or less wt% Pb. In general, Zn–Pb and Zn slags contain less Pb than Pb slags with maximum values of 6.2 wt% and 1.5 wt%, respectively, and averages of approximately 1 wt% for both types. Excluding the two groups of Pb slags with anomalously high Pb contents,[19,58,63] the remaining Pb slags have generally comparable concentrations of Pb to those found in Zn–Pb and Zn slags. Similar to Pb, the concentrations of Zn in primary Pb, Zn–Pb, and Zn slags all have similar averages at 5.0 wt%, 4.0 wt%, and 7.0 wt%, respectively; secondary Pb slag contains less, with an average of 0.5 wt% (Table 3.2). The studies that report Zn content in slags at greater than 10 wt% for most samples from a site are for primary Pb blast furnace slags from France[61,64] and Australia.[29] Obviously, Zn was not targeted for recovery during the processing that generated these slags. The higher amount of Zn in some Pb slags compared to Zn–Pb slags for which both Zn and Pb may have been recovered, in particular those generated in an ISF, was also mentioned in Chapter 2.

Other trace elements that are enriched in Pb and Zn slags and that may exceed soil guidelines include As, Cd, Cu, Ni, Sb, and V (Table 3.2 and Figure 3.4). Arsenic was reported at around 1 wt% for one anomalous Zn slag (Pb–Zn ore) and one Pb–Zn slag for suites of samples collected at two different locations in Poland.[15,62] Additionally, Pb slags generated from refining Pb bullion from Poland consistently contain approximately 1 wt% As, as well as the highest Cd and Cu reported in Pb and Zn–Pb slags with concentrations ranging from 1600 to 3500 mg kg^{-1} for Cd and from 10.8 to 16.5 wt% for Cu;[63] Sb is also high in these slags at up to 4400 mg kg^{-1}. Interestingly, the slags were disposed of as waste according to Nowinska.[63] Excluding the Pb ladle slags reported by Nowinska,[63] Cu and Ni are both, on average, higher in the secondary Pb slag (0.4 wt% Cu and 564 mg kg^{-1} Ni) compared to the Pb and Zn–Pb slags complied in Table 3.2 and shown in Figure 3.4. Lead-acid batteries can contain a core of Cu in the composite grids in the battery, and therefore could possibly be the source of Cu in the secondary Pb slag. Excluding the anomalous concentrations in Pb ladle slag, Cd, Ni, and Sb are frequently present at hundreds of mg kg^{-1} or less and Cu at thousands of mg kg^{-1} or less in Pb and Zn–Pb slags (Figure 3.4). Another example of anomalous trace element concentrations in slag is Sb, which ranges from 599 to 5490 mg kg^{-1} for 14th century Pb–Ag slags from the Czech Republic.[19]

3.3.3.3 Ni Slags

Nickel slags are generated from either laterite (silicate-oxide) or sulfide ores to produce ferronickel alloy to be used mainly in stainless steel or to produce refined Ni metal to be used in applications such as catalysis or chemicals. In Chapter 2, the metallurgical processes that generate these types of slag are discussed under ferroalloy production as well as under non-ferrous metal production. For this chapter, they are included under base metal slags.

Pyrometallurgical extraction of Ni from Ni sulfide ores utilizes flash, electric, or reverberatory furnaces to smelt concentrates that may have been dried or roasted. Nickel laterite ores are commonly calcined, partially reduced, and then smelted and refined in electric furnaces. Smelting of sulfide ores generates a Ni-rich, Fe-poor matte, whereas the smelting of laterite ores generates a FeNi alloy.

The major element chemistry of the Ni slag generated from sulfides is different from those generated from laterites in Table 3.2 and Figure 3.2. The major element chemistry for Ni slags is also reported in Table 2.13 compiled from several studies referenced in Chapter 2 with compositional values comparable to those presented in this chapter. On average, Ni slag from sulfide ores contains higher amounts of Fe and lower amounts of Al, Ca, Mg, and Mn compared to Ni slag from laterite ores. Remarkably, the Mg content in Ni laterite ore slag reaches up to 31.2 wt% MgO from the mafic to ultramafic minerals that weathered to generate the laterite ores (Table 3.2 and Figure 3.3). Silica compositions are similar for both slag types, averaging approximately 40 wt% SiO_2. In general, Ni laterite ore slags are composed of Al_2O_3, CaO, FeO, MgO, and SiO_2, whereas Ni sulfide ore slags are mainly FeO and SiO_2 (Figures 3.2 and 3.3). Additionally, S is lower (less than 0.2 wt%) in Ni laterite-generated slag compared to 0.9 to 1.8 wt% in Ni sulfide-generated slag due to the lack of sulfide minerals in the ores (Table 3.2).

Trace elements such as Co, Cr, and Cu are enriched in some Ni slags, commonly exceeding soil guidelines (Figure 3.4). The amount of Co present in Ni sulfide ore slag is consistently higher than in Ni laterite ore slag and averages 1699 (±623) mg kg^{-1}, compared to 199 (±254) mg kg^{-1} in the latter. Copper is particularly enriched reaching over 1 wt% in most Ni slag samples from processing Cu–Ni–PGM ores at Sudbury, Ontario, Canada.[65] The slags from Sudbury also contain approximately 1 wt% Ni and over 1000 mg kg^{-1} Zn. The high trace element content of the Ni sulfide ore slags varies widely and depends on the sulfide mineralogy of the ores. The potential to recover Co, Cu, Ni, and Zn from the enriched slags from Sudbury was investigated by Li *et al.*[65] with favorable results. Similar to Co and Cu enrichment in Ni slag from sulfide ores, Cr is anomalously high in Ni slags from laterite ores due to Cr being high in the weathered mafic and ultramafic minerals in the ores; Cr reaches over 1 wt% in these slags (Table 3.2). Other trace elements, such as Cd and Pb, are not notably enriched in Ni slags due to them not being present in the ores (Table 3.2).

3.3.3.4 *Other Types of Non-ferrous Slag*

Some types of base metal slag do not have an abundance of literature describing their geochemistry and information regarding their chemical composition may not be readily available in the literature. Some types are merely not generated in large quantities or are mostly recycled within the processing plant, resulting in a minimal footprint in the environment. This section highlights the compositions of several slag types that are less typically the focus of environmental or resource recovery related research. They include slags generated from the pyrometallurgical processing of natural ores or from the recycling of secondary materials.

Platinum-group metals are either recovered as a by-product of Ni–Cu deposits or are recovered as the main commodity with Ni and Cu and other elements as by-products (see Chapter 2, Section 2.4.6). Slags generated from deposits where Ni and Cu are the dominant commodities, such as Sudbury, Ontario, Canada, are discussed in Section 3.3.3.3. Slags generated from the smelting of PGM-dominated ores, such as the Bushveld Igneous Complex in South Africa, are included here. Other elements recovered from the Bushveld ores include Au, Cr, Cu, Ni, and V. Jones[66] noted compositions of slags generated from processing ore from several mines in the Bushveld Complex with the table reproduced as Table 2.14 in Chapter 2. The smelter furnace slags are dominated by SiO_2 followed by FeO and MgO and then CaO and Al_2O_3 in abundance, whereas converter slags contain FeO in higher concentrations than SiO_2 and only a few weight percent or less of Al_2O_3, CaO, and MgO (Table 2.14). The major element compositions of the slags from the Bushveld are generally slightly different to those reported for slags generated from other Ni sulfide-bearing ores (Table 3.2). However, the uniqueness of these slags is the amount of Cr they contain, with concentrations ranging from 0.4 to 2.8 wt% Cr_2O_3; these values are more similar to the average Cr reported in Ni laterite ore slags of 1.0 wt% Cr_2O_3 (0.7 wt% Cr).

Another type of slag, referred to as salt slag or salt cake, is produced as a waste product generated during recovery of Al from Al scrap in either a rotary or reverberatory furnace. This type of slag is unique in that it is composed mostly of oxides, nitrides, and chlorides, instead of oxides and silicates like previously discussed base metal and ferrous slags. In a review on salt slag, Tsakiridis[67] referred to several studies from around the world; in summary, salt slags are predominantly composed of Al (20 to 40 wt% Al_2O_3), followed by Cl and Na in abundance (approximately 10 to 20 wt%). Nitrogen and K can vary from a percent to up to almost 10 wt%, with lesser amounts of F, Mg, Si,

Fe, Ca, C, and P, all usually present at a few weight percent or less. Stated in terms of chemical compounds, the salt slag is 15 to 30 wt% Al_2O_3, 30 to 55 wt% NaCl, 15 to 30 wt% KCl, and 5 to 7 wt% metallic Al and impurities such as carbides, nitrides, sulfides, and phosphates.[67] In another study, Jody *et al.*[68] stated that salt slag from the USA was approximately composed of the following compounds: 3 to 5 wt% Al, 15 to 30 wt% Al_2O_3, 30 to 40 wt% NaCl, and 20 to 30 wt% KCl, generally consistent with Tsakiridis.[67] Trace elements present include Pb, Ti, V, and Zn, among others, depending on the furnace charge.[67] Currently, salt slag is commonly treated to recover any entrained metals, then leached and crystallized to reform the salts to be used again as flux in the furnaces. However, historically it was disposed of as waste, despite its toxic and hazardous characteristics.[67]

Other types of slags are tin (Sn) slags produced during pyrometallurgical processing of cassiterite (SnO_2) ores. Allain *et al.*[69] examined Sn slags from Asia, Africa, and Europe attributing the variations in bulk chemistry to the differences in the geochemistry of the deposits and smelting conditions. The compositions of these slags are as follows: 11.9 to 41.9 wt% SiO_2, 11.6 to 29.2 wt% CaO, 4.7 to 11.2 Al_2O_3, 0.9 to 27.7 wt% FeO, 1.3 to 17.8 wt% TiO_2, 0.5 to 8.4 wt% ZrO_2, and 0.4 to 3.7 wt% MnO, and less than 1.4 wt% WO_3.[69] Gaballah *et al.*[70] reported the bulk composition of an Sn slag (sample origin not specified) within the ranges reported by Allain *et al.*,[69] with that slag mainly composed of SiO_2 (41.9 wt%), CaO (11.6 wt%) and Al_2O_3 (11.2 wt%), with lesser amounts of MnO, FeO, K_2O, and TiO_2 ranging from 1 to 4 wt% (listed in order of decreasing abundance). The concentration of Sn in the slag reported by Gaballah *et al.*[70] is 0.7 wt% SnO_2, whereas slags reported by Allain *et al.*[69] contained between 0.1 wt% and 1.6 wt% SnO_2. Interestingly, Nb and Ta are concentrated in the Sn slags with concentrations of 5.2 wt% Nb_2O_5 and 7.5 wt% Ta_2O_5 reported by Gaballah *et al.*[70] Furthermore, the Sn slags from Asia, Africa, and Europe contain from 1 to 25 wt% $(Ta + Nb)_2O_5$.[69] Research exists on the recovery of these strategic elements from Sn slags.[69–71] Additionally, the reader is referred to Chirikure *et al.*[72] for the chemistry of Sn slags from archaeological sites in South Africa.

Lastly, the recycling of postconsumer waste, including batteries, magnets, and computer circuit boards, may involve pyrometallurgical processing and generate slags. In addition, municipal solid waste may be thermally degraded, sometimes to recover heat or electrical energy. A bottom slag may be generated in this process. These types of slags from recycling or waste treatment have been studied as materials to target for element recovery or use in construction. A thorough review of their diverse chemical compositions is beyond the scope of this chapter.

3.3.4 Chemical Diversity of Slags

The various types of ferrous and non-ferrous slag have diverse chemical compositions. However, some types stand out as being more enriched in certain elements when compared to other slag types. Moreover, some major differences in composition exist for non-ferrous slags compared to ferrous slags. In general, modern BF slags are dominated by Ca and Si, steel slags are dominated by Ca, Si, and Fe, and base metal slags, most produced from processing sulfide ores, contain predominantly Fe and Si (Figure 3.2). This generalization does not consider the highly variable composition of slags and that other elements are also present, sometimes in significant concentrations (Tables 3.1 and 3.2). Nonetheless, base metal slags in Table 3.2 generally contain higher concentrations of Fe (averages range from 16.2 to 46.0 wt% FeO) than modern BF slags (average 2.6 wt% FeO) and ladle slags (4.3 wt% FeO) (Table 3.1). In addition, the more Fe-rich steel slags contain similar ranges of Fe concentrations to base metal slags (averages between 19.6 and 27.8 wt% FeO) (Table 3.1). In general, Ni sulfide ore slags, Cu slags, and secondary Pb slags contain the highest average concentrations of Fe among non-ferrous slags (Table 3.2).

The average SiO_2 contents for the individual base metal slag types (excluding secondary Pb slags) and modern BF slags are similar and range from 23.1 to 39.7 wt%. In comparison, the SiO_2 contents of steel slags are lower, containing, on average, 12.0 to 21.1 wt% SiO_2. In contrast, the concentration of Ca is generally higher in ferrous slags, especially steel slags, compared to most base metal slag types (Tables 3.1 and 3.2, Figures 3.2 and 3.3). The average concentration of CaO is 18.3 wt% or less for the base metal slag types (Table 3.2). In contrast, modern BF and steel slags generally contain more Ca and are similar to each other with average concentrations of CaO ranging from 31.0 to 46.6 wt%. The BOF and ladle slags commonly contain more Ca than EAF slags. The next most abundant elements after Si, Ca, and Fe for base metal slags and ferrous slags are Al and Mg. Maximum concentrations of Al are 41.2 wt% Al_2O_3 for modern BF slags and 35.3 wt% Al_2O_3 for ladle slags, and average concentrations are 12.5 wt% Al_2O_3 (modern BF slags) and between 3.6 and 6.9 wt% Al_2O_3 (BOF, EAF, and steel and legacy slag types) (Table 3.1). Ladle slags are the most Al-rich of the ferrous slags and contain an average of 18.3 wt% Al_2O_3 (Table 3.1).

For base metal slags, averages are similar and range from 4.2 wt% to 10.9 wt% Al_2O_3 for the various types, with the highest maximum concentration reported for Zn–Pb slags (21.9 wt%) (Table 3.2). Magnesium is generally higher in ferrous slags (averaging from 6.3 to 8.2 wt% MgO for modern BF and steel slags) than most base metal slags (averaging from 1.5 to 5.4 wt% MgO for all types except Ni laterite ore slag) (Tables 3.1 and 3.2). The exception is Ni laterite ore slags, which can contain up to 31.2 wt% MgO and average 16.5 wt% MgO; Ni laterite ore is derived from the weathering of mafic and ultramafic rocks, which contain abundant Mg.

Since base metals are typically pyrometallurgically recovered from sulfide-mineral ores, compared to Fe-oxide ores for Fe recovery, there is generally more S present in base metal furnaces compared to ironmaking and steelmaking furnaces. The reducing conditions in the Fe blast furnace and some steel refining furnaces facilitate entrapment of S in the slag to avoid its presence in the Fe and steel (*i.e.*, desulfurization). Due to relatively low S in the furnace charge compared to base metal slags, ferrous slags are low in S with averages between 0.1 and 1.0 wt% with the highest maximum and average concentrations found in modern BF slags and ladle slags in Table 3.1. In comparison, the average S contents in base metal slags in Table 3.2 are from 1.1 to 5.6 wt% S, with the exception being Ni laterite ore slags, which contain an average of 0.1 wt% S. Maximum S concentrations reached 22.0 wt% for Pb slags from the processing of Cu processing wastes due to the entrapment of matte in the slag.[73] The entrapment of matte likely explains the high S reported in various other studies on base metal slags.[58,63] The high S that reached 6.4 wt% in 13th century C.E. Cu slags in Italy was attributed to uncomplete roasting according to Manasse *et al.*[74]

Another element that can be found in major amounts in slag is Mn, which is generally higher in steel slags (Table 3.1; average approximately 4 to 5 wt% and reaches up to 21.5 wt% MnO) than in other slag types. However, ferroalloy slags from the production of FeMn contain the highest amounts of Mn (average 42.2 wt% MnO for duplex slag, Table 2.6). Some Pb slags can contain Mn, reaching up to 9.4 wt% MnO (Table 3.2). Slags that are consistently enriched in Mn (5.6 to 9.4 wt% MnO) are from the Zeehan historical smelter site in Tasmania, Australia, which processed Ag-, Pb-, Zn-, and Cu-rich ores from replacement, vein, skarn, and volcanic-hosted massive sulfide deposits.[29] Besides the major elements, minor elements, such as K, Na, and Ti, are found in similarly low concentrations in ferrous slags as in base metal slags (usually less than 1 wt% on average).

As for trace elements, ferrous slags are commonly higher in certain elements when compared to non-ferrous slags. BF slags and steel slags contain higher amounts of Cr and Mn and lower amounts of As, Cd, Co, Mn, Ni, Pb, and Zn than most non-ferrous slags (Tables 3.1 and 3.2, Figure 3.4). However, the concentrations of Ni and Zn in steel slags reach those of some base metal slags. These elements, along with Cr and Mn, are likely to be from the alloys and steel scrap added during steelmaking. Nickel laterite ore slags contain similar amounts of Cr to steel slags due to the Cr present in the ores. Similarly, Cr is particularly enriched in FeCr ferroalloy slags. As shown in Figure 3.3, base metal slags are generally enriched in the element(s) targeted during extraction because they are the most abundant trace elements present and regularly concentrated in pre-smelting processing steps (Figure 3.4). On average, Cu slags contain the highest Cu (2.0 (±4.6) wt%) and Ni slags contain the highest average Ni (0.6 (±0.4) wt% sulfide ores and 0.2 (±0.2) wt% laterite ores), primary and secondary Pb slags contain the highest Pb (8.9 (±8.7) wt% for primary and 8.9 (±9.7) wt% for secondary), and Zn slags contain the highest Zn (7.0 wt%) (Table 3.2). However, some overlap exists for Pb and Zn content, as shown in Figure 3.4, where their concentrations reach similarly high values in Cu, Pb, Zn–Pb, and Zn slags. Cadmium reaches the highest concentrations in Cu and Pb slags and Co reaches the highest in Cu slags.

Vanadium, although not consistently reported in slag studies, is enriched in some ferrous slags, in particular steel slags, compared to most base metal slags. BOF slags have an average of 0.5 (±0.8) wt% V in Table 3.1 and several slags from Sweden contain greater than 1.0 wt% V.[75,76] In addition, V is noted to be particularly high in one base metal slag study and averages 0.6 wt% V in historical Pb–Zn slags in Namibia from processing Zn–Pb–V-rich ores.[77] Other elements that are not consistently reported in slag studies but may be enriched in slags include As (averaging over 1000 mg kg^{-1} in Cu, Pb, Zn–Pb, and Zn slags), Au (averaging 345 and 403 mg kg^{-1} in Cu and Pb slags, respectively), Sb (averaging 1538, 1424, and 785 mg kg^{-1} in primary Pb, secondary Pb, and Cu slags, respectively), and Se (averaging over 100 mg kg^{-1} in primary Pb slag). Indium is present at concentrations between 310 and 480 mg kg^{-1} in Pb slag from the refining of Pb bullion;[63] for the few other slag studies that report In, it may occur at 6 or 7 mg kg^{-1}[24,78] but is more commonly less than 1 mg kg^{-1}.

Some elements, such as rare earth elements (REEs), are occasionally reported whereas others, such as PGMs, are rarely, if ever, reported. The sum of the REEs reaches up to 140 mg kg^{-1} in Pb slags,[79,80] up to 350 mg kg^{-1} in Cu slags,[79,81] up to 420 mg kg^{-1} in steel slag,[49,82–84] and up to 440 mg kg^{-1} in BF slags.[49,82,85] Concentrations of the PGMs were reported by Lottermoser[86] as 4 mg kg^{-1} or less for palladium (Pd) and 1 mg kg^{-1} for platinum (Pt) for several base metal slags from Australia.

3.4 Mineralogical Characterization

The slag chemistry and the rate and method of cooling after the slag is tapped from the smelting and refining furnaces determine the crystalline and amorphous phases that form and their textures. This section focuses on common phases found in slag (Figure 3.5) and their textures (Figure 3.6).

3.4.1 Textures

Many slag textures are analogous to igneous rock textures because they form from the cooling and solidification of a melt. As was mentioned in Chapter 2, molten slag may be air-cooled, granulated, expanded, or pelletized.

Air-cooled slags are commonly crystalline silicate and oxide phases with interstitial glass and fine-grained immiscible sulfide and/or metallic blebs. Glassy rinds can be found on the edges of slag pots or on the exterior portions of slag dumps where cooling occurred rapidly. Crystallinity increases away from these chilled margins toward the center of the mass. Flow textures may be evident in slag deposited in the molten state and allowed to cool under atmospheric conditions.

Olivine-group phases are abundant in most types of air-cooled slags, both ferrous and base metal; they frequently display a spinifex texture composed of large dendritic crystals, either oriented randomly or parallel (Figure 3.6A–C). Spinifex texture is frequently found in komatiites, a type of ultramafic mantle-derived rock that crystallized at high temperature from an Mg-rich melt typically of Archaean age. According to Shore and Fowler,[88] the characteristic features of komatiites are spinifex olivine, millimeters to decimeters

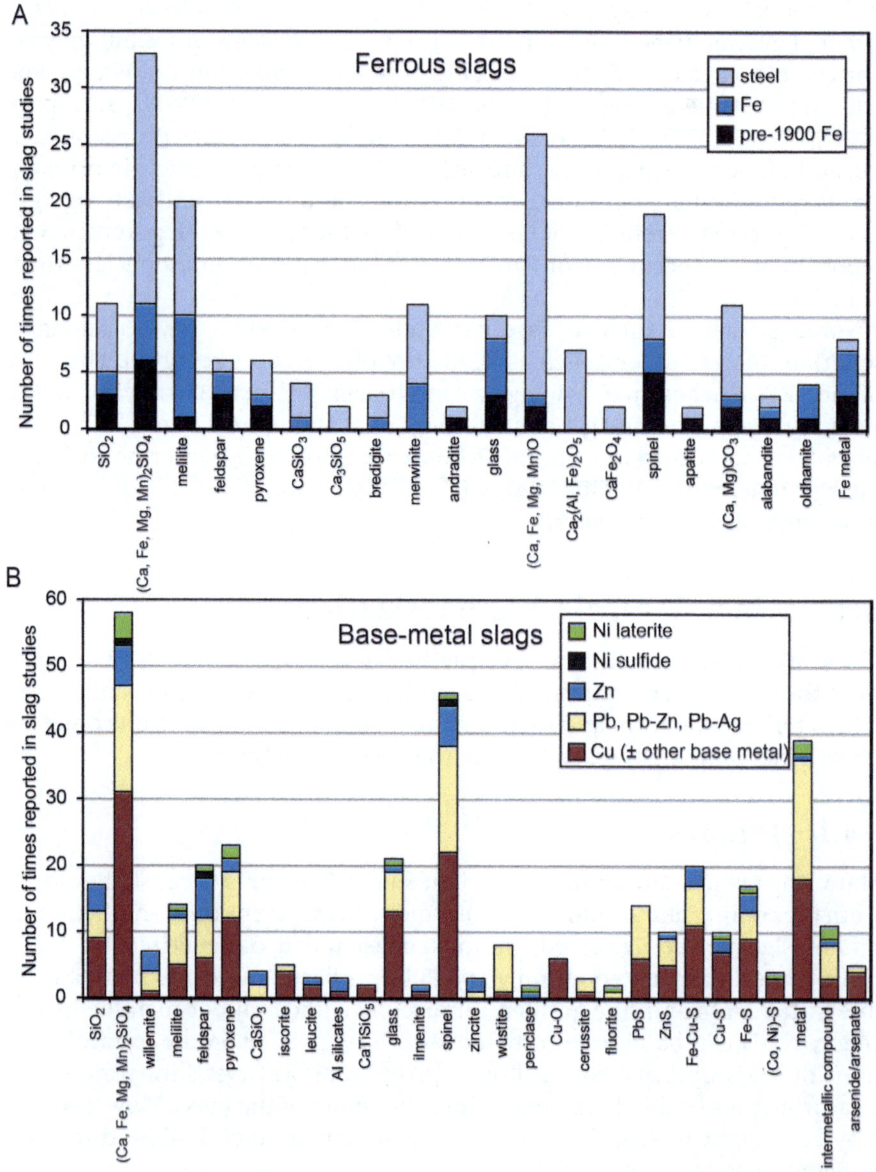

Figure 3.5 Number of times primary phases were identified in (A) ferrous slags and (B) non-ferrous slags. Only phases that were identified repeatedly (at least twice) were included in the figure. References are provided in Piatak *et al.*[1] Reproduced from ref. 1 with permission from Elsevier, Copyright 2014.

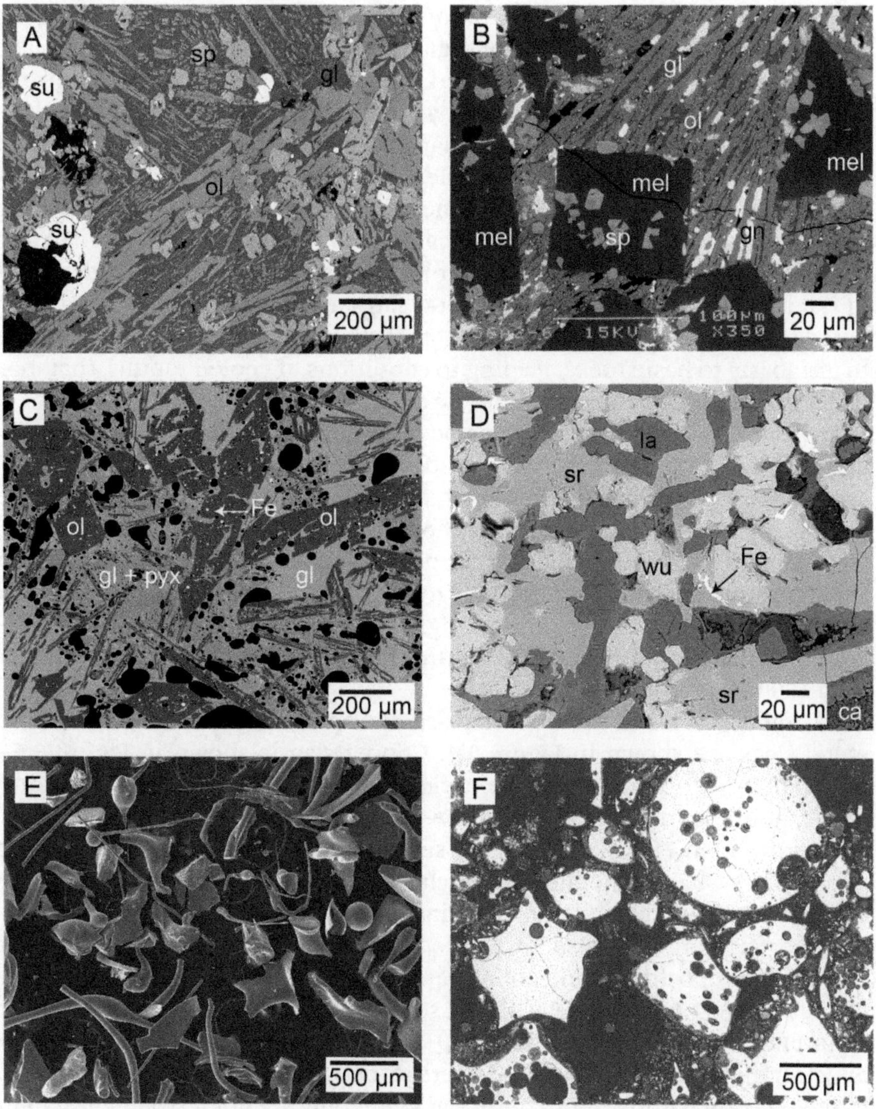

Figure 3.6 Backscattered scanning electron and secondary electron photomicrographs of slag. (A) Cu slag from the Elizabeth mine, Vermont, USA. Image shows randomly oriented elongate skeletal laths of fayalite olivine (ol), a euhedral spinel (sp)-group phase, and fine to coarse-sized Cu–Fe–sulfide blebs (su) set in a glass (gl) matrix. Reproduced from ref. 24 with permission from Elsevier, Copyright 2004. (B) Pb–Zn slag from Příbram, Czech Republic with prismatic crystals of melilite (mel) including euhedral crystals of a zoned spinel (sp)-group phase, accompanied by long needles of olivine (ol) in the matrix glass (gl), with galena inclusions (gn). Reproduced from ref. 87 with permission from the Mineralogical Association of Canada, Copyright 2001. (C) Pre-1900 blast furnace Fe slag from the Hopewell Furnace National Historic Site in Pennsylvania, USA. Image shows skeletal forsterite olivine (ol) laths

in length, in a fine-grained matrix of spherulitic clinopyroxene, dendritic spinel, of the chromite variety, and altered glass. This mineral assemblage and textural description are similar to air-cooled slag, likely due to analogous cooling conditions with a large thermal gradient and commonly mafic/ultramafic melt compositions, in particular for base metal slags and some steel slags. However, there are several differences in the characteristics of slags and komatiites. First, pyroxenes are not always present in slags and melilite is frequently reported. According to Piatak *et al.*,[1] pyroxene is reported in less than half the number of studies that olivine and spinel phases are identified in (Figure 3.5). In addition, base metal slags are more enriched in Fe compared to Mg, unlike komatiites, yet both bulk compositions are comparable to ultrabasic to basic rocks, leading to conditions, if cooled rapidly, that may be favorable for developing spinifex texture.

The morphology of crystals has been frequently used to experimentally determine the solidification and cooling rates of basaltic melts and other volcanic rocks.[89,90] Similar approaches have also been adopted to estimate the cooling regimes in slags, especially those issued from (pre)historic smelting technologies.[19,74,91] Prismatic and "hopper" olivine crystals in slag indicate a cooling rate corresponding to lower hundreds of $°C\ h^{-1}$, chain-like skeletal crystals formed at a cooling rate of higher hundreds of $°C\ h^{-1}$, and "herringbone" textures and fine dendrites formed during rapid quenching at greater than 1500 $°C\ h^{-1}$.[19]

Several examples of the type of air-cooled spinifex textures found in base metal slags are shown in Figure 3.6. Copper slag is shown in Figure 3.6A with randomly oriented elongate skeletal laths of fayalite olivine, euhedral spinel, and fine to coarse-sized Cu–Fe–sulfide blebs within a glass matrix.[24] Furthermore, in Figure 3.6B, Zn–Pb slag from the Czech Republic contains prismatic crystals of melilite with inclusions of euhedral spinels, which are accompanied by skeletal laths of olivines in a glass matrix.[87]

Some ferrous slags also display spinifex textures depending on their chemical composition and cooling rate. Modern BF slags are generally low in Fe (average 2.6 wt% FeO) whereas steel slags contain significantly higher Fe (averages of 19.6 to 27.8 wt% FeO) (Table 3.1). Each of these slags contains similar ranges of Mg that average between 6.3 and 8.2 wt% MgO. Based on composition, modern BF slags generally are not ferromagnesian in

Figure 3.6 *(continued)* and Fe metal blebs (Fe) within a background of glass only (gl) and areas of glass and dendritic and skeletal pyroxenes (pyx). Also note abundant vesicles (black circular shapes) Reproduced from ref. 23 with permission from Elsevier, Copyright 2012. (D) Legacy ferrous slag from the Chicago area, Illinois, USA. Intergrown srebrodolskite (sr), larnite (la), and wüstite (wu) with Fe metal (Fe) blebs. Also labeled is secondary calcite (ca) in a crack. Reproduced from ref. 2 with permission from Elsevier, Copyright 2018. (E) Granulated Cu slag from Tsumeb, Namibia, showing glassy fragments resulting from rapid quenching. (F) Granulated Fe slag from the Chicago area, Illinois, USA. Image shows glassy fragments containing abundant vesicles mounted in epoxy resin (black).

composition, whereas steel slags and some historical BF slags may be, influencing which melts are conducive to forming spinifex textures upon rapid cooling. For example, pre-1900 BF slags from the USA contain, on average 15.0 wt% MgO and 3.1 wt% FeO, and display skeletal forsterite olivine laths and Fe blebs within a matrix of glass only or glass and dendritic pyroxenes (Figure 3.6C).[23]

Besides spinifex textures, air-cooled slags can display euhedral and anhedral crystal growth without the dendrites or elongate laths. For example, the legacy ferrous slag in Figure 3.6D shows subhedral to anhedral larnite (Ca_2SiO_4), srebrodolskite (Ca_2FeO_5), wüstite (FeO), and metallic Fe. The metallurgical origin of this sample is unknown due to sample collection at a legacy disposal site. However, the textures and mineralogy are similar to those reported for a slowly cooled BOF slag that also displayed subhedral to anhedral grains of larnite and srebrodolskite, but contained periclase (MgO), not wüstite.[92]

A unique air-cooled slag texture was reported by Warchulski *et al.*[93] for historical Zn slags in Poland. Coarse-grained (up to 2.5 cm in diameter) segregations (up to 40 cm long) with fine-grained (generally less than 1 mm in diameter) rims were documented in crystalline slags lacking a glass matrix. The pegmatite-like segregations, some massive and some occurring in cavities, are composed of subhedral crystals of melilite, leucite, spinel-group phases, clinopyroxene, and hematite, whereas the rims consist of olivine, spinel-group phases, melilite, leucite, nepheline, and sulfides.[93] According to Warchulski *et al.*,[93] the pegmatite-like areas formed from a volatile-rich slag melt overlain by a vesicular slag layer that prevented rapid cooling and the escape of the volatiles. These conditions were compared to igneous rock conditions that formed layered or tabular intrusions with pegmatoidal structures.[93]

Expanded or pelletized slags are lightweight porous materials. Expanded slags are generated by using a spray of water that contacts a molten slag, forming steam, which expands the slag. Alternatively, the molten slag may be agitated in a confined space with a controlled amount of water, entrapping the steam and causing expansion. Expanded slags are like pumice in their porosity and are crushed to produce an angular, roughly cubical in shape material.[94] According to Ramachandra Rao,[94] the bulk relative density of expanded slag is approximately 70% of that of air-cooled slag. Unlike expanded slags, pelletized slags are formed by water and air quenching in a spinning drum. During pelletization, the molten slag is thrown into the air during cooling for a long enough time for surface tension to form pellets. Round and smooth pellets are formed from the gas bubbles projected through a water spray. Pellet sizes range from less than 0.1 mm to 13 mm and have a smooth texture and rounded shape.[94] Coarser pellets (tens of millimeters in diameter) are more crystalline whereas finer pellets are more similar to vitrified slags.[95] Therefore, pelletized and expanded slags, depending on their pellet size, slag composition, and cooling rate, may contain crystalline phases similar to air-cooled slags or be composed predominantly of amorphous glass, or a mixture of both crystalline and glass phases. Often, the glass is most abundant with only minor or trace amounts of crystalline phases.[96,97]

Granulated slags are glassy materials produced from quenching molten slag with high-volume, high-pressure sprays of water. The cooling process prevents crystallization and generates vitrified sand-sized fragments that may be further ground. Figure 3.6E and F showing granulated glassy Cu and Fe slags, respectively, illustrate the angular as well as rounded grains that may form during the rapid cooling. Granulated slags may contain minor to trace amounts of crystalline phases, such as silicates, oxides, or intermetallic and sulfidic compounds, depending on the cooling rate and its composition. Slags rich in certain elements, such as SiO_2, vitrify when rapidly cooled; however, some slag compositions have been shown to form crystalline structures despite being cooled rapidly. In addition, the temperature at which initial crystallization occurs from a cooling melt depends on the melt composition. Kang *et al.*[98] found that as the basicity (*i.e.*, CaO to SiO_2 ratio) of a slag melt increased, the temperature at which phases precipitated increased; thus, less basic melts may be more suitable for preparing amorphous slag granules. Another study by Tossavainen *et al.*[75] found that rapidly cooled melted ladle slag with abundant Al_2O_3, MgO, and SiO_2 (12.6 to 22.9 wt% of each oxide) formed a nearly completely amorphous material (98 wt% glass) with trace crystalline periclase. However, in the same study, the melted and then water granulated BOF and EAF slags (with less than 7.0 wt% Al_2O_3, and SiO_2 varying from 11.1 to 32.2 wt%) formed crystalline materials containing ferrous slag phases (*i.e.*, larnite, merwinite ($Ca_3Mg(SiO_4)_2$), hatrurite (Ca_3SiO_5), (Fe,Mn,Mg)O, and spinel). The BOF and EAF slags contain 1 to 17 wt% glass, with the highest amount of glass found in the most Si-rich slag. In general, BF and ladle slags, as well as some base metal slags such as Cu and Pb slags, have been effectively granulated to form a vitrified product for reuse, whereas others, such as low-Si steel slags, may not produce a vitrified material even if rapidly cooled.

3.4.2 Mineralogy

During the fusion processes in pyrometallurgical furnaces, the melt stratifies according to the density (see also Chapter 2). For instance, four liquids are generally formed in Pb shaft furnace smelting: (i) Pb bullion at the bottom of the furnace, (ii) a layer of speiss (arsenides and antimonides of Cu and Fe), (iii) a layer of matte (metal sulfides), and (iv) a silicate slag floating on the top.[115] Metallic/sulfide melts are immiscible with the silicate melts. The viscosity of the molten slags may hinder settling and the metallic/sulfide droplets that do not have time to settle towards the furnace bottom by gravity will, upon slag crystallization, form droplets entrapped within the silicate matrix (Figure 3.6A–D). Ferrous and non-ferrous slags exhibit chemical variability in their crystalline and amorphous phases and sulfides, as discussed in the following sections.

3.4.2.1 Ferrous Slags

Figure 3.5 provides a compilation from the literature of the frequency of individual phases found in slags. Table 3.3 lists phases found in ferrous slags that may be enriched in metal(loid)s, which are considered potential environmental contaminants, or valuable constituents for possible recovery, or both.

There are several groups of silicates found in ferrous slags, some are analogous to minerals abundant in nature (*e.g.*, olivine, melilite, pyroxene, and plagioclase ($(Na,Ca)(Si,Al)_4O_8$)), whereas others are less typical in the natural environment (*e.g.*, hatrurite and kirschsteinite ($CaFe^{2+}SiO_4$)). The most frequently reported primary silicate phases in ferrous slag are olivine-group phases, melilite, merwinite, crystalline SiO_2 (quartz or cristobalite), and silicate glass, with olivine-group phases being the most common by far (Figure 3.5). Larnite (Ca_2SiO_4) is the most typical composition of the olivine-group phases found in ferrous slag, followed in frequency by monticellite ($CaMgSiO_4$). Larnite is the moderate-temperature monoclinic β-Ca_2SiO_4 polymorph of dicalcium silicate, which may also be present in slags as α-Ca_2SiO_4 (high temperature, orthorhombic bredigite $Ca_7Mg(SiO_4)_4$) and γ-Ca_2SiO_4 (low temperature, orthorhombic belite).[106] Forsterite (Mg_2SiO_4), fayalite (Fe_2SiO_4), and kirschsteinite–glaucochroite ($CaFeSiO_4$–$CaMnSiO_4$) may also form, but are less regularly reported, and their formation depends on the bulk chemistry of the slag melt and the cooling conditions. Olivine-group phases may contain trace to minor amounts of Al, Mn, or Ti in ferrous slags. In air-cooled slags, olivine crystals are euhedral and subhedral in shape or found as skeletal laths (Figure 3.6C and D).

Melilite-group phases are the next most frequently reported phases in ferrous slags, with compositions typically between gehlenite ($Ca_2Al_2SiO_7$) and åkermanite ($Ca_2MgSi_2O_7$). Sodium, Fe, K, Mn, S, and Ti can be present in some melilites in minor amounts. Melilite phases may occur as fine-grained anhedral to subhedral and dendritic crystals in the glass matrix of pre-1900 BF slags from the USA.[23] Merwinite, crystalline SiO_2, and silicate glass are the next most frequently reported silicates and are found in both BF and steel slags (Figure 3.5). Both quartz and, less typically, the high-temperature polymorph cristobalite, occur in these slags. Residual quartz as fractured grains in unmelted inclusions has been documented by Piatak and Seal.[23] Glass is likely ubiquitous, although it is not always reported and is a challenge to quantify in slags.[18] Nearly all of a sample may be composed of glass, in particular for granulated slags (Figure 3.6F), or it may be present in only trace amounts as the matrix for crystalline phases (Figure 3.6C) and at quickly chilled margins of air-cooled slags. The composition of glass in ferrous slags is similar to the bulk chemical composition of the slags in that it is usually dominated by SiO_2, CaO, FeO, Al_2O_3, and MgO, but their amounts vary widely.[1] Other major elements, such as K_2O, MnO, Na_2O, S, and TiO_2, are commonly less abundant and present at a few weight percent or less.[1]

Table 3.3 Examples of main metal(loid)-hosting phases in ferrous slags.

Group	Phase	Formula	Slag type	Contaminants and/or valuable elements (wt%)[a]	Reference
Silicates	Larnite	β-Ca_2SiO_4	BOF	V (0.44 ave.)	Hobson *et al.*[30]
	Forsterite	Mg_2SiO_4	FeCr	Cr (1.6 one value)	Tanskanen and Makkonen[99]
	Glaucochroite	$CaMnSiO_4$	FeMn	Mn (29.4 formula)	Groot *et al.*[100]
	Pyroxene	$(Ca,Na)(Mg,Fe,Al)(Al,Si)_2O_6$	FeCr	Cr (2.24 max)	Tanskanen and Makkonen[99]
	Glass	Complex compositions	BF	Mn (0.88 max), Ti (1.9 max)	Piatak and Seal[23]
	Glass	Complex compositions	FeCr	Cr (5.02 max)	Tanskanen and Makkonen[99]
Oxides	Chromite	$MgCr_2O_4$	EAF, FeCr converter	Cr (46.5 formula)	Horckmans *et al.*,[101] Nakasuga *et al.*[102]
	Fe–V spinel	FeV_2O_4	Converter	V (45.9 formula)	Xiang *et al.*[103]
	Mg–Cr–Al spinel	$Mg(Cr,Al)_2O_4$	FeCr	Cr (36.03 max)	Tanskanen and Makkonen[99]
	Hercynite–magnetite	$FeAl_2O_4$–Fe_3O_4	BF	Cr (1.8 max)	Severin *et al.*[46]
	Calcium vanadium oxide	CaV_2O_7	Converter	V (40.1 formula)	Aarabi-Karasgani *et al.*[104]
	Manganosite	MnO	FeMn	Mn (77.5 formula)	Groot *et al.*[100]
	Brownmillerite	$Ca_2(Fe,Al)_2O_5$	BOF	V (main host), Cr (0.42 ave.), V (1.14 ave.), Mn (0.81 ave.)	Chaurand *et al.*,[12] Hobson *et al.*[30]
	Hausmannite	Mn_3O_4	EAF	Mn (72.0 formula)	Chiang and Pan[105]
	Groutellite	MnO_2	EAF	Mn (63.2 formula)	Chiang and Pan[105]
Metals and sulfides	Metal	Fe	BF	V (0.17 max), P (8.0 max), Mn (0.84 max), Co (0.66 max), Ni (1.07 max), Cu (4.73 max)	Scott *et al.*,[97] Piatak and Seal[23]
	Oldhamite	CaS	BF	Mn (1.25 max)	Scott *et al.*,[97] Piatak and Seal[23]
	Alabandite	MnS	BF	Mn (58.55 max), Ti (1.03 max), Cu (0.25 max)	Piatak and Seal[23]
	Ferrochrome	FeCr	FeCr	Cr (48.2 formula)	Horckmans *et al.*[101]

[a]average (ave.), maximum (max.) or calculated on formula (formula), one value reported (one value).

In slowly cooled slags, such as air-cooled, the glass is enriched in elements not compatible with the crystallizing phases, whereas glass generated by rapid cooling (*e.g.*, granulation) may resemble the original melt composition if crystallization does not occur. This chemical variation in glass can also occur on the millimeter and centimeter scale, as discussed by Piatak and Seal,[18] where darker-colored glass areas associated with pyroxenes contained less Ca and Mg compared to crystallite-free colorless areas of glass.

Pyroxenes are less frequently identified in ferrous slags than the previously discussed silicate phases. For BF slags, pyroxenes containing Ca in greater abundance than Mg, some with appreciable Al and minor Ti, and all with very little Fe have been reported.[23,96] In pre-1900 BF air-cooled slags from the USA, pyroxene is found as fine-grained dendrites within a glass matrix (Figure 3.6C) or as coarse-grained euhedral to subhedral skeletal crystals or laths.[23] The pyroxenoid wollastonite ($CaSiO_3$) can be found in BF and steel slags (Figure 3.5). Other silicates found in ferrous slags, but only occasionally reported, include hatrurite and bredigite (mostly in steel slags), and feldspar and andradite ($Ca_3Fe_2(SiO_4)_3$) (Figure 3.5). Hatrurite may be unique to steel slags.

The most common types of oxides found in ferrous slags are simple oxides of the form XO where X = Ca, Fe, Mn, or Mg, and spinel-group (AB_2O_4, where A = Fe^{2+}, Mg, or Mn, and B = Al, Cr, or Fe^{3+}) phases (Figure 3.5). Calcium oxide (CaO, lime), MgO (periclase), and FeO (wüstite), and less frequently MnO, are common in steel slags but only rarely reported in BF slags. Figure 3.6D shows subhedral to anhedral wüstite grains tens of micrometers in diameter in a legacy ferrous slag from the USA, which is likely a steel slag. Compared to ladle and modern BF slags, the higher Fe content of BOF and EAF slags results in wüstite typically being observed as a main phase in these types of steel slag.[7] The compositions of spinel-group phases include spinel ($MgAl_2O_4$), hercynite ($FeAl_2O_4$), magnetite (Fe_3O_4), magnesioferrite ($MgFe_2O_4$), and galaxite ($MnAl_2O_4$). Portlandite ($Ca(OH)_2$) is a hydration product of lime that is also observed in steel slags. Spinel crystals in ferrous slags may occur as coarse-to fine-grained euhedral to subhedral crystals. Maghemite (γFe_2O_3), formed by weathering of spinels containing Fe, has also been found in ferrous slags. Srebrodolskite, also known as dicalcium ferrite, and calcium ferrite ($CaFe_2O_4$), which does not have a natural mineral analog, are found in steel slags (Figure 3.5). Srebrodolskite can accommodate Al in the structure and, in some slags, is reported as brownmillerite ($Ca_2(Al,Fe^{3+})_2O_5$). The occurrence of srebrodolskite in a legacy slag is shown in Figure 3.6D. Lastly, perovskite ($CaTiO_3$) was identified in steel slags from direct iron reduction.[107]

Silicates and oxides may host several elements that are of interest because of their environmental impact or their potential for recovery. Table 3.3 includes examples of phases found in ferrous slag and concentrations of elements of potential interest reported in those phases. For example, Cr, Mn, and V are usually present at only a few weight percent or less in crystalline silicates and glass. However, end-member phases containing these elements have been reported, such as the occurrence of glauchchroite

($CaMnSiO_4$) in FeMn slag.[100] Oxide phases may contain these elements as end-member compositions or as substitutions for the primary elements. Spinel phases, particularly in ladle slags, may host significant Cr and V (Table 3.3). Xiang *et al.*[103] reported that the host of V in a steelmaking converter slag was a spinel of the composition FeV_2O_4; however, V can also occur as a component of the spinel, not only an end member (*e.g.*, (Mn, Fe)(V,Cr)$_2$O$_4$). As for Cr, Horckmans *et al.*[101] examined several types of ferrous slag (EAF, ladle, and FeCr) and found that Cr was often present in small-zoned spinels with variable compositions. Table 3.3 also includes oxides other than spinels that are end members in composition and contain elements of potential interest, including calcium vanadium oxides (CaV_2O_7) and hausmannite (Mn_3O_4), among others.

Unlike base metal slags, carbonates are reported to occur in weathered or legacy ferrous slags. The most frequently reported are calcite ($CaCO_3$) and dolomite (Figure 3.5), which are often found in steel slags. Calcite filling in a void in a weathering larnite grain in a ferrous slag sample is shown in the lower right corner of Figure 3.6D. Piatak *et al.*[108] identified three polymorphs of $CaCO_3$ (*i.e.*, calcite, aragonite, and vaterite) in modern and legacy ferrous slags from the Chicago area of the USA. Aragonite was identified in modern steel slags, vaterite in modern BF and steel slags and at legacy sites, and calcite was present in nearly all samples. On the other hand, dolomite was only identified at a few legacy sites. These carbonates form from atmospheric carbonation of primary phases, such as lime and periclase.

Sulfides are only found in trace amounts in some types of ferrous slags and include CaS (oldhamite) and MnS (alabandite) (Figure 3.5 and Table 3.3). Subhedral to anhedral grains of alabandite with up to 10 wt% Ca associated spatially with metallic Fe forming in cracks were reported by Piatak and Seal,[23] as shown in Figure 3.7A; these authors also found fine-grained oldhamite (<1 μm in diameter) within the glass phase. The solid solution between CaS–MnS phases is temperature dependent with an increase in alabandite solid solution with respect to oldhamite as temperature increases. Therefore, the amount of Ca in the oldhamite may reflect the smelting furnace's maximum temperature. Iron may also be present in minor amounts in the CaS–MnS phases.[23] During refining steps in modern operations, the hot metal from the BF or the steel from steelmaking furnaces is treated to remove S by adding reagents (*e.g.*, CaO, CaF_2) forming sulfides, which create a layer on the slag. Due to the presence of oxygen in the steelmaking furnaces, this desulfurization cannot occur during steelmaking.

The only metallic phase common in ferrous slag is Fe, usually occurring as fine-grained blebs or elongated sinuous forms (a few micrometers in diameter) (Figures 3.6C,D and 3.7B) and may reach up to hundreds of micrometers in diameter (Figure 4.4B,C, Chapter 4). The immiscible iron usually forms isolated blebs but has also been documented within silicate phases along crystal growth planes. Scott *et al.*[97] described Fe as being "squeezed out" of melilite as it crystallizes. The metallic Fe may contain up to a few percent of other elements, such as Ca, Mg, Mn, P, Si, and V, and trace amounts of Cu,

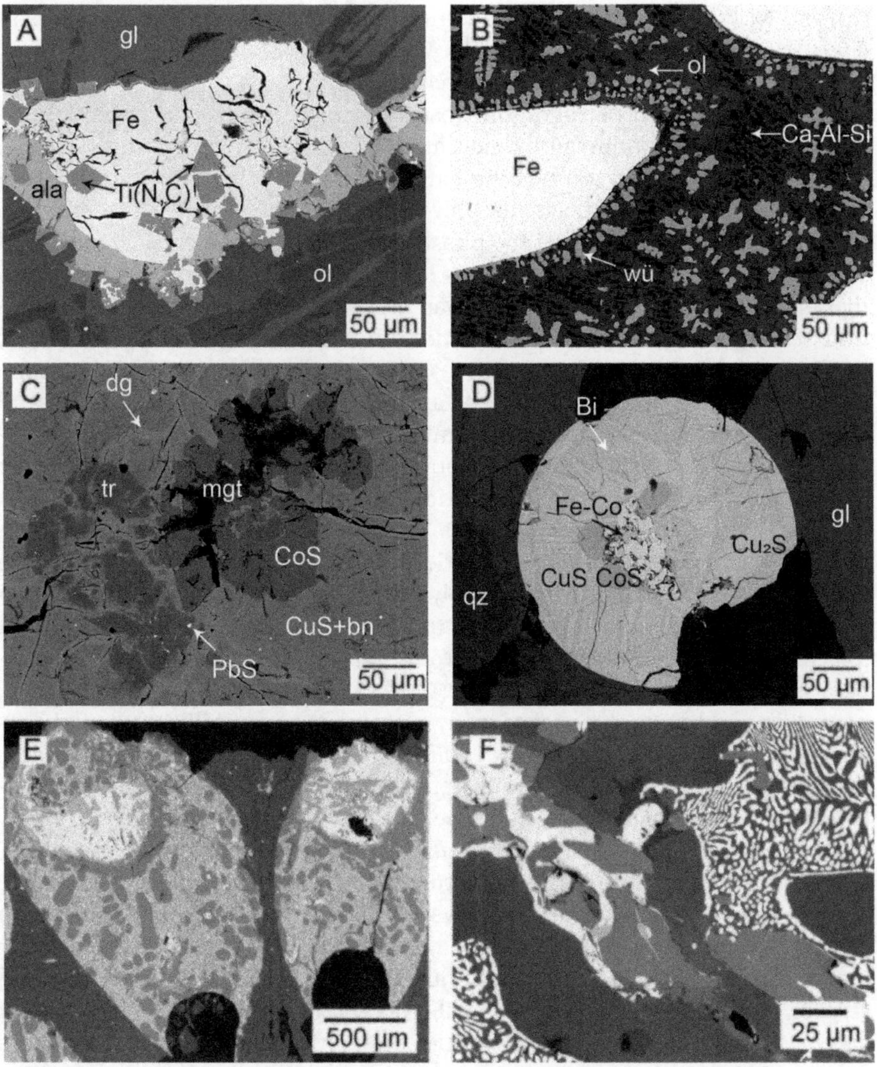

Figure 3.7 Backscatter scanning electron photomicrographs of slag. (A) Pre-1900 blast furnace slag from the Hopewell Furnace National Historic Site in Pennsylvania, USA. Image shows anhedral Fe metal (Fe), alabandite (ala), and euhedral Ti(N,C) vein within skeletal olivine-group phase (ol) and glass (gl). Reproduced from ref. 23 with permission from Elsevier, Copyright 2012. (B) Legacy ferrous slag from the Chicago area, Illinois, USA. Anhedral Fe metal blebs (Fe) surrounded by a matrix of dendritic wüstite (wü) and subhedral grains of an olivine (ol)-group phase and prismatic Ca–Al–silicate phase (Ca–Al–Si). (C) Copper slag from the Zambian Copperbelt Province. Image shows troilite (tr) and CoS crystals associated with CuS-bornite (bn) intergrowth with trace digenite (dg) and galena (PbS). Magnetite (mgt) is associated with CoS. Reproduced from ref. 26 with permission from Cambridge University Press, Copyright 2010. (D) Copper slag from the Zambian Copperbelt

Co, and Ni.[23,97] In ladle slags, it is common to have immiscible inclusions of a ferroalloy such as FeCr or FeNi.[109]

The crystallization sequences of phases from ferrous slag melts have been examined both from petrographic observations of cooled slag, from laboratory experiments conducted by melting existing slag or generating synthetic melts mimicking their composition, and from simulating phase transformations. The cooling rate affects not only the crystal size and structure but also the distribution and purity of the phases that form.[92] Zhao *et al.*[110] compared simulated crystallization of stainless steel slags to laboratory-based experiments during which a slag sample was melted and then cooled. Crystallization sequences by computer simulations predicted spinel to precipitate at the highest temperature (above 1600 °C), then, with decreasing temperature, α-dicalcium silicate to form next. At approximately 1400 °C, merwinite and melilite phases form and the dicalcium silicate may undergo crystal transformation to the β polymorph. The laboratory melting and cooling results were generally consistent with the predictions; dicalcium silicate did not persist when temperates decreased from 1600 to 1300 °C. Petrographic examination of steel slags from direct iron reduction documented the occurrence of the different high to low-temperature polymorphs of dicalcium silicate (*i.e.*, α, α high, α low, and β) dependent on the temperature; other phases are reported to crystallize in the following order from high (1420 °C) to low (500 °C) temperature: melilite, calcium aluminate ($CaAl_2O_4$), merwinite, wüstite, periclase, perovskite, and glass.[107] Another study examined the crystallization of phases during rapid, moderate, and slow cooling of a melted BOF slag.[92] The rapidly cooled slag contained haturite and srebrodolskite with some chemical substitutions and a small amount of periclase. The phase evolution of the less rapidly cooled slags was examined using *in situ* XRD in experiments with increasing temperatures. At the highest temperature (above 1400 °C), α-dicalcium silicate was the only phase observed; the lower temperature polymorphs along with wüstite, spinel, srebrodolskite, lime, and periclase were observed below that temperature.[92] As for BF slags, computer simulations and laboratory-based studies by Kang *et al.*[98] revealed melilite as the first phase to crystalize with less basic melts (*i.e.*, lower Ca to Si ratio) then forming anorthite ($CaAl_2Si_2O_8$), pyroxene, and wollastonite, whereas the more basic melts (*i.e.*, higher Ca to Si ratios) precipitated spinel. Although a wide range of phases are reported in ferrous slags and it is a challenge to generalize crystallization sequences, these

Figure 3.7 (*continued*) Province. A rounded inclusion composed of Cu sulfides (CuS, Cu_2S) and Co sulfide (CoS), native Cu, Bi, and Fe–Co alloy with a matrix of quartz (qz) and glass (gl). Reproduced from ref. 26 with permission from Cambridge University Press, Copyright 2010. (E) Pb–Ag slag from the Clayton smelter, USA. Polymineralic particles displaying myrmekitic texture composed of galena, bornite, sphalerite/wurtzite, and metallic and intermetallic phases containing Fe, Cu, As, Pb, and Pb enclosed by silicate phases. (F) Close up of polymineralic particle shown in E. (E and F) Reproduced from ref. 24 with permission from Elsevier, Copyright 2004.

studies suggest that olivine-group, specifically dicalcium silicate, and melilite-group phases commonly are the first to crystallize as slag melts cool, followed by other Ca silicates, such as merwinite or pyroxene, and then oxides such as spinel, wüstite, or srebrodolskite, among others. However, similar to the non-ferrous slags discussed in the next section, spinel-group oxides may form first or early in the sequence. For example, Tanskanen and Makkoneh[99] examined FeCr slags and found spinels enclosed in glass for granulated slags, and spinel, forsterite, and sometimes pyroxene enclosed in glass for air-cooled slag. Based on petrographic observations, the spinel appears to crystallize first from the melt. In air-cooled, the forsterite is the next phase to crystallize followed by pyroxene, if present.

3.4.2.2 Non-ferrous Slags

Table 3.4 summarizes the list of phases most frequently observed in non-ferrous slags and Figure 3.5 illustrates their reporting frequency in slag studies. The chemical composition of the slags and crystallization conditions (temperature, cooling rate) determine the mineralogical composition. The Ca–Fe aluminosilicates, silicate glass, and oxides forming most of the slag volume are the most frequently reported primary phases (Figure 3.5). In addition, metal-bearing sulfides, intermetallic phases, and metals shown in Figure 3.5 and listed in Table 3.4 are also observed; most frequently they occur as spherical inclusions of variable size (<1 μm to hundreds of μm) entrapped within the silicate matrix (Figures 3.6A,B and 3.7C–F).

Olivine-group phases are very common in crystalline slags (Figure 3.6A and B). The predominance of individual end members in the slag is controlled by the bulk chemistry of the slag melt and the temperature. Whereas fayalite (Fe_2SiO_4) and kirschsteinite ($CaFeSiO_4$) were found to be dominant in many Pb–Zn and Cu slags;[24,26,58,87] forsterite (Mg_2SiO_4) and monticellite ($CaMgSiO_4$) were more abundant in Zn and some Ni slags.[62,77,111,112]

Pyroxenes are abundant in slags and their chemistry varies mainly between hedenbergite ($CaFeSi_2O_6$)[26,37,41] and diopside ($CaMgSi_2O_6$),[15,41] although Ca-poorer pyroxenes with augite ($(Ca,Na)(Mg,Fe,Al,Ti)(Si,Al)_2O_6$) compositions[15,41,111] and pigeonite ($(Ca,Mg,Fe)(Mg,Fe)Si_2O_6$)[41] or clustering near the clinoferrosilite ($Fe_2Si_2O_6$)[37] or clinoenstatite ($Mg_2Si_2O_6$)[112] compositions have also been observed. Kądziołka *et al.*[41] also reported the presence of esseneite ($CaFeAlSiO_6$), a pyroxene containing trivalent Fe. For a more detailed compilation of the data see Piatak *et al.*[1]

Melilite-group phases with the general formula $(Ca,Na)_2(Al,Mg,Fe)(Al,Si)_2O_7$ crystallize mostly from the Ca-rich slag melts (Figure 3.6B). They exhibit variable compositions ranging from åkermanite ($Ca_2MgSi_2O_7$), ferroåkermanite ($Ca_2FeSi_2O_7$), and gehlenite ($Ca_2Al_2SiO_7$) to soda melilite ($CaNaAlSi_2O_7$) but can contain a generally high proportion of a Zn-rich end member, hardystonite ($Ca_2ZnSi_2O_7$). For instance, Ettler *et al.*[58] found up to 40 mol% of hardystonite end member in melilites from Cu–Pb slags from Namibia and Zn smelting slags from Poland and France often contained almost pure hardystonite.[32,62]

Silicate glass is an almost ubiquitous phase in slags and is found in larger proportions especially in granulated slags (Figures 3.5 and 3.6). It is a very important hosting phase for metal(loid) contaminants and, given its generally lower stability during the weathering process in comparison with crystalline phases, it is responsible for the release of the contained elements when exposed to water[87] (see also Chapter 4).

Spinel-group oxides are relatively frequent phases in air-cooled non-ferrous smelting slags. Their composition with the general formula (Zn,Mg,Fe^{2+})(Fe^{3+},Al,Cr)$_2$O$_4$ is rather variable owing to numerous substitutions in their crystal structure. They represent the key carriers for Zn and Cr in the majority of slags produced by non-ferrous metal smelting (Table 3.4).[15,24,32,37,58,62,87,111]

Many of the silicates and oxides can incorporate metal(loid)s, which are important in terms of the environmental impact of slag (see also Chapters 4, 5, and 6) and/or potential recovery of valuable elements from slags (see also Chapter 9). As a result, Table 3.4 also indicates the maximum reported concentrations of metal(loid)s within individual phases as obtained by EPMA. It has been suggested that there is a common replacement in the crystalline structure of silicates, silicate glass, and oxides due to similar ionic radii in octahedral coordination, where Co (75 pm), Cu (77 pm), Zn (74 pm), and Ni (69 pm) substitute for common elements like Fe (78 pm) and Mg (72 pm).[26,87,111] Similarly, if the structural positions are sufficiently large in the given phase structure, Pb (119 pm) can substitute for Na (102 pm), K (138 pm), and/or Ca (100 pm) (*e.g.*, in melilite or feldspar);[19,32,58,62,87] (note: the effective ionic radii were taken from Shannon).[114] Tyszka *et al.*[28] used EPMA and LA-ICP-MS to determine the Cd distribution between individual phases in Pb–Zn slags and found that this metal does not readily substitute for other elements (according to Shannon,[114] its ionic radius is 95 pm in octahedral coordination), and is incompatible and enriched in the late melts. As a result, it is entrapped in Pb-bearing hardystonite rims crystallizing toward the end of slag solidification, in late Pb-bearing silicates (compositionally close to alamosite, $PbSiO_3$) and in rare phosphates.[28]

Despite their low volumetric proportion in slags, sulfides and intermetallic and metallic phases are key metal(loid)-hosting phases. As mentioned, they typically form inclusions of variable size corresponding to the immiscible mattes or metallic melts (Figure 3.7). The richness of species is also demonstrated in Table 3.4. Iron sulfides, especially pyrrhotite ($Fe_{1-x}S$) or troilite (FeS), are almost ubiquitous for all non-ferrous slags. Triolite-enriched Cu and Co is shown in Figure 3.7C for a Cu slag from the Zambian Copperbelt.[26] Copper slags generally contain many Cu–Fe sulfides, such as digenite ($Cu_{9+x}S_5$), bornite (Cu_5FeS_4), chalcocite (Cu_2S), covellite (CuS), cubanite ($CuFe_2S_3$), chalcopyrite ($CuFeS_2$), and intermediate solid solution formed by quenching in the system Cu–Fe–S[24,26,58,73,113] (Figure 3.7C and D). If Cu slags are rich in Co (like in the Zambian Copperbelt), Co-bearing sulfides such as Co-pentlandite (Co_9S_8) and CoS also occur (Figure 3.7D).[26] Galena (PbS) and sphalerite/wurtzite (ZnS) are typical for Pb–Zn and Zn slags.[15,19,37,87] Pentlandite ($(Fe,Ni)_9S_8$) and heazlewoodite (Ni_3S_2) have been detected in laterite

Table 3.4 Examples of main metal(loid)-hosting phases in non-ferrous slags.

Group	Phase	Formula	Slag type	Contaminants and/or valuable elements (up to, wt%)	Reference
Silicates	Olivine group	$(Ca,Mg,Mn,Fe)_2SiO_4$	Cu	Cu: 0.34, Zn: 2.73, Pb: 0.1,2 Co: 5.62	Potysz *et al.*,[73] Vítková *et al.*,[26] Manasse *et al.*[74]
			Cu–Pb	Zn: 7.02	Ettler *et al.*[58]
			Pb	Zn: 4.22	Piatak *et al.*,[24] Ettler *et al.*[19]
			Pb–Zn	Zn: 5.05	Ettler *et al.*[87]
			Zn–Pb–V	Zn: 14.6, V: 0.37	Ettler *et al.*[77]
			Zn	Zn: 2.27	Puziewicz *et al.*[62]
			Ni	Ni: 1.37, Cr: 0.47	Kierczak *et al.*,[111] Ettler *et al.*[112]
	Melilite group	$(Ca,Na)_2(Al,Mg,Fe)(Al,Si)_2O_7$	Cu–Pb	Zn: 9.35, Pb: 0.13	Ettler *et al.*[58]
			Pb–Zn	Pb: 1.46, Zn: 8.44	Ettler *et al.*,[87] Warchulski *et al.*[15]
			Zn–Pb–V	Zn: 13.04, V: 0.24	Ettler *et al.*[77]
			Zn	Pb: 1.87, Zn: 18.49	Vanaecker *et al.*,[32] Piatak and Seal,[37] Puziewicz *et al.*[62]
	Pyroxene	$Ca(Mg,Fe,Al)(Al,Si)_2O_6$	Cu	Zn: 0.82, Pb: 0.42, Cu: 0.18, Co: 1.4	Potysz *et al.*,[73] Vítková *et al.*[26]
			Pb	Zn: 3.54	Piatak *et al.*,[24] Ettler *et al.*[19]
			Pb–Zn	Zn: 14.41	Warchulski *et al.*,[15] Ettler *et al.*[87]
			Zn–Pb–V	Zn: 3.38, V: 3.40	Ettler *et al.*[77]
			Zn	Zn: 14.40	Vanaecker *et al.*,[32] Puziewicz *et al.*[62]
			Ni	Cr: 1.03	Ettler *et al.*[112]
	Glass	Complex compositions	Cu	Co: 0.72, Cu: 5.51, Zn: 1.21, Pb: 4.43	Vítková *et al.*,[26] Potysz *et al.*[73]
			Cu–Pb	Pb: 10.95, Zn: 13.02, Cu: 1.42, Cd 0.09, As 2.46	Ettler *et al.*[58]
			Pb–Zn	Pb: 3.45, Zn: 7.87	Ettler *et al.*[87]
			Ni	Cr: 0.88	Ettler *et al.*[112]
	Willemite	Zn_2SiO_4	Zn	Zn: 55.97	Piatak and Seal,[37] Vanaecker *et al.*,[32] Puziewicz *et al.*[62]

(continued)

Table 3.4 (*continued*)

Group	Phase	Formula	Slag type	Contaminants and/or valuable elements (up to, wt%)	Reference
	Feldspars	$NaAlSi_3O_8$–$CaAl_2Si_2O_8$–$KAlSi_3O_8$	Cu–Pb	Pb: 31.56	Ettler *et al.*[58]
			Pb	Pb: 14.25	Ettler *et al.*[19]
			Pb–Zn	Pb: 3.42	Warchulski *et al.*[15]
			Zn	Pb: 6.36, Zn: 0.13	Puziewicz *et al.*,[62] Piatak and Seal[37]
	Leucite	$KAlSi_2O_6$	Zn	Pb: 12.72, Zn: 0.37	Puziewicz *et al.*[62]
Oxides	Spinel group	$(Zn,Mg,Fe^{2+})(Fe^{3+},Al,Cr)_2O_4$	Cu	Co: 4.03, Cu: 3.29, Zn: 0.80	Vítková *et al.*[26]
			Cu–Pb	Zn: 20.41, Cr: 36.04	Ettler *et al.*[58]
			Pb–Zn	Zn: 16.00, Cr: 4.76	Ettler *et al.*[87]
			Zn	Zn: 32.19	Piatak and Seal,[37] Puziewicz *et al.*,[62] Vanaecker *et al.*[32]
			Ni	Ni: 0.13, Cr: 47.48	Kierczak *et al.*[111]
	Zincite	ZnO	Zn	Zn: 78.17	Vanaecker *et al.*,[32] Piatak and Seal[37]
Sulfides	Cu–Fe–sulfides	$Cu_{9+x}S_5$, Cu_5FeS_4, Cu_2S, CuS, $CuFe_2S_3$, $CuFeS_2$, iss[a]	Pb–Zn, Zn, Cu		Ettler *et al.*,[87] Piatak and Seal,[37] Veselská and Majzlan,[113] Potysz *et al.*,[73] Ettler *et al.*[58]
	Pb sulfides	PbS	Pb–Zn		Ettler *et al.*,[87] Warchulski *et al.*[15]
	Fe sulfides	$Fe_{1-x}S$, FeS	Pb–Zn, Zn, Cu		Ettler *et al.*,[87] Warchulski *et al.*,[15] Piatak and Seal,[37] Vítková *et al.*[26]
	Zn sulfides	(Zn,Fe)S	Pb–Zn, Zn		Piatak *et al.*,[24] Warchulski *et al.*,[15] Piatak and Seal[37]
	Ni sulfides	$(Fe,Ni)_9S_8$, Ni_3S_2	Ni		Kierzcak *et al.*[111]
	Co sulfides	Co_9S_8, Co_4S_3, CoS	Cu		Vítková *et al.*[26]
Interme-tallic	Arsenides	$Cu_{5-x}As_2$, Cu_3As, $FeAs_2$, Fe_2As, FeAs, $(Co,Fe)_2As$	Pb–Zn, Cu–Pb, Cu		Ettler *et al.*,[87] Ettler *et al.*,[58] Vítková *et al.*[26]
	Antimonides	Cu_5Sb, $Cu_3(Sn,Sb)$, Cu_2Sb, NiSb, Ag_3Sb	Cu–Pb, Pb		Ettler *et al.*,[58] Ettler *et al.*[19]
	Others	Ni_3Fe, Cu_6Sn_5	All types		Kierczak *et al.*[111]
Metals	Fe	Fe, Pb, Cu, Sb	All types		Kierczak *et al.*,[111] Ettler *et al.*,[87] Veselská and Majzlan,[113] Potysz *et al.*,[73] Ettler *et al.*[58]

[a]iss = intermediate solid solution formed by quenching in the ternary system Cu–Fe–S.

Ni slags; however Ni in these slags is also bound in various Ni–Fe intermetallic compounds.[111,112] Large numbers of binary intermetallic compounds and alloys have been reported from slags (some of them having mineral analogues and mineral names in nature). Table 3.4 lists only some of them; other compounds (including pure metals and unknown ternary or quaternary intermetallic phases) similar to those observed in mattes and speiss[115] have also been described in slags. Complex admixtures of sulfides and metallic and intermetallic compounds displaying myrmekitic or symplectic textures are not uncommon in base metal slags, as shown in Figure 3.7E and F.

Many phase assemblages have been identified for different non-ferrous smelting slags; microscopic observations have then been used to determine the crystallization sequences in the slag melts.[15,37,62,87] Rapidly quenched slags (*e.g.*, granulated) have very simple phase assemblages with predominantly glass, and possibly dendritic olivines and minor sulfide/metallic droplets. In contrast, for the air-cooled slags, the phase assemblages are much more complex, but the crystallization sequences generally start with the formation of the spinel-group oxides followed by the crystallization of olivines, pyroxenes, and melilites (in variable order and dependent on the melt compositions and cooling rates) and finally by the solidification of glass at the end. Feldspars and leucite were also observed in these slags and crystallized at various positions in the crystallization sequences.[15,26,37,62,93,111] Other silicates, such as garnet ($(Fe^{2+},Ca,Mg,Mn)_3(Fe^{3+},Ti,Al)_2Si_3O_{12}$) or hatrurite, are rare and were observed in Ni slags only.[111] In Zn-rich slags, willemite and zincite (ZnO) commonly form at the beginning of the crystallization sequences.[37,62,77] In contrast, local grains of quartz and/or cristobalite correspond to residues of non-melted gangue from the furnace charge.

The mineralogy of salt slag from the pyrometallurgical processing of Al scrap to recover Al is distinct from the mineralogy of other non-ferrous slags discussed previously in that it contains an abundance of soluble salts. Based on Tsakiridis'[67] review of salt slag from several studies, the water-soluble compounds identified were halite ($NaCl$), sylvite (KCl), aluminum nitride (AlN), aluminum carbide (Al_4C_3), AlC_3, and aluminum sulfide (Al_2S_3). Halite and sylvite were the most abundant and may account for 50 wt% of the slag. Non-soluble compounds in water included corundum (Al_2O_3), spinel ($MgAl_2O_4$), quartz, calcium silicate (β-$CaSiO_3$), sodalite ($Na_8Al_6Si_6O_{24}Cl_2$), Al_2S, aluminum oxide nitride (Al_5O_6N), gibbsite ($Al(OH)_3$), diaoyudaoite ($NaAl_{11}O_{17}$), cryolite (Na_3AlF_6), elpasolite (K_2NaAlF_6), aluminum calcium ($AlCa$), fluorite (CaF_2), villiaumite (NaF), iron (Fe), silicon (Si), and aluminum (Al).[67]

3.5 Summary and Outlook

This chapter has provided an overview of the diverse chemistry and mineralogy of slags. Ferrous and non-ferrous slags are subdivided based on the metal targeted for extraction and, sometimes, the step in the extraction process. In general, modern BF slags are chemically dominated by Ca and Si, steel slags are composed mostly of Ca, Si, and Fe, and base metal slags contain predominantly Fe and Si. The next most abundant elements after Si, Ca, and Fe for base metal slags and ferrous slags are Al and Mg. Ferrous and many non-ferrous slags contain only trace amounts of S. As for other elements, slags may be enriched in the element targeted for extraction due to its heightened concentration in the ore, concentrate, or furnace feed and further enrichment during processing. For example, Cu slags contain the highest amount of Cu. The trace elements enriched in ferrous slags include Cr, Mn, and V.

As slag melts cool, the phases that form and their texture and crystallization sequence all depend on the chemical composition of the melt and the cooling rate and process. Quenched or granulated slags are commonly glassy with minor to trace amounts of crystalline phases and/or remnants of the immiscible matte or metals. Some melt compositions do not form glassy products even when cooled rapidly. Moderately cooled slags, such as those that have been pelletized or expanded to generate a lightweight porous product may be glassy and/or crystalline. Lastly, air-cooled slags are usually crystalline with some amount of intestinal glass and may have glassy chilled margins at their exterior portions. Air-cooled slags frequently display spinifex textures, usually olivine, reminiscent of komatiites. Fine-grained dendritic silicates or oxides in a matrix of glass are typical, yet the phases may also form euhedral to anhedral crystals depending on the conditions.

There are various phases with unique chemistries that exist in the different slag types; some are abundant in nature whereas others are rare or not found in rocks. For ferrous slags, including BF, BOF, EAF, and ladle slags, the most frequently reported silicate phases include olivine-group, melilite, merwinite, crystalline SiO_2, and glass. Spinel and simple oxides (*i.e.*, XO) are variable in composition but are usually a part of the mineralogical suite, as well as immiscible iron metal. Periodically, carbonates are also reported in weathered ferrous slags. For ferrous slags, sulfides are not common or abundant. However, desulfurization in Fe blast furnaces and some steel refining furnaces results in the presence of alabandite and oldhamite, and a solid solution between the two, in the slags.

For non-ferrous slags, the focus in this chapter was mainly on base metal slags. Their mineralogical composition is similar to the suite of crystalline phases reported in ferrous slags, but their chemical composition is usually different. Olivine-group phases are typical but may contain appreciable Fe or Mg compared to Ca-rich olivine in ferrous slags. Pyroxene, melilite, glass, and spinel are also common in base metal slags, with diverse chemical compositions usually dissimilar to their compositions reported in ferrous slags. Sulfides, intermetallic compounds, and metallic phases occur in many base metal slags and are of interest due to the elements they host.

Crystallization sequences are different for different slags. For ferrous slags, olivine or melilite phases have been reported to form first from cooling melts, followed by Ca silicates such as merwinite or pyroxene, and by spinel or other oxides. The remaining melt solidifies to form interstitial glass containing some immiscible metal inclusions. However, this sequence is not always the case. For base metal slags, the crystallization sequences generally start with the formation of the spinel-group oxides followed by the crystallization of olivine, pyroxenes and melilites (in variable order, depending on the melt composition and cooling rate), and finally by the solidification of glass at the end.

A wide range of chemical and mineralogical variability exists within and among slag types; however, some similarities do exist. This chapter has provided an overview of the major, minor, and trace elements found in different types of slag along with their mineralogical and textural characteristics. The chemistry and mineralogy of a slag set the framework for its behavior in the environment, including its susceptibility to weathering and the formation of secondary weathering products, as discussed in Chapter 4, and the potential to leach elements into the ecosystem including water and soil, as discussed in Chapters 5 and 6. As will be discussed, base metal slags are generally more of an environmental concern than slags in the ferrous category due to their high trace element content. Additionally, the chemical, physical, and mineralogical characteristics of slags determine their potential for use, for example, as a material to add to cement or use as road aggregate, topics discussed in Chapter 7. For example, the high Ca content in ferrous slags makes them a potential substitute for Portland cement in concrete. Lastly, a slag's characteristics dictate its value for use in environmental applications (Chapter 8), such as the use of highly basic ferrous slags to treat acid mine drainage. Furthermore, these characteristics influence the potential for reprocessing to extract residual elements (Chapter 9), such as reprocessing of historical Cu slag to recover Co. More detail on these topics will be provided in subsequent chapters.

Appendix

Summary of slag characterization references used to generate Tables 3.1 and 3.2 and Figures 3.2–3.4.

Slag type	Reference	Location information
BF	Butler, 1977[96]	United Kingdom
	Cravotta *et al.*, 2005[85]	Johnstown, Pennsylvania, USA
	Das *et al.*, 2007[116]	Location not provided
	Douglas & Zerbino, 1986[117]	Hamilton and Sault Ste. Marie, Ontario, Sparrow's Point, Maryland, USA
	EUROSLAG, 2003[118]	Location not provided
	Fällman & Hartlén, 1994[119]	Sweden
	Goodarzi *et al.*, 2015[120]	Iran
	Hadjsadok *et al.*, 2012[121]	Algeria
	Hoppe *et al.*, 2020[49]	Gary, Indiana, USA
	Lewis, 1982[122]	Location not provided
	Mahieux *et al.*, 2009[123]	Location not provided
	Muhmood *et al.*, 2009[124]	India
	Poh *et al.*, 2006[125]	Newport, United Kingdom
	Proctor *et al.*, 2000[36]	USA and Canada
	Scott *et al.*, 1986[97]	United Kingdom and France
	Singh *et al.*, 2008[126]	India
	Svirenko *et al.*, 2003[127]	Ukraine, Russia and Latvia
	Wang, 2016[128]	Location not provided
	Wendling *et al.*, 2013[82]	Jiangxi Province, China

Slag type	Reference	Location information
Pre-1900 BF	Eekelers *et al.*, 2016 [48]	Hellenistic to Byzantine city of Sagalassos, Turkey
	Mahé-Le Carlier *et al.*, 2000 [47]	France, Sweden, and Italy
	Piatak & Seal, 2012 [23]	Pennsylvania, USA
	Severin *et al.*, 2011 [46]	Ethiopia
	West, 1902 [45]	Location not provided
BOF	Barca *et al.*, 2012 [129]	France, Luxembourg, and Germany
	Chaurand *et al.*, 2006 [130]	Location not provided
	Chaurand *et al.*, 2007 [12]	Austria
	Das *et al.*, 2007 [116]	Location not provided
	Gahan *et al.*, 2009 [76]	Sweden
	Gautier *et al.*, 2013 [92]	France
	Goodarzi *et al.*, 2015 [120]	Iran
	Hobson *et al.*, 2017 [30]	Scunthorpe, United Kingdom
	Hobson *et al.*, 2018 [131]	Scunthorpe, United Kingdom
	Juckes, 2013 [132]	United Kingdom
	Mahieux *et al.*, 2009 [123]	Location not provided
	Poh *et al.*, 2006 [125]	United Kingdom
	Proctor *et al.*, 2000 [36]	USA and Canada
	Shen *et al.*, 2009 [133]	Location not provided
	Shi, 2004 [134]	Location not provided
	Tossavainen *et al.*, 2007 [75]	Sweden
	Waligora *et al.*, 2010 [135]	Location not provided
	Xuequan *et al.*, 1999 [136]	Taiyuan, China
	Yildirim & Prezzi, 2011 [7]	Indiana, USA
EAF	Barca *et al.*, 2012 [129]	Spain, France, Luxembourg, and Germany
	Barra *et al.*, 2001 [137]	Location not provided
	Drizo *et al.*, 2002 [138]	Canada
	Drizo *et al.*, 2006 [139]	Canada
	Fällman & Hartlén, 1994 [119]	Sweden
	Gahan *et al.*, 2009 [76]	Sweden
	Luxán *et al.*, 2000 [140]	Location not provided
	Manso *et al.*, 2006 [141]	Location not provided
	Milacic *et al.*, 2011 [142]	Location not provided
	Muhmood *et al.*, 2009 [124]	Location not provided
	Neuhold *et al.*, 2019 [143]	Location not provided
	Proctor *et al.*, 2000 [36]	USA and Canada
	Qian *et al.*, 2002 [144]	Location not provided
	Shi, 2004 [134]	Location not provided
	Suer *et al.*, 2009 [51]	Sweden
	Tossavainen *et al.*, 2007 [75]	Sweden
	Tsakiridis *et al.*, 2008 [145]	Greece
	Yildirim & Prezzi 2011 [7]	Indiana, USA
Steel and legacy sites	Douglas *et al.*, 2012 [84]	Kwinana, Western Australia
	Hoppe *et al.*, 2020 [49]	Illinois and Indiana, USA
	Kourounis *et al.*, 2007 [146]	Greece
	Navarro *et al.*, 2010 [147]	Spain
	Oh *et al.*, 2012 [148]	Korea
	Roadcap *et al.*, 2005 [149]	Chicago, USA
	Shen *et al.*, 2009 [133]	China
	Svirenko *et al.*, 2003 [127]	Ukraine, Russia and Latvia

Slag type	Reference	Location information
	US Department of Transportation, 2008[150]	Location not provided
	Wendling *et al.*, 2010[83]	Australia
	Wendling *et al.*, 2013[82]	Jiangxi Province, China
	Xuequan *et al.*, 1999[136]	Taiyuan, China
	Ziemkiewicz & Skousen, 1999[151]	USA
Ladle	Gahan *et al.*, 2009[76]	Sweden
	Nicolae *et al.*, 2007[52]	Romania
	Qian *et al.*, 2002[144]	Location not provided
	Setién *et al.*, 2009[152]	Location not provided
	Shi, 2004[134]	Location not provided
	Tossavainen *et al.*, 2007[75]	Sweden
Cu	Ali *et al.*, 2013[153]	India
	Álvarez-Valero *et al.*, 2008[154]	Portugal
	Ettler *et al.*, 2009[58]	Tsumeb, Namibia
	Jarošíková *et al.*, 2017[155]	Tsumeb, Namibia
	Kądziołka *et al.*, 2020[41]	Old Copper Basin, Poland
	Kaksonen *et al.*, 2017[156]	Harjavalta, Finland
	Khorasanipour & Esmaeilzadeh, 2016[56]	Sarcheshmeh, Iran
	Kierczak & Pietranik, 2011[40]	Poland
	Kinnunen *et al.*, 2020[157]	Location not provided
	Lottermoser, 2002[158]	Australia
	Lottermoser, 2005[86]	Rio Tinto, Spain
	Mahé-Le Carlier *et al.*, 2000[47]	Location not provided
	Manasse *et al.*, 2001[74]	Italy
	Mikoda *et al.*, 2018[79]	Poland
	Parsons *et al.*, 2001[59]	California, USA
	Pérez-López *et al.*, 2010[81]	Iberian Pyrite Belt, Portugal
	Piatak *et al.*, 2004[24]	USA
	Potysz *et al.*, 2016[73]	Rudawy Janowickie Mountains, Poland
	Potysz *et al.*, 2018[159]	Leszczyna, Poland
	Sáez *et al.*, 2003[57]	Iberian Pyrite Belt, Spain
	Sarfo *et al.*, 2017[160]	Anaconda, Montana
	Tatarinov, 2002[161]	Russia
	Urosevic *et al.*, 2015[162]	Serbia
	US Department of Transportation, 2008[150]	Location not provided
	Vítková *et al.*, 2010[26]	Copperbelt Province, Zambia
	Zuo *et al.*, 2020[163]	Gansu Province, China
Ni (laterite ore)	Ettler *et al.*, 2016[112]	Niquelandia, Brazil
	Kierczak *et al.*, 2009[111]	Poland
	Prasetyo *et al.*, 2020[164]	Morowali, Central Sulawesi, Indonesia
	Ratié *et al.*, 2016[165]	Niquelandia and Barro Alto, Brazil
	Soultana *et al.*, 2019[166]	Larymna, Greece
Ni (sulfide ore)	Gbor *et al.*, 2000[167]	Thompson Manitoba, Canada
	Li *et al.*, 2008[65]	Location not provided
	Li *et al.*, 2020[168]	Jinchuan, China
	Souter & Watmough, 2017[169]	Coniston (east of Sudbury), Ontario, Canada
	US Department of Transportation, 2008[150]	Location not provided

Slag type	Reference	Location information
Pb (primary)	Barna *et al.*, 2004 [61]	France
	Bosso & Enzweiler, 2008 [170]	Brazil
	Chaudhuri *et al.*, 1993 [171]	Harz Mountain, Germany
	de Andrade Lima & Bernardez, 2011 [80]	Santo Amaro, Brazil
	Ettler *et al.*, 2003 [172]	Příbram, Czech Republic
	Ettler *et al.*, 2004 [173]	Příbram, Czech Republic,
	Ettler *et al.*, 2009 [19]	Příbram, Czech Republic
	Ettler & Johan, 2014 [174]	Příbram, Czech Republic
	Lottermoser, 2002 [158]	Montalbion, Australia
	Mikoda *et al.*, 2018 [79]	Poland
	Navarro *et al.*, 2008 [175]	Spain
	Nowinska, 2020 [63]	Poland
	Parbhakar *et al.*, 2019 [29]	Tasmania, Australia
	Parsons *et al.*, 2001 [59]	Belledune, Canada
	Piatak *et al.*, 2004 [24]	Idaho, USA
	Potysz *et al.*, 2016 [73]	Rudawy Janowickie Mountains, Poland
	Saikia *et al.*, 2012 [176]	Location not provided
	Sobanska *et al.*, 2000 [64]	France
	US Department of Transportation, 2008 [150]	Location not provided
	Yin *et al.*, 2016 [177]	Nord-Pas-de-Calais, France
Pb (secondary)	De Angelis & Medici, 2012 [178]	Location not provided
	Ettler & Johan, 2014 [174]	Příbram, Czech Republic
	Forte *et al.*, 2017 [179]	Location not provided
	Kim *et al.*, 2017 [180]	Location not provided
	Knežević *et al.*, 2010 [181]	Location not provided
	Penpolcharoen, Metta, 2005 [182]	Location not provided
	Shirinbayan *et al.*, 2014 [183]	Ghaniabad Tehran, Iran
Zn	Barna *et al.*, 2004 [61]	France
	Yin *et al.*, 2016 [177]	Nord-Pas-de-Calais, France
Zn–Pb	De Angelis & Medici, 2012 [178]	Location not provided
	Ettler *et al.*, 2020 [184]	Kabwe, Zambia
	Ettler *et al.*, 2020 [77]	Berg Aukas, Namibia
	Kucha *et al.*, 1996 [185]	Belgium
	Piatak & Seal, 2010 [37]	Illinois, USA
	Puziewicz *et al.*, 2007 [62]	Upper Silesia, Poland
	Tyszka *et al.*, 2014 [186]	Upper Silesia, Poland
	US Department of Transportation, 2008 [150]	Location not provided
	Warchulski *et al.*, 2015 [15]	Piekary Slaskie, Poland
	Warchulski *et al.*, 2016 [39]	Upper Silesia, Poland

Acknowledgements

Vojtěch Ettler was supported by the Czech Science Foundation (GAČR) project no. 19-18513S and institutional funding from the Centre for Geosphere Dynamics (UNCE/SCI/006). The mentorship provided by Robert (Bob) Seal (US Geological Survey) to Nadine Piatak over the years is greatly appreciated.

The valuable comments from Jakub Kierczak (University of Wrocław), Elias Matinde (MINTEK), and Robert Seal (US Geological Survey) helped to significantly improve the original version of the manuscript. Any use of trade, firm, or product names is for descriptive purposes only and does not imply endorsement by the US Government.

References

1. N. M. Piatak, M. B. Parsons and R. R. Seal II, *Appl. Geochem.*, 2015, **57**, 236.
2. N. M. Piatak, in *Environmental Geochemistry: Site Characterization, Data Analysis and Case Histories*, ed. B. De Vivo, H. E. Belkin and A. Lima, Elsevier, 2 edn, 2018, ch. 19, pp. 487–519.
3. A. L. Riley, J. M. MacDonald, I. T. Burke, P. Renforth, A. P. Jarvis, K. A. Hudson-Edwards, J. McKie and W. M. Mayes, *J. Geochem. Explor.*, 2020, **219**, 106630.
4. A. Potysz, E. D. van Hullebusch, J. Kierczak, M. Grybos, P. N. L. Lens and G. Guibaud, *Crit. Rev. Environ. Sci. Technol.*, 2015, **45**, 2424.
5. T. S. Naidu, C. M. Sheridan and L. D. van Dyk, *Miner. Eng.*, 2020, **149**, 106234.
6. D. Pan, L. Li, X. Tian, Y. Wu, N. Cheng and H. Yu, *Resour., Conserv. Recycl.*, 2019, **146**, 140.
7. I. Z. Yildirim and M. Prezzi, *Adv. Civ. Eng.*, 2011, **2011**, 463638.
8. D. Harvey, *Modern Analytical Chemistry*, McGraw-Hill Companies, 2.1 electronic version edn, 2016.
9. R. Tessadri, Analytical techniques for elemental analysis of minerals, *Encyclopedia of Life Support Systems (EOLSS)*, Oxford, UK, 2009.
10. A. B. M. Helaluddin, R. S. Khalid, M. Alaama and S. A. Abbas, *Trop. J. Pharm. Res.*, 2016, **15**, 427.
11. S. Calvin, *XAFS for Everyone*, CRC Press, 1 edn, 2013.
12. P. Chaurand, J. Rose, V. Briois, L. Olivi, J. L. Hazemann, O. Proux, J. Domas and J. Y. Bottero, *J. Hazard. Mater.*, 2007, **139**, 537.
13. A. Roy, *Cem. Concr. Res.*, 2009, **39**, 659.
14. V. Ettler, *Appl. Geochem.*, 2016, **64**, 56.
15. R. Warchulski, A. Gawęda, M. Kądziołka-Gaweł and K. Szopa, *Mineral. Mag.*, 2015, **79**, 459.
16. B. Lavina, P. Dera and R. T. Downs, *Rev. Mineral. Geochem.*, 2014, **78**, 1.
17. G. Will, *Powder Diffraction: The Rietveld Method and the Two Stage Method to Determine and Refine Crystal Structures from Powder Diffraction Data*, Springer-Verlag, Berlin, Germany, 1 edn, 2006.
18. N. M. Piatak and R. R. Seal II, in *Sampling and Monitoring for the Mine-Life Cycle*, ed. V. T. McLemore, K. S. Smith and C. C. Russell, Appendix 5—Case Studies of Sampling and Monitoring, Society for Mining, Metallurgy, and Exploration, Inc., Eaglewood, Colorado, 2014.
19. V. Ettler, R. Červinka and Z. Johan, *Archaeometry*, 2009, **51**, 987.

20. M. Tuhý, T. Hrstka and V. Ettler, *Environ. Pollut.*, 2020, **266**, 115118.
21. A. Pietranik, J. Kierczak, R. Tyszka and B. Schulz, *Minerals*, 2018, **8**, 513.
22. A. L. Morrison, Z. Swierczek and B. L. Gulson, *Environ. Pollut.*, 2016, **210**, 271.
23. N. M. Piatak and R. R. Seal II, *Appl. Geochem.*, 2012, **27**, 623.
24. N. M. Piatak, R. R. Seal II and J. M. Hammarstrom, *Appl. Geochem.*, 2004, **19**, 1039.
25. V. Ettler, Z. Johan, M. Vítková, R. Skála, M. Kotrlý, G. Habler and M. Klementová, *J. Hazard. Mater.*, 2012, **221–222**, 298.
26. M. Vítková, V. Ettler, Z. Johan, B. Kříbek, O. Šebek and M. Mihaljevič, *Mineral. Mag.*, 2010, **74**, 581.
27. R. Wirth, *Chem. Geol.*, 2009, **261**, 217.
28. R. Tyszka, A. Pietranik, J. Kierczak, G. Zieliński and J. Darling, *J. Geochem. Explor.*, 2018, **186**, 215.
29. A. Parbhakar-Fox, S. Gilmour, N. Fox and P. Olin, *Minerals*, 2019, **9**, 415.
30. A. J. Hobson, D. I. Stewart, A. W. Bray, R. J. G. Mortimer, W. M. Mayes, M. Rogerson and I. T. Burke, *Environ. Sci. Technol.*, 2017, **51**, 7823.
31. A. Courtin-Nomade, O. Rakotoarisoa, H. Bril, M. Grybos, L. Forestier, F. Foucher and M. Kunz, *Geochemistry*, 2012, **72**, 29.
32. M. Vanaecker, A. Courtin-Nomade, H. Bril, J. Laureyns and J.-F. Lenain, *J. Geochem. Explor.*, 2014, **136**, 1.
33. S. Sobanska, D. Deneele, J. Barbillat and B. Ledésert, *Appl. Geochem.*, 2016, **64**, 107.
34. S. S. Potgieter-Vermaak, J. H. Potgieter, M. Belleil, F. DeWeerdt and R. Van Grieken, *Cem. Concr. Res.*, 2006, **36**, 663.
35. A. Riboldi, L. Borgese, I. Vassalini, G. Cornacchia, M. Gelfi, M. V. Boniardi, A. Casaroli and L. E. Depero, *Appl. Sci.*, 2020, **10**, 4115.
36. D. M. Proctor, K. A. Fehling, E. C. Shay, J. L. Wittenborn, J. J. Green, C. Avent, R. D. Bigham, M. Connolly, B. Lee, T. O. Shepker and M. A. Zak, *Environ. Sci. Technol.*, 2000, **34**, 1576.
37. N. M. Piatak and R. R. Seal II, *Appl. Geochem.*, 2010, **25**, 302.
38. A. Vartiainen and M. Kytö, *Scand. J. Metall.*, 2002, **31**, 298.
39. R. Warchulski, *Eur. J. Mineral.*, 2016, **28**, 375.
40. J. Kierczak and A. Pietranik, *Can. Mineral.*, 2011, **49**, 1281.
41. K. Kądziołka, A. Pietranik, J. Kierczak, A. Potysz and T. Stolarczyk, *J. Archaeol. Sci.*, 2020, **118**, 105142.
42. E. M. Levin, C. R. Robbins and H. F. McMurdie, *Phase Diagrams for Ceramists*, The American Ceramic Society, Columbus, Ohio, 1964.
43. USEPA, Regional Screening Levels (RSLs) – Summary Table, https://www.epa.gov/risk/regional-screening-levels-rsls-generic-tables, accessed February 2021.
44. CCME, *Canadian Soil Quality Guidelines for the Protection of Environmental and Human Health: Chromium (Total 1997) (VI 1999)*, Canadian environmental quality guidelines, Canadian Council of Ministers of the Environment, Winnipeg, 1999.

45. T. D. West, *Metallurgy of Cast Iron*, The Cleveland Printing and Publishing Company, Cleveland, Ohio, USA, 7th edn, 1902.
46. T. Severin, T. Rehren and H. Schleicher, *Eur. J. Mineral.*, 2011, **23**, 981.
47. C. Mahé-Le Carlier, C. Le Carlier de Veslud, A. Ploquin and J.-J. Royer, *C. R. Acad. Sci., Ser. IIa: Sci. Terre Planetes*, 2000, **330**, 179.
48. K. Eekelers, P. Degryse and P. Muchez, *Am. Mineral.*, 2016, **101**, 1072.
49. D. A. Hoppe, N. M. Piatak, R. R. Seal II and C. J. Green, *Geochemical Characterization, Acid Neutralization Potential, and Phosphate Removal Capacity of Modern and Legacy Iron and Steel Slag from the Chicago-Gary Area of Illinois and Indiana*, US Geological Survey data release, USA, 2020, DOI: 10.5066/P9X7SPIK.
50. S. Ndlovu, G. S. Simate and E. Matinde, *Waste Production and Utilization in the Metal Extraction Industry*, CRC Press, Boca Raton, Florida, 2017.
51. P. Suer, J. E. Lindqvist, M. Arm and P. Frogner-Kockum, *Sci. Total Environ.*, 2009, **407**, 5110.
52. M. Nicolae, V. Irina and F. Zaman, *Sci. Bull. - Univ. "Politeh." Bucharest, Ser. B*, 2007, **69**, 99.
53. H. I. Gomes, V. Funari, W. M. Mayes, M. Rogerson and T. J. Prior, *J. Environ. Manage.*, 2018, **222**, 30.
54. H. Shen and E. Forssberg, *Waste Manage.*, 2003, **23**, 933.
55. M. E. Schlesinger, K. C. Sole and W. G. Davenport, *Extractive Metallurgy of Copper*, Elsevier, Oxford, UK, 5 edn, 2011.
56. M. Khorasanipour and E. Esmaeilzadeh, *J. Geochem. Explor.*, 2016, **166**, 1.
57. R. Sáez, F. Nocete, J. M. Nieto, Á. Capitán and S. Rovira, *Can. Mineral.*, 2003, **41**, 627.
58. V. Ettler, Z. Johan, B. Kříbek, O. Šebek and M. Mihaljevič, *Appl. Geochem.*, 2009, **24**, 1.
59. M. B. Parsons, D. K. Bird, M. T. Einaudi and C. N. Alpers, *Appl. Geochem.*, 2001, **16**, 1567.
60. A. M. Álvarez-Valero, R. Pérez-López and J. M. Nieto, *Am. Mineral.*, 2009, **94**, 1417.
61. R. Barna, P. Moszkowicz and C. Gervais, *Waste Manage.*, 2004, **24**, 945.
62. J. Puziewicz, K. Zainoun and H. Bril, *Can. Mineral.*, 2007, **45**, 1189.
63. K. Nowinska, *Minerals*, 2020, **10**, 371.
64. S. Sobanska, B. Ledésert, D. Deneele and A. Laboudigue, *C. R. Acad. Sci., Ser. IIa: Sci. Terre Planetes*, 2000, **331**, 271.
65. Y. Li, I. Perederiy and V. G. Papangelakis, *J. Hazard. Mater.*, 2008, **152**, 607.
66. R. T. Jones, *S. Afr. J. Sci.*, 1999, **95**, 525.
67. P. E. Tsakiridis, *J. Hazard. Mater.*, 2012, **217–218**, 1.
68. B. J. Jody, E. J. Daniels, P. V. Bonsignore and D. E. Karvelas, presented in part at the 7th International Conference on Solid Waste Management and Secondary Materials, Philadelphia, PA USA, December, 1991.
69. E. Allain, N. Kanari, F. Diot and J. Yvon, *Miner. Eng.*, 2019, **134**, 97.

70. I. Gaballah, E. Allain and M. Djona, *Metall. Mater. Trans. B*, 1997, **28b**, 359.
71. O. Rodríguez, F. J. Alguacil, E. E. Baquero, I. García-Díaz, P. Fernández, B. Sotillo and F. A. López, *RSC Adv.*, 2020, **10**, 21406.
72. S. Chirikure, R. B. Heimann and D. Killick, *J. Archaeol. Sci.*, 2010, **37**, 1656.
73. A. Potysz, J. Kierczak, Y. Fuchs, M. Grybos, G. Guibaud, P. N. L. Lens and E. D. van Hullebusch, *J. Geochem. Explor.*, 2016, **160**, 1.
74. A. Manasse, M. Mellini and C. Viti, *Eur. J. Mineral.*, 2001, **13**, 949.
75. M. Tossavainen, F. Engstrom, Q. Yang, N. Menad, M. Lidstrom Larsson and B. Bjorkman, *Waste Manage.*, 2007, **27**, 1335.
76. C. S. Gahan, M. L. Cunha and Å. Sandström, *Hydrometallurgy*, 2009, **95**, 190.
77. V. Ettler, M. Mihaljevič, A. Jarošíková, A. Culka, B. Kříbek, V. Majer, A. Vaněk, V. Penížek, O. Sracek, B. Mapani and F. Kamona, *Appl. Geochem.*, 2020, **114**, 104473.
78. L. R. P. de Andrade Lima and L. A. Bernardez, *Environ. Earth Sci.*, 2013, **68**, 1033.
79. B. Mikoda, H. Kucha, A. Potysz and E. Kmiecik, *Appl. Geochem.*, 2018, **98**, 459.
80. L. R. Lima and L. A. Bernardez, *J. Hazard. Mater.*, 2011, **189**, 692.
81. R. Pérez-López, J. Delgado, J. M. Nieto and B. Márquez-García, *Chem. Geol.*, 2010, **276**, 29.
82. L. A. Wendling, M. T. Binet, Z. Yuan, F. Gissi, D. J. Koppel and M. S. Adams, *Environ. Toxicol. Chem.*, 2013, **32**, 2602.
83. L. A. Wendling, G. B. Douglas, S. Coleman and Z. Yuan, *Assessment of the Ability of Low-Cost Materials to Remove Metals and Attenuate Acidity in Contaminated Waters*, CSIRO, Water for a Healthy Country National Research Flagship, Australia, 2010.
84. G. B. Douglas, L. A. Wendling and S. Coleman, *Miner. Eng.*, 2012, **35**, 49.
85. C. A. I. Cravotta, *Assessment of Characteristics and Remedial Alternatives for Abandoned Mine Drainage: Case Study at Staple Bend Tunnel Unit of Allegheny Portage Railroad National Historic Site, Cambria County, Pennsylvania, 2004*, Report 2005-1283, US Geological Survey, Reston, Virginia, 2005.
86. B. G. Lottermoser, *Neues Jahrb. Mineral., Geol. Palaeontol., Abh., Abt. A*, 2005, **181**, 183.
87. V. Ettler, O. Legendre, F. Bodénan and J. C. Touray, *Can. Mineral.*, 2001, **39**, 873.
88. M. Shore and A. D. Fowler, *Nature*, 1999, **397**, 691.
89. C. H. Donaldson, *Contrib. Mineral. Petrol.*, 1976, **57**, 187.
90. L. Giuliani, G. Iezzi, T. Hippeli, M. Davis, A. Elbrecht, F. Vetere, M. Nazzari and S. Mollo, *Front. Earth Sci.*, 2020, **8**, 337.
91. A. Addis, I. Angelini, P. Nimis and G. Artioli, *Archaeometry*, 2016, **58**, 96.
92. M. Gautier, J. Poirier, F. Bodénan, G. Franceschini and E. Véron, *Int. J. Miner. Process.*, 2013, **123**, 94.

93. R. Warchulski, A. Gawęda, J. Janeczek and M. Kądziołka-Gaweł, *Mineral. Petrol.*, 2016, **110**, 681.
94. S. Ramachandra Rao, in *Resource Recovery and Recycling from Metallurgical Wastes*, Elsevier, Oxford, UK, 1 edn, 2006, vol. 7, ch. 8, p. 269.
95. G. C. Wang, *The Utilization of Slag in Civil Infrastructure Construction*, Woodhead Publishing, Duxford, UK, 2016.
96. B. C. M. Butler, *Mineral. Mag.*, 1977, **41**, 493.
97. P. W. Scott, S. R. Critchley and F. C. F. Wilkinson, *Mineral. Mag.*, 1986, **50**, 141.
98. Y. Kang, C. Liu, Y. Zhang and H. Xing, *Crystals*, 2020, **10**, 30.
99. P. Tanskanen and H. Makkoneh, *Global Slag Magazine*, 2006, **5**, 16.
100. D. R. Groot, D. M. Kazadi, H. Pollmann, J. P. R. de Villiers, T. Redtmann and J. Steenkamp, presented in part at the Proceedings of the 13th International Ferroalloys Congress, Almaty, Kazakhstan, January, 2013, p. 1051.
101. L. Horckmans, R. Möckel, P. Nielsen, F. Kukurugya, C. Vanhoof, A. Morillon and D. Algermissen, *Minerals*, 2019, **9**, 646.
102. T. Nakasuga, K. Nakashima and K. Mori, *ISIJ Int.*, 2004, **44**, 665.
103. J. Xiang, Q. Huang, X. Lv and C. Bai, *J. Cleaner Prod.*, 2018, **170**, 1089.
104. M. Aarabi-Karasgani, F. Rashchi, N. Mostoufi and E. Vahidi, *Hydrometallurgy*, 2010, **102**, 14.
105. P.-C. Chiang and S.-Y. Pan, in *Carbon Dioxide Mineralization and Utilization*, Springer, Singapore, 2017, ch. 11, p. 233, DOI: 10.1007/978-981-10-3268-4_11.
106. N. A. Yamnova, N. V. Zubkova, N. N. Eremin, A. E. Zadov and V. M. Gazeev, *Crystallogr. Rep.*, 2011, **56**, 210.
107. N. A. Wessey, *Int. J. Mater. Sci. Appl.*, 2017, **6**, 99.
108. N. M. Piatak, R. R. Seal II, D. A. Hoppe, C. J. Green and P. M. Buszka, *Minerals*, 2019, **9**, 468.
109. E. Matinde, G. S. Simate and S. Ndlovu, *J. South. Afr. Inst. Min. Metall.*, 2018, **118**, 825.
110. Q. Zhao, C. Liu, L. Cao, X. Zheng and M. Jiang, *Minerals*, 2018, **8**, 445.
111. J. Kierczak, C. Neel, J. Puziewicz and H. Bril, *Can. Mineral.*, 2009, **47**, 557.
112. V. Ettler, J. Kvapil, O. Šebek, Z. Johan, M. Mihaljevič, G. Ratié, J. Garnier and C. Quantin, *Appl. Geochem.*, 2016, **64**, 118.
113. V. Veselská and J. Majzlan, *Environ. Sci. Pollut. Res.*, 2016, **223**, 7308.
114. R. D. Shannon, *Acta Crystallogr., Sect. A: Found. Adv.*, 1976, **32**, 751.
115. V. Ettler, Z. Johan, P. Bezdička, M. Drábek and O. Šebek, *Eur. J. Mineral.*, 2009, **21**, 837.
116. B. Das, S. Prakash, P. S. R. Reddy and V. N. Misra, *Resour., Conserv. Recycl.*, 2007, **50**, 40.
117. E. Douglas and R. Zerbino, *Cem. Concr. Res.*, 1986, **16**, 662.
118. EUROSLAG, Granulated Blastfurnace Slag, Technical Leaflet No. 1, https://www.euroslag.com/wp-content/uploads/2018/12/LeafletGBS.pdf, accessed September 2020.

119. A.-M. Fällman and J. Hartlén, *Stud. Environ. Sci.*, 1994, **60**, 39.
120. A. R. Goodarzi and M. Salimi, *Appl. Clay Sci.*, 2015, **108**, 61.
121. A. Hadjsadok, S. Kenai, L. Courard, F. Michel and J. Khatib, *Cem. Concr. Compos.*, 2012, **34**, 671.
122. D. W. Lewis, presented in part at the Symposium on Slag National Institute for Transport and Road Research, South Africa, February, 1982.
123. P. Y. Mahieux, J. E. Aubert and G. Escadeillas, *Constr. Build. Mater.*, 2009, **23**, 742.
124. L. Muhmood, S. Vitta and D. Venkateswaran, *Cem. Concr. Res.*, 2009, **39**, 102.
125. H. Y. Poh, G. S. Ghataora and N. Ghazireh, *J. Mater. Civ. Eng.*, 2006, **18**, 229.
126. S. P. Singh, D. P. Tripathy and P. G. Ranjith, *Waste Manage.*, 2008, **28**, 1331.
127. L. Svirenko, J. Vergeles and O. Spirin, in *Approaches to Handling Environmental Problems in the Mining and Metallurgical Regions*, ed. W. Leal Filho and I. Butorina, Springer, Netherlands, 1 edn, 2003, vol. 20, ch. 22, p. 211.
128. G. C. Wang, in *The Utilization of Slag in Civil Infrastructure Construction*, Elsevier, Duxford, UK, 1 edn, 2016, ch. 7, p. 131, DOI: 10.106/B987-0-08-100381-7.00007-0.
129. C. Barca, C. Gerente, D. Meyer, F. Chazarenc and Y. Andres, *Water Res.*, 2012, **46**, 2376.
130. P. Chaurand, J. Rose, J. Domas and J.-Y. Bottero, *J. Geochem. Explor.*, 2006, **88**, 10.
131. A. J. Hobson, D. I. Stewart, R. J. G. Mortimer, W. M. Mayes, M. Rogerson and I. T. Burke, *Waste Manage.*, 2018, **81**, 1.
132. L. M. Juckes, *Miner. Process. Extr. Metall.*, 2013, **112**, 177.
133. D.-H. Shen, C.-M. Wu and J.-C. Du, *Constr. Build. Mater.*, 2009, **23**, 453.
134. C. Shi, *J. Mater. Civ. Eng.*, 2004, **16**, 230.
135. J. Waligora, D. Bulteel, P. Degrugilliers, D. Damidot, J. L. Potdevin and M. Measson, *Mater. Charact.*, 2010, **61**, 39.
136. W. Xuequana, Z. Honga, H. Xinkaib and L. Husenb, *Cem. Concr. Res.*, 1999, **29**, 1103.
137. M. Barra, E. V. Ramonich and M. A. Munoz, presented in part at the Beneficial Use of Recycled Materials in Transportation Applications, Arlington, Virginia, November, 2001, p. 423.
138. A. Drizo, Y. Comeau, C. Forget and R. P. Chapuis, *Environ. Sci. Technol.*, 2002, **36**, 4642.
139. A. Drizo, C. Forget, R. P. Chapuis and Y. Comeau, *Water Res.*, 2006, **40**, 1547.
140. M. P. Luxán, R. Sotolongo, F. Dorrego and E. Herrero, *Cem. Concr. Res.*, 2000, **30**, 517.
141. J. M. Manso, J. A. Polanco, M. Losañez and J. J. González, *Cem. Concr. Compos.*, 2006, **28**, 528.
142. R. Milacic, T. Zuliani, T. Oblak, A. Mladenovic and J. S. Ancar, *J. Environ. Qual.*, 2011, **40**, 1153.

143. S. Neuhold, A. van Zomeren, J. J. Dijkstra, H. A. van der Sloot, P. Drissen, D. Algermissen, D. Mudersbach, S. Schüler, T. Griessacher, J. G. Raith, R. Pomberger and D. Vollprecht, *Minerals*, 2019, **9**, 525.
144. G. R. Qian, D. D. Sun, J. H. Tay and Z. Y. Lai, *Br. Ceram. Trans.*, 2002, **101**, 159.
145. P. E. Tsakiridis, G. D. Papadimitriou, S. Tsivilis and C. Koroneos, *J. Hazard. Mater.*, 2008, **152**, 805.
146. S. Kourounis, S. Tsivilis, P. E. Tsakiridis, G. D. Papadimitriou and Z. Tsibouki, *Cem. Concr. Res.*, 2007, **37**, 815.
147. C. Navarro, M. Díaz and M. A. Villa-García, *Environ. Sci. Technol.*, 2010, **44**, 5383.
148. C. Oh, S. Rhee, M. Oh and J. Park, *J. Hazard. Mater.*, 2012, **213–214**, 147.
149. G. S. Roadcap, W. R. Kelly and C. M. Bethke, *Ground Water*, 2005, **43**, 806.
150. US Department of Transportation, *User Guidelines for Waste and Byproduct Materials in Pavement Construction*, Federal Highway Administration, Report FHWA-RD-97–148, McLean, VA, USA, 1997.
151. P. Ziemkiewicz and J. Skousen, presented in part at the National Meeting of the America Society for Surface Mining and Reclamation, Phoenix, Arizona, USA, August 13–19 1999, p. 651.
152. J. Setién, D. Hernández and J. J. González, *Constr. Build. Mater.*, 2009, **23**, 1788.
153. M. M. Ali, S. K. Agarwal and A. Pahuja, *Adv. Cem. Res.*, 2013, **25**, 208.
154. A. M. Álvarez-Valero, R. Pérez-López, J. Matos, M. A. Capitán, J. M. Nieto, R. Sáez, J. Delgado and M. Caraballo, *Environ. Geol.*, 2008, **55**, 1797.
155. A. Jarošíková, V. Ettler, M. Mihaljevič, B. Kříbek and B. Mapani, *J. Environ. Manage.*, 2017, **187**, 178.
156. A. H. Kaksonen, S. Särkijärvi, J. A. Puhakka, E. Peuraniemi, S. Junnikkala and O. H. Tuovinen, *Miner. Eng.*, 2017, **106**, 97.
157. P. Kinnunen, J. Mäkinen, M. Salo, R. Soth and K. Komnitsas, *Minerals*, 2020, **10**, 654.
158. B. G. Lottermoser, *Mineral. Mag.*, 2002, **66**, 475.
159. A. Potysz, J. Kierczak, A. Pietranik and M. Kądziołka-Gaweł, *Appl. Geochem.*, 2018, **98**, 22.
160. P. Sarfo, G. Wyss, G. Ma, A. Das and C. Young, *Miner. Eng.*, 2017, **107**, 8.
161. A. V. Tatarinov, *Geochem. Int.*, 2002, **40**, 1075.
162. D. M. Urosevic, M. D. Dimitrijevic, Z. D. Jankovic and D. V. Antic, *Physicochem. Probl. Miner. Process.*, 2015, **51**, 73.
163. Z. Zuo, Q. Yu, S. Luo, J. Zhang and E. Zhou, *Energy Fuels*, 2020, **34**, 491.
164. A. B. Prasetyo, R. Darmawansyah, W. Mayangsari, E. Febriana, S. Permana, A. Maksum, S. Oediyani, F. Firdiyono and J. W. Soedarsono, *East.-Eur. J. Enterp. Technol.*, 2020, **1**, 6.
165. G. Ratié, C. Quantin, D. Jouvin, D. Calmels, V. Ettler, Y. Sivry, L. C. Vieira, E. Ponzevera and J. Garnier, *Appl. Geochem.*, 2016, **64**, 136.
166. A. Soultana, A. Valouma, G. Bartzas and K. Komnitsas, *Minerals*, 2019, **9**, 551.

167. P. K. Gbor, I. B. Ahmed and C. Q. Jia, *Hydrometallurgy*, 2000, **57**, 13.
168. X. Li, X. Zhang, X. Zang and X. Xing, *Minerals*, 2020, 10, 313.
169. L. Souter and S. A. Watmough, *Sci. Total Environ.*, 2017, **605–606**, 461.
170. S. T. Bosso and J. Enzweiler, *Environ. Geochem. Health*, 2008, **30**, 219.
171. J. N. Bhadra Chaudhuri and H. Newesely, *Can. Metall. Q.*, 1993, **32**, 1.
172. V. Ettler, P. Piantone and J. C. Touray, *Mineral. Mag.*, 2003, **67**, 1269.
173. V. Ettler, M. Komárková, J. Jehlička, P. Coufal, D. Hradil, V. Machovič and F. Delorme, *Chemosphere*, 2004, **57**, 567.
174. V. Ettler and Z. Johan, *Appl. Geochem.*, 2014, **40**, 97.
175. A. Navarro, E. Cardellach, J. L. Mendoza, M. Corbella and L. M. Domènech, *Appl. Geochem.*, 2008, **23**, 895.
176. N. Saikia, G. Cornelis, Ö. Cizer, C. Vandecasteele, D. Van Gemert, K. Van Balen and T. Van Gerven, *J. Mater. Cycles Waste Manage.*, 2012, **14**, 102.
177. N. H. Yin, Y. Sivry, F. Guyot, P. N. Lens and E. D. van Hullebusch, *J. Environ. Manage.*, 2016, **180**, 310.
178. G. De Angelis and F. Medici, *J. Solid Waste Technol. Manage.*, 2012, **38**, 117.
179. F. Forte, L. Horckmans, K. Broos, E. Kim, F. Kukurugya and K. Binnemans, *RSC Adv.*, 2017, **7**, 49999.
180. E. Kim, J. Roosen, L. Horckmans, J. Spooren, K. Broos, K. Binnemans, K. C. Vrancken and M. Quaghebeur, *Hydrometallurgy*, 2017, **169**, 589.
181. M. Knežević, M. Korac, Ž. Kamberović and M. Ristić, *Metall. Mater. Eng.*, 2010, **16**, 195.
182. M. Penpolcharoen, *Cem. Concr. Res.*, 2005, **35**, 1050.
183. M. Shirinbayan, H. Razavizadeh and M. R. Aboutalebi, *Russ. J. Non-Ferrous Metals*, 2014, **55**, 538.
184. V. Ettler, D. Štěpánek, M. Mihaljevič, P. Drahota, R. Jedlicka, B. Kříbek, A. Vaněk, V. Penížek, O. Sracek and I. Nyambe, *Chemosphere*, 2020, **260**, 127642.
185. H. Kucha, A. Martens, R. Ottenburgs, W. De Vos and W. Viaene, *Environ. Geol.*, 1996, **27**, 1.
186. R. Tyszka, J. Kierczak, A. Pietranik, V. Ettler and M. Mihaljevič, *Appl. Geochem.*, 2014, **40**, 70.

CHAPTER 4

Weathering of Slags

JAKUB KIERCZAK*[a], ANNA PIETRANIK[a] AND NADINE M. PIATAK[b]

[a]Institute of Geological Sciences, University of Wrocław, Pl. M. Borna 9, 50-204 Wrocław, Poland; [b]US Geological Survey, 954 National Center, Reston, Virginia, 20192, USA
*E-mail: jakub.kierczak@uwr.edu.pl

4.1 Introduction

Weathering of metallurgical slags must be discussed and understood as it is a geological process affecting both rocks and human-made materials. The classic approach from the Earth sciences defines weathering as a process by which minerals, rocks, and related materials break down under near-surface conditions (low pressure and temperature and the presence of water and oxygen); these conditions are different to those under which the rocks were formed. Weathering, together with volcanism, metamorphism, and diagenesis, are considered fundamental parts of the rock cycle.[1] This process directly affects life on our planet because it is responsible for soil formation and essential nutrient supply. Weathering is, therefore, a friendly operation of nature, without which the functioning of ecosystems would become impossible. However, weathering not only affects natural materials, such as rocks and minerals, but can also impact anthropogenic materials. Metallurgical slags are an excellent example of human-made materials that are structurally analogous to naturally occurring volcanic rocks.[2] The detailed physical, chemical, and mineralogical characteristics of metallurgical slags are presented in the previous chapter of this book and our intention in this chapter is to discuss what happens when they undergo natural weathering. The feature

Chemistry in the Environment Series No. 2
Metallurgical Slags: Environmental Geochemistry and Resource Potential
Edited by Nadine M. Piatak and Vojtěch Ettler

Published by the Royal Society of Chemistry, www.rsc.org

that distinguishes metallurgical slags from volcanic rocks is the metal and metalloid content, especially for slag generated from base metal production (*i.e.*, base metal slags). This is an important difference when weathering is considered because potentially hazardous substances (*i.e.*, metal(loid)s) may be mobilized; weathering turns from a process supporting life into one that is a major environmental threat. In contrast, the release of metalloids as a result of weathering may not be as significant in the case of slags generated from steelmaking (*i.e.*, ferrous slags) compared to base metal slags. However, the weathering of ferrous slags may generate alkaline to hyperalkaline drainage water, which may be hazardous to aquatic and ecosystem health.

Interestingly, published studies on the weathering of base metal slags and ferrous slags generally have very different focuses. A large knowledge base exists on the environmental behavior of base metal slags enriched in potentially toxic metal(loid)s.[3] All over the world, there are examples of the leaching of deleterious metal(loid)s from base metal slags disposed of in waste piles or used as aggregate and in building materials causing human and ecosystem impacts. In contrast, the majority of research on ferrous slags has been on their physical and mechanical properties due to their wide use in construction and cement, as well as their environmental applications.[3] Base metals slags are also used for these purposes, but the literature on ferrous slag applications is overwhelmingly more abundant in comparison. Steelmaking slags are intentionally weathered or activated before use in construction to hydrate the lime (CaO) to prevent expansion, which would create instability or swelling when the slag is exposed to the atmosphere in the final product.[3] The generation of secondary phases, such as portlandite ($Ca(OH)_2$) and calcite ($CaCO_3$), is intentional and studies focus on how to facilitate and expedite it.[3] Additionally, environmental applications of ferrous slag, such as in treating nutrient-rich or metal-rich wastewater or agricultural runoff, neutralizing acid mine drainage, or sequestering carbon dioxide gas cause the dissolution of phases and precipitation of secondary phases; laboratory and field application studies abound, particularly for nutrient removal. Overall, the perspectives from which ferrous and base metal slag weathering studies are approached are distinct.

Regardless of the type of slag or the reason for the investigation, slag weathering studies usually focus on two stages of the process. The first stage includes the incipient weathering occurring on a microscale and is limited to the places that are generally the most susceptible (*e.g.*, cracks and/or fissures), during which weathering solutions are mostly affected by the dissolution of nearby primary slag phases. As an example, studies exist in which the authors observed the weathering of primary phases and their replacement by closely related secondary phases (*e.g.*, the hydrated form of willemite (Zn_2SiO_4) and hemimorphite ($Zn_4Si_2O_7(OH)_2 \cdot H_2O$) replace primary willemite), while the majority of slag phases as well as the texture remain unaffected.[4] Perhaps the best example of rapid slag weathering is Figure 4.1, which shows images of thin sections prepared from fresh slag that were left under indoor atmospheric conditions for several weeks. Clearly, room

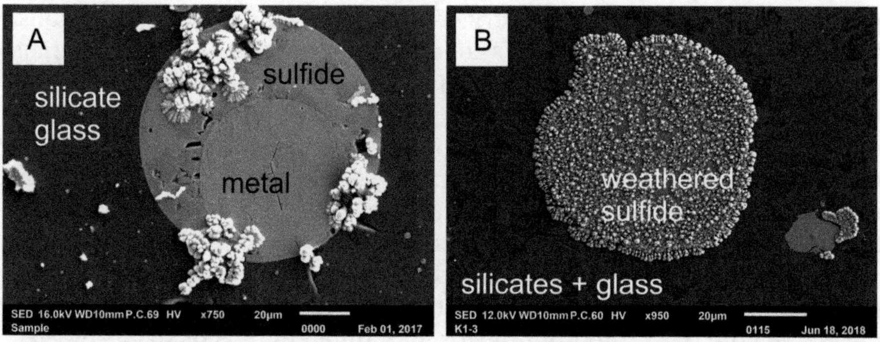

Figure 4.1 Incipient weathering of Cu sulfides observed after a thin section of Cu slag (Old Copper Basin site, Lower Silesia, Poland) was prepared and stored for several months at room temperature (secondary electron microscope image). (A) Cu–Fe sulfide with inclusion of metal phase and (B) pure Cu–Fe sulfide.

temperature and contact with the open air were enough to trigger extensive sulfide oxidation and formation of secondary phases (probably sulfates).

The second situation, *i.e.*, the second stage of slag weathering, is far more complicated because, as weathering progresses, the solutions circulating within the (often disintegrated) slag material represent the cumulative effect of different dissolution reactions occurring within the slag heap.[5] The whole process depends on many variables, such as the physical and chemical properties of drainage water and microbial activity. Therefore, the types of secondary phases that form in this case are related to the whole assemblage of phases susceptible to weathering but also are affected by the surrounding environment.

In this chapter, we summarize factors that affect rates of slag weathering and characterize the two stages of weathering as observed in a wide variety of individual phases and slags. Weathered slags require specific analytical approaches, which are also outlined in this chapter. We also discuss whether slag weathering and the weathering of rocks are similar processes.

4.2 Factors Affecting the Weathering of Slags

Weathering is a very complex phenomenon influenced by many variables. Factors that affect the weathering of metallurgical slags can be divided into two groups: (i) internal and (ii) external. Internal factors are directly related to the properties of the material undergoing weathering and comprise the mineralogy and geochemistry of slags (especially the stability of slag phases) as well as their grain size and texture, which in turn is often related to the conditions in which the slags are formed (mainly with crystallization and cooling rates). The second group of weathering factors, external factors, includes climatic and hydrologic conditions as well as site-specific environmental parameters (*e.g.*, pH and chemistry of circulating fluids, redox conditions, microbial activity, organic matter content, the presence of other materials with different resistance to weathering) that may affect slag weathering. Internal and external factors interact with each other, resulting in a slow-down or acceleration of the dissolution processes for individual slag components. As the dissolution begins, chemically variable fluids are formed that may be saturated with respect to secondary phases, and the precipitation of these phases is yet another factor affecting slag weathering.

It is difficult to determine which is the most important factor influencing slag weathering. Additionally, the most important factor for one type of slag may not be the most important factor for a different type of slag. However, some slag features show similar responses to weathering, regardless of slag type. In particular, smaller grain size and/or increased slag porosity accelerate weathering. For example, small particles of Pb slags and Ni slags present within soil are characterized by more pronounced weathering features than larger particles.[6,7] Schmukat *et al.*[8] showed that the available specific surface area of Cu slags, directly related to grain-size distribution, is positively correlated with the release of hazardous substances. Furthermore, larger amounts of secondary phases, direct evidence of slag weathering, have been observed in areas with higher porosity slags.[4,9] However, the phase and chemical composition of slags are strongly site-specific and these characteristics result in different rates of weathering among slags of different compositions. This is particularly pronounced when comparing weathering of base metal and ferrous slags.

Studies on the geochemical processes that facilitate the dissolution of phases found in base metal slags have revealed the importance of sulfides, intermetallic compounds, and glass weathering to metal(loid) release.[10–12] These metal(loid)s could subsequently be sequestered by the precipitation of secondary phases, which have the potential to release the aforementioned metal(loid)s upon further weathering.[4,13,14] Conversely, studies of ferrous slag commonly focus on the breakdown of the Ca and Mg silicates and oxides, which influences the efficiency of the material to be used in a variety of applications. Weathering of these same phases leads to the environmental issues associated with ferrous slag, predominantly the generation of alkaline drainage but also the release of potentially hazardous elements, such as Cr and V. A focus on silicates and oxides in ferrous slag is in stark contrast to the sulfides, intermetallic compounds, and glass phases that draw attention in base metal slag studies. Additionally, the sulfides (*i.e.*, oldhamite (CaS) and alabandite (MnS)) that occur in ferrous slag commonly do not host potentially hazardous metal(loid)s. Furthermore, the metallic component of ferrous slags is commonly Fe metal, unlike the complex intermetallic phases reported in base metal slags; studies on metallic Fe focus on reprocessing to recover the metal instead of environmental issues associated with its oxidation.

Slags of different chemical and phase compositions may be deposited in different environmental conditions (external factors), resulting in specific rates of weathering for slag constituents. Undoubtedly, the basic factor that determines slag stability is the pH of the environment. The phase composition of slags and pH are strongly related to each other; slags may cause acidification (base metals slags) or alkalization of the environment,[9,15,16] which in turn may enhance further secondary transformations of the slags themselves, but also of materials (including rocks) occurring in their immediate vicinity. The general rule observed by many researchers from slag leaching experiments is that slags are most stable under close to neutral pH conditions. Additionally, individual phases within the slags may exhibit contrasting pH-dependent reactivity. For example, the study by Potysz *et al.*[17] revealed that crystalline constituents of Cu slag are preferentially dissolved under acidic conditions, in contrast to glass, which is preferentially dissolved under alkaline conditions. The rates of dissolution for laboratory-synthesized silicates and oxides that commonly occur in ferrous slags were generally slower under alkaline compared to acidic conditions.[18] Among other factors that influence slag stability, the presence of organic acids and microbial activity should also be noted. As shown by Potysz *et al.*[19] glass dissolution is enhanced in the presence of rhizosphere-like organic acids. Furthermore, another bio-weathering study of slags indicated that microbial activity also accelerates the disintegration of slags.[20] Observations of natural assemblages of secondary minerals in slag heaps also provide information on the environmental conditions prevailing at disposal sites, which is complementary information to that obtained from leaching tests. The types of secondary phases that form depends on both the factors related directly to the properties of the slag and on external factors (such as the pH of the environment). For example, the presence of jarosite ($KFe^{3+}{}_3(OH)_6(SO_4)_2$), a product of sulfide weathering from Cu slags, indicates acidic conditions with pH below 4.2.[21] However, such straightforward interpretation of pH conditions based on the secondary assemblage is rare. Actually, secondary phases that should be stable in a variety of environmental conditions (*e.g.*, sulfates, oxides or carbonates) are often present together at a single site, which in turn could indicate dynamically changing conditions.[13,22] This is understandable as both seasonal fluctuations and alternating dry and rainy periods may lead to the formation and subsequent dissolution of secondary phases, changing conditions within a slag heap.[9,11,23] This shows that climate is another important factor affecting slag weathering, with more dynamic conditions and more advanced weathering expected in warm to temperate and humid climates as compared to dry ones.[24]

4.3 Approaches to Analyzing Weathering Processes in Slags

Because weathering processes control the environmental stability of slags, the main aim of numerous studies is to characterize the mobility of toxic elements during slag weathering. Three common approaches are: (i) leaching studies and experimental simulations, (ii) analyses of natural solutions and geochemical modeling, and (iii) observations of naturally occurring secondary assemblages.

Leaching experiments are used to determine a slag's stability both in the conditions it is deposited, but also in the conditions in which it could be potentially placed (reviewed in Chapter 5). However, here we would like to mention that leaching experiments can be used as a laboratory approach to reconstruct slag weathering. For example, the impact of changing pH conditions is simulated through pH_{stat} leaching experiments, consisting of leaching of the crushed material at several different pH values with the same liquid:solid ratio.[17,25–27] These experiments may also determine the acid-neutralizing capacity of the slags. In recent years, scientists have also used leaching experiments to investigate the influence of other factors (*e.g.*, soil organic matter or microbial activity) on slag stability.[19,20,28–30]

Geochemical modeling is another way of reproducing the conditions of slag weathering depending on both internal and external factors. Modeling requires reliable thermodynamic data for all slag phases. Rigorous thermodynamic data complemented by detailed characterization of slag material (including field observations as well as identification of primary and secondary phases) will improve the accuracy of the modeling results.[16] For example, Parsons *et al.*[31] used geochemical modeling to reveal that glass is more susceptible to dissolution than other silicate phases; this was confirmed by several studies focused on slag weathering using direct mineralogical observations of slags.[7,32] Additionally, geochemical modeling may predict how slag weathering could change over time in response to changing environmental conditions.

Finally, another way is to *directly analyze secondary phases* that can be collected from their accumulation points, for example, along cracks or from pores or shielded surfaces. The phases can be identified using different approaches, usually scanning electron microscopy with energy dispersive spectroscopy (SEM-EDS), X-ray diffraction, or Raman spectrometry (Figure 4.2). Numerous phases similar to naturally occurring minerals were identified this way (Table 4.1), providing information on elements that were mobilized from slag and sequestered in secondary phases. However, valuable information comes also from analyzing weathering of primary slag phases *in situ*. This approach provides information on the primary phases that were dissolved so that we know exactly which elements were released. This is important because not all the mobilized elements are sequestered in secondary phases; some, such as Zn, may be continuously washed away in solution.[33] *In situ* techniques also include SEM-EDS[34] and Raman spectrometry,[35] with the latter being able to identify complex Fe (hydro)oxide intergrowths. An emerging field is the use of automated-SEM analyses (Mineral Liberation Analyses – MLA or Quantitative Evaluation of Mineralogy by a SCANning electron microscope – QEMSCAN), which can provide quantitative information on phases that are being weathered. For example, Morrison *et al.*[36] showed that 80% of Pb-enriched phases are accessible to circulating fluids and

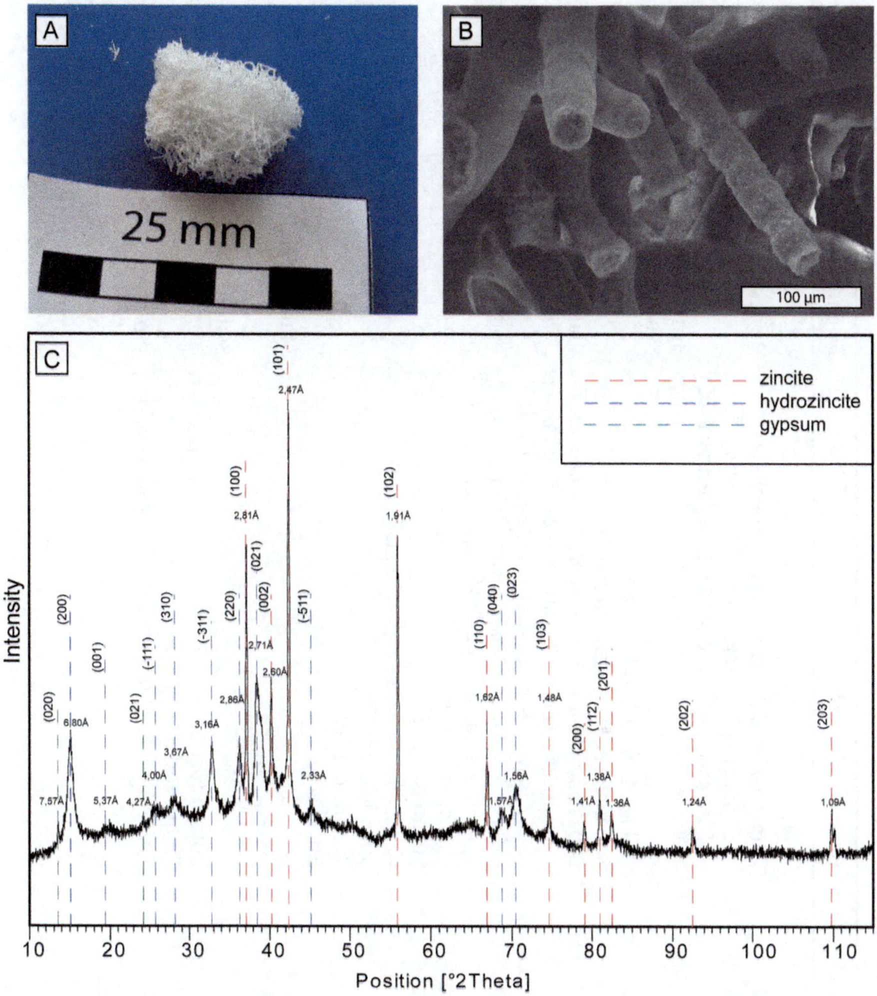

Figure 4.2 Direct analyses of secondary phases issued from Zn–Pb slag (Katowice-Wełnowiec site, Upper Silesia, Poland) weathering with (A) macroscopic sample, (B) secondary electron microscope image, and (C) powder X-ray diffraction pattern.

identified them as prone to leaching. Using this method for strongly weathered slags is challenging because such slags are extremely complex materials with numerous phases not being recognized due to the too low resolution of SEM techniques.[14] However, the grain-size distribution of individual phases or the proportion of different phases in mineral assemblages are great tools for distinguishing primary and secondary phases, and their relations and distribution of toxic elements among them.[14,37]

Table 4.1 Secondary phases identified in base metal and ferrous slags.

Secondary phases	Theoretical formula	Base metal slags	Fe and steel slags
Oxides and hydroxides			
HFO *undistinguished*	Not identified	Zdravković *et al.*,[5] Kierczak *et al.*,[7] Sueoka and Sakakibara,[12] Kierczak *et al.*,[22] Ettler *et al.*,[28] Parsons *et al.*,[31] Ettler and Johan,[34] Piatak *et al.*[41]	Bayless and Schultz,[15] Svirenko *et al.*[66]
Al hydro-oxides	$Al(OH)_3$	Piatak *et al.*,[16] Sobanska *et al.*[35]	Svirenko *et al.*,[66] Navarro *et al.*[67]
Brucite	$Mg(OH)_2$		Svirenko *et al.*[66]
Cuprite	Cu_2O	Bril *et al.*,[13] Kądziołka *et al.*[68]	
Ferrihydrite	$Fe^{3+}{}_{10}O_{14}(OH)_2$	Sobanska *et al.*,[35] Navarro *et al.*[69]	
Goethite	α-FeOOH	Bril *et al.*,[13] Alvarez-Valero *et al.*,[21] Ettler *et al.*,[24] Cappuyns *et al.*,[25] Sobanska *et al.*,[35] Navarro *et al.*,[69] Alvarez-Valero *et al.*,[70] Piatak and Seal[71]	Piatak and Seal,[55] Svirenko *et al.*,[66] Costagliola *et al.*[72]
Hematite	Fe_2O_3	Bril *et al.*,[13] Alvarez-Valero *et al.*,[21] Cappuyns *et al.*,[25] Sobanska *et al.*,[35] Alvarez-Valero *et al.*,[70] Piatak and Seal[71]	Piatak and Seal,[55] Piatak *et al.*[73]
Lepidocrocite	γ-FeO(OH)	Sobanska *et al.*[35]	
Litharge	PbO	Gee *et al.*,[6] Ettler *et al.*,[11] Sobanska *et al.*[35]	
Maghemite	γFe_2O_3	Sobanska *et al.*[35]	
Magnetite	Fe_3O_4	Sobanska *et al.*[35]	
Mcconnellite	$CuCrO_2$	Ettler *et al.*[11]	
Portlandite	$Ca(OH)_2$		Mahieux *et al.*,[47] Navarro *et al.*[67]
Carbonates			
Ankerite	$Ca(Fe,Mg)(CO_3)_2$		Svirenko *et al.*[66]
Azurite	$Cu_3(CO_3)_2(OH)_2$	Lottermoser[43]	
Brianyoungite	$Zn_3(CO_3)(OH)_4$	Piatak and Seal[71]	
Calcite/aragonite/vaterite	$CaCO_3$	Vitková *et al.*,[23] Sobanska *et al.*,[35] Warchulski *et al.*[74]	Bayless and Schultz,[15] Mahieux *et al.*,[47] Piatak and Seal,[55] Svirenko *et al.*,[66] Navarro *et al.*,[67] Costagliola *et al.*,[72] Cravotta[75]

Cerussite, hydrocerussite	$PbCO_3$, $Pb_3(CO_3)_2(OH)_2$	Gee *et al.*,[6] Ettler *et al.*,[11] Bril *et al.*,[13] Parsons *et al.*,[31] Tyszka *et al.*,[34] Lottermoser,[43] Navarro *et al.*[69]	
Dolomite	$CaMg(CO_3)_2$		Bayless and Schultz,[15] Svirenko *et al.*,[66] Navarro *et al.*[67]
Hydrozincite	$Zn_5(CO_3)_2(OH)_6$	Kucha *et al.*,[10] Parsons *et al.*,[31] Tyszka *et al.*[34]	
Magnesite	$MgCO_3$		Svirenko *et al.*[66]
Malachite	$Cu_2CO_3(OH)_2$	Kierczak *et al.*,[22] Vitková *et al.*,[23] Parsons *et al.*,[31] Sobanska *et al.*,[35] Lottermoser[43]	
Monohydrocalcite	$CaCO_3{\cdot}H_2O$		Navarro *et al.*[67]
Smithsonite	$ZnCO_3$	Vanaecker *et al.*[4]	
Sphaerocobaltite	$CoCO_3$	Vitková *et al.*[23]	
Siderite	$FeCO_3$		Svirenko *et al.*[66]
Sulfates			
Alum-(K)	$KAl(SO_4)_2{\cdot}12H_2O$	Ashelford and Gore[76]	
Anglesite	$PbSO_4$	Gee *et al.*,[6] Ettler *et al.*,[11] Bril *et al.*,[13] Navarro *et al.*[69]	
Barite	$BaSO_4$	Bril *et al.*,[13] Parsons *et al.*[31]	
Beaverite	$PbCu(Fe,Al)_2(SO_4)_2(OH)_6$	Zdravković *et al.*[5]	
Bechererite	$Zn_7Cu(OH)_{13}[SiO(OH)_3(SO_4)]$	Piatak and Seal[71]	
Bianchite	$(Zn,Fe)SO_4{\cdot}6H_2O$	Bril *et al.*[13]	
Bloedite	$Na_2Mg(SO_4)_2{\cdot}4H_2O$	Lottermoser[9]	
Brochantite	$Cu_4(SO_4)(OH)_6$	Zdravković *et al.*,[5] Bril *et al.*,[13] Vitková *et al.*,[23] Piatak *et al.*[41]	
Chalcanthite	$CuSO_4{\cdot}5H_2O$	Parsons *et al.*,[31] Piatak *et al.*,[41] Navarro *et al.*[69]	
Copiapite	$Fe^{2+}Fe^{3+}{}_4(SO_4)_6(OH)_2{\cdot}20(H_2O)$	Lottermoser,[9] Navarro *et al.*[69]	
Epsomite	$MgSO_4{\cdot}7H_2O$	Lottermoser[9]	
Ettringite	$Ca_6Al_2(SO_4)_3(OH)_{12}{\cdot}26H_2O$		Cravotta[75]
Eugsterite	$Na_4Ca(SO_4)_3{\cdot}2H_2O$	Lottermoser[43]	
Gypsum	$CaSO_4{\cdot}2H_2O$	Vanaecker *et al.*,[4] Lottermoser,[9] Ettler *et al.*,[11] Bril *et al.*,[13] Tyszka *et al.*,[34] Piatak *et al.*,[41] Lottermoser,[43] Navarro *et al.*,[69] Piatak and Seal,[71] Warchulski *et al.*,[74] Ashelford and Gore,[76] Warchulski *et al.*[77]	Bayless and Schultz,[15] Svirenko *et al.*,[66] Costagliola *et al.*,[72] Cravotta[75]
Goslarite	$ZnSO_4{\cdot}7H_2O$	Vanaecker *et al.*,[4] Navarro *et al.*[69]	

(*continued*)

Table 4.1 (*continued*)

Secondary phases	Theoretical formula	Base metal slags	Fe and steel slags
Halotrichite	$Fe^{2+}Al_2(SO_4)_4 \cdot 22H_2O$	Navarro *et al.*[69]	
Hexahydrite	$MgSO_4 \cdot 6H_2O$	Lottermoser,[9] Bril *et al.*,[13] Warchulski *et al.*[74]	
Jarosite	$KFe^{3+}_3(OH)_6(SO_4)_2$	Bril *et al.*,[13] Alvarez-Valero *et al.*,[21] Piatak *et al.*,[41] Navarro *et al.*,[69] Alvarez-Valero *et al.*[70]	
Langite	$Cu_4(SO_4)(OH)_6 \cdot 2H_2O$	Kierczak *et al.*[22]	
Leadhillite	$Pb_4SO_4(CO_3)_2(OH)_2$	Gee *et al.*[6]	
Linarite	$Pb,Cu(SO_4)(OH)_2$	Zdravković *et al.*[5]	
Poitevinite	$(Cu,Fe)SO_4 \cdot H_2O$	Lottermoser[9]	
Rapidcreekite	$Ca_2(SO_4)(CO_3) \cdot 4H_2O$	Warchulski *et al.*[77]	
Roemerite	$Fe^{2+}Fe^{3+}_2(SO_4)_4 \cdot 14H_2O$	Lottermoser[9]	
Siderotil	$(Fe,Cu)SO_4 \cdot 5H_2O$	Piatak *et al.*[41]	
Syngenite	$K_2Ca(SO_4)_2 \cdot H_2O$	Lottermoser,[43] Warchulski *et al.*[74]	
Szomolnokite	$FeSO_4 \cdot H_2O$	Navarro *et al.*[69]	
Arsenates and phosphates			
Apatite	$Ca_5(PO_4)_3(F,Cl,OH)$	Warchulski *et al.*[77]	
Bayldonite	$PbCu_3(AsO_4)_2(OH)_2$	Ettler *et al.*[11]	
Bobierrite	$Mg_3(PO_4)_2 \cdot 8H_2O$		Svirenko *et al.*[66]
Brushite	$CaHPO_4 \cdot 2H_2O$		Svirenko *et al.*[66]
Lammerite	$Cu_3(AsO_4)_2$	Ettler *et al.*[11]	
Lavendulan	$NaCaCu_5(AsO_4)_4Cl \cdot 5H_2O$	Ettler *et al.*[11]	
Mimetite	$Pb_5(AsO_4)_3Cl$	Bril *et al.*[13]	
Olivenite	Cu_2AsO_4OH	Ettler *et al.*[11]	
Pyromorphite	$Pb_5(PO_4)_3Cl$	Gee *et al.*[6]	
Scorodite	$FeAsO_4 \cdot 2H_2O$	Lottermoser[43]	
Whitlockite	$Ca_9Mg(PO_4)_6(PO_3OH)$		Svirenko *et al.*[66]
Other			
Atacamite	$Cu_2Cl(OH)_3$	Lottermoser[43]	
Clays	Variable composition	Lottermoser[9]	Bayless and Schultz[15]
Cotunnite	$PbCl_2$	Navarro *et al.*[69]	
Hemimorphite	$Zn_4Si_2O_7(OH)_2 \cdot H_2O$	Vanaecker *et al.*,[4] Kucha *et al.*,[10] Tyszka *et al.*[34]	

4.4 Incipient Weathering

The weathering of different slag components (amorphous and/or crystalline, being synthetic equivalents of minerals) controls the release and mobility of metal(loid)s from all slag types and the generation of hyperalkaline drainage from ferrous slags.

Research over many years has focused on determining the susceptibility of different slag phases to weathering and their rates of dissolution. In the following subsections, we present an overview of major slag components and how each of them contributes to weathering.

4.4.1 Sulfides

Sulfides are much more abundant and diverse in base metal slags than in Fe and steel slags. Base metal slags are most commonly generated from the smelting of sulfide ores whereas Fe ores consist almost exclusively of Fe oxides or oxy-hydroxides. The principle of base metal smelting is aimed at the separation of two melts: a silicate one (in solid form, it is a slag) and a liquid metal (*i.e.*, matte, consisting mainly of sulfides, metals, and various intermetallic compounds). In practice, however, no process is 100 percent effective, therefore matte fragments normally can be found in base metal slags and most often they occur in the form of sulfides. Although constituting a relatively low percentage by volume in the whole phase composition, sulfides are very important components of base metal slags because they are the main source of some environmentally important elements, for example, Pb, Cu, As, and Ni. On the other hand, sulfides found in ferrous slags include trace amounts of CaS and MnS (oldhamite and alabandite). There is a sparsity of information regarding the weathering of these sulfides in ferrous slags, likely due to the lack of environmental issues related to their dissolution as a result of their low abundance and the absence of potentially deleterious trace elements, unlike sulfides found in base metal slags.

There is general agreement that sulfides are the slag components that weather first (see Figure 4.1). This is because they generally form under a reducing environment and, after the slag is removed and deposited outside the furnace, the conditions become oxidizing. The initial weathering of sulfides is sometimes barely noticeable and manifests itself only by increased porosity at the contact with vesicles present in slags.[7] However, such traces of sulfide decomposition may be observed even in slags that at first glance seem to be completely unweathered.[7] On the other end of the weathering spectrum, the original sulfide may be partly or completely replaced by hydrous ferric oxides (HFO) and other secondary phases, as is the case in Zn–Pb slags from Upper Silesia (Figure 4.3).

The general reactions controlling sulfide weathering are similar to those observed in ore deposits and mine waste.[38,39] As sulfides in slag are often Fe-rich (pyrite (FeS_2), pyrrhotite ($Fe_{1-x}S$), chalcopyrite ($CuFeS_2$), bornite (Cu_5FeS_4), and Fe-rich sphalerite ((Zn,Fe)S)), the oxidation of the sulfides is

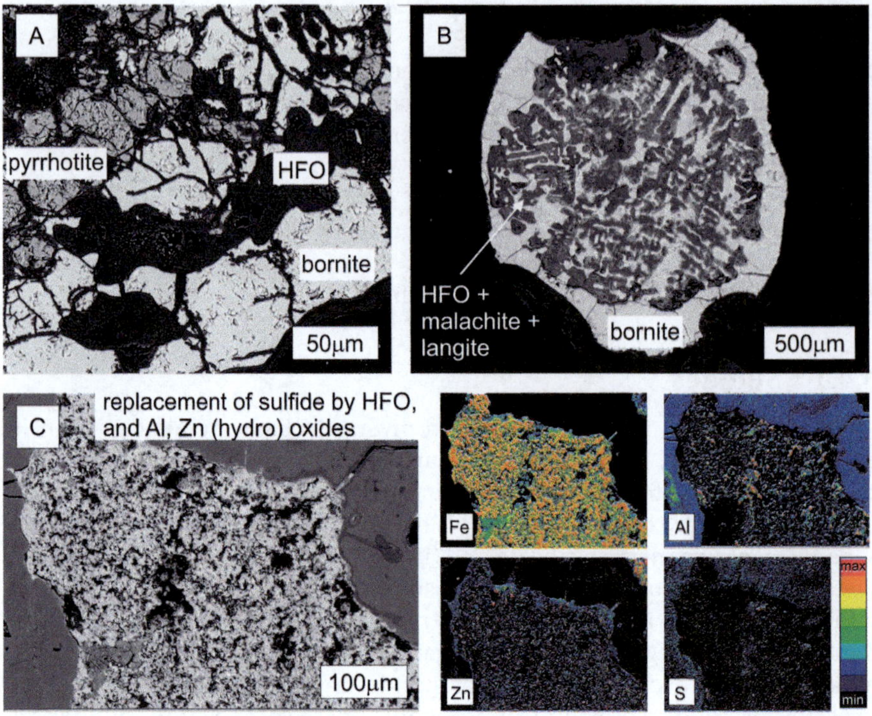

Figure 4.3 Early-stage weathering (back-scattered electron images) of different slag phases. (A) Intergrowths of bornite and pyrrhotite in Cu slag replaced by hydrous ferric oxides (HFO). (B) Complex internal structure of bornite and its weathering products in Cu slag (malachite and langite). The presence of HFO suggests complete weathering of pyrrhotite (Rudawy Janowickie site, Lower Silesia, Poland); Reproduced from ref. 20 with permission from Elsevier, Copyright 2012. (C) Complete weathering of sulfide, which was replaced by HFO and other phases, as indicated in microprobe elemental mappings. The false color scale shows relative enrichment in a given element (Świętochłowice site, Upper Silesia, Poland).

an acid-generating reaction associated with precipitation of Fe oxyhydroxides;[39] for example, pyrrhotite and bornite oxidation can be described as follows (eqns (4.1) and (4.2)):

$$Fe_{0.9}S_{(s)} + 2.175O_{2(g)} + 2.35H_2O_{(l)} \rightarrow 0.9Fe(OH)_{3(s)} + SO_4{}^{2-}{}_{(aq)} + 2H^+{}_{(aq)} \quad (4.1)$$

$$Cu_3FeS_{4(s)} + 7.75O_{2(g)} + 3.5H_2O_{(l)} \rightarrow Fe(OH)_{3(s)} + 4SO_4{}^{2-}{}_{(aq)} + 3Cu^{2+}{}_{(aq)} + 4H^+{}_{(aq)} \quad (4.2)$$

The rate of sulfide weathering depends on several factors, the most important being the type of sulfide itself. Also important are the texture and size as well as admixtures of some trace elements.[38] The reconstruction of sulfide weathering in slags is often difficult because sulfides are very diverse, their composition is affected by substitutions, and they often have a complicated internal structure with symplectic intergrowths.[40,41] Despite that, it is sometimes possible, based on observations of weathered sulfides, to determine which of them dissolves faster than the others. For example, in the historical Cu slags studied by Kierczak *et al.*,[22] the most common sulfides are bornite and pyrrhotite intergrown with each other, with pyrrhotite being more affected by weathering than bornite (Figure 4.3A). This is consistent with pyrrhotite being the least resistant to weathering among naturally occurring ore sulfides.[38] In slags, pyrrhotite weathering is manifested by the presence of HFO as a major component among the secondary phases (Figure 4.3A and B). The oxides may include goethite (FeO(OH)), hematite (Fe_2O_3), maghemite (γ-Fe_2O_3), lepidocrocite (FeO(OH)), and ferrihydrite ($Fe^{3+}{}_{10}O_{14}(OH)_2$).[35] Other secondary phases may form when base metal sulfides are also weathered, for example, malachite ($Cu_2(CO_3)(OH)_2$) and langite ($Cu_4(SO_4)(OH)_6 \cdot 2H_2O$) as the products of bornite weathering (Figure 4.3B). In Figure 4.3B only bornite is present, but this is probably because pyrrhotite has been completely dissolved and only its decomposition products are left.

An interesting feature indicating that sulfides are less stable than other slag phases can be observed at slag surfaces affected by glass weathering. For example, it was noted that sulfides are quickly dissolved at the surface of slags.[42] It seems to be the rule that the outer parts of slightly weathered slags are sulfide-free (Figure 4.4A) or if present they are much more weathered than the sulfides occurring in the inner part of slags (Figure 4.4A).

4.4.2 Metallic Compounds

Metallic compounds, which include pure metals, intermetallic compounds, and metal alloys, are important carriers of toxic metal(loid)s and are readily weathered. The rate of weathering depends on the conditions and the compositions of the phases, which are strongly variable from slag site to slag site. The main process that triggers weathering is oxidation and a metal cation is formed that may be leached away from the reaction surface or may be precipitated as a secondary phase. The major secondary minerals are oxides and hydroxides, for example, litharge or massicot (α-PbO and β-PbO) from Pb metallic droplets,[11,35] Zn hydroxide from metallic Zn,[10] or HFO from Fe metal.[24] In general, the first reaction step for Fe metal weathering is as follows:

$$2Fe^0{}_{(s)} + 1.5O_{2(g)} + 3H_2O_{(l)} \rightarrow 2Fe(OH)_{3(s)} \tag{4.3}$$

Sulfates are also common weathering products of metallic compounds if sulfate ions are present in the circulating fluids. This is often the case if sulfides occur in the slag. Carbonates may also be formed if CO_2 is present in the fluids. Different minerals may precipitate depending on the pH and fluid composition. A Raman spectroscopy study of Pb droplets by Sobanska *et al.*[35]

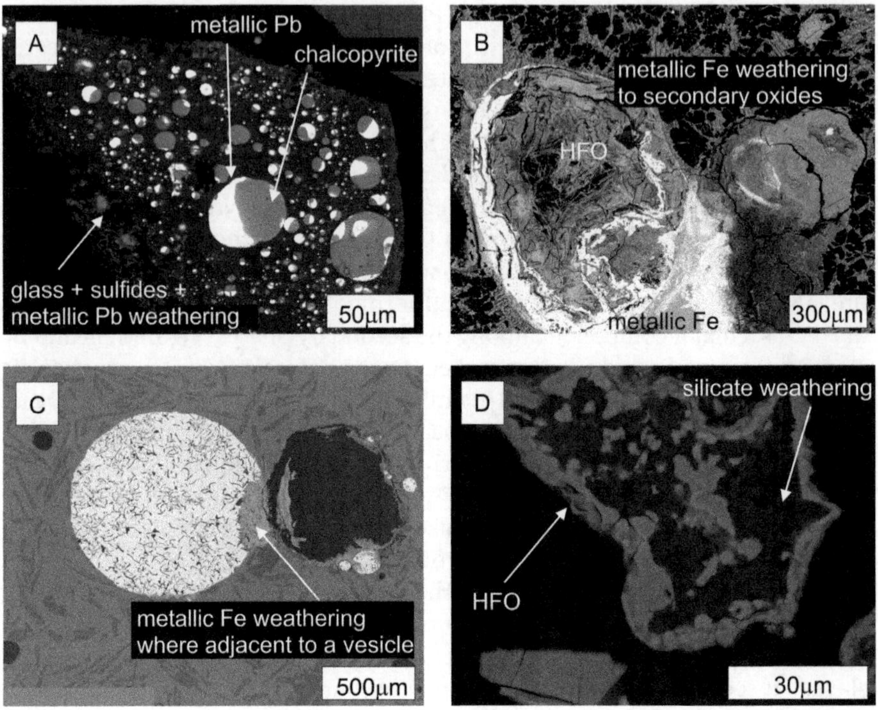

Figure 4.4 Early-stage weathering (back-scattered electron images) of different slag phases. (A) Glass weathering observed in external parts of a Cu-slag fragment; please note that the weathered part is porous and devoid of metal/sulfide phases (Legnica site, Lower Silesia, Poland); Reproduced from ref. 42, https://doi.org/10.1007/s11356-016-6100-z, under the terms of the CC BY 4.0 license, http://creativecommons.org/licenses/by/4.0/. (B) Metallic iron (bright white) inclusion weathering to secondary Fe oxides. The slag surrounding the weathered inclusion is subhedral to euhedral cristobalite (dark gray/black) within a matrix of fine-grained dendritic olivine (Fe- and Mn-rich) and glass (steel slag from legacy slag dump from Chicago area, Illinois, USA). (C) Metallic iron inclusion (white) weathering to secondary Fe oxide (light gray). Note weathering of inclusion occurs where the surface is adjacent to a vesicle. Surrounding slag is composed of euhedral prismatic spinel (medium gray) and long laths of skeletal olivine (similar medium gray as spinel) within a glass matrix (pre-1900 Fe slag from Hopewell Furnace, Pennsylvania, USA). (D) Silicate weathering in Zn–Pb slag with HFO precipitation (Świętochłowice site, Upper Silesia, Poland).

showed that several secondary phases may precipitate within a single weathered Pb droplet (litharge, massicot, basic Pb sulfates, cerussite ($PbCO_3$), hydrocerussite ($Pb_3(CO_3)_2(OH)_2$), and malachite), suggesting complex conditions during the precipitation.

For base metal slags, the release of metals from metallic particles is strongly controlled by their structure as they often form complex intergrowths of several metallic or metallic–sulfide phases[43] (Figures 4.1A and 4.3B). Direct weathering of a complex metallic compound was shown by Ettler *et al.*[11] with Pb metal being altered to litharge; the intermetallic compound $CuAs_3$ that occurred as inclusions in the Pb metal was less affected. Kucha *et al.*[10] also noted that pure Zn metal particles weather more rapidly than metal alloys. In contrast to the complex intergrowths of metals found in base metal slags, the metal found in ferrous slags is most commonly Fe, the amount of which generally depends on the efficiency of the smelting process. Metallic Fe in ferrous slag weathers to iron oxides, including HFO and goethite (Figure 4.4B and C). Another aspect that affects the weathering rates of metallic phases is their size. They are usually small and therefore often effectively armored by surrounding silicate minerals and glass, which would postpone their weathering until the silicate phase has weathered. This is seen in Figure 4.4A where chalcopyrite/Pb droplets are not weathered inside the particle but the metal is weathered away from its rim.[44]

4.4.3 Silicates

Silicates in slag are a chemically and structurally variable group with the presence of phases that are both similar to common natural ones (*e.g.*, olivine ($(Fe,Mg)_2SiO_4$), melilite ($(Ca,Na)_2(Al,Mg,Fe^{2+})(Si,Al)_2O_7$), pyroxene ($(Mg,Fe^{2+},Ca)Si_2O_6$), plagioclase ($(Na,Ca)(Si,Al)_4O_8$), and leucite ($KAlSi_2O_6$)) and those found only in very specific rocks (*e.g.*, kirschsteinite ($CaFe^{2+}SiO_4$), willemite, and alamosite ($PbSiO_3$)). Generally, silicate weathering is controlled by acid-consuming reactions that may proceed as congruent (*e.g.*, pyroxene dissolution in eqn (4.4)) or incongruent dissolution (*e.g.*, formation of kaolinite ($Al_2Si_2O_5(OH)_4$) after feldspar weathering, as in eqn (4.5)):

$$(Mg,Fe,Ca)^{2+}SiO_{3(s)} + 2H^+_{(aq)} + H_2O_{(l)} \rightarrow (Mg,Fe,Ca)^{2+}_{(aq)} + H_4SiO_{4(aq)} \quad (4.4)$$

$$2(Na,K)AlSiO_{4(s)} + 2H^+_{(aq)} + H_2O_{(l)} \rightarrow (Na,K)^+_{(aq)} + Al_2Si_2O_5(OH)_{4(s)} \quad (4.5)$$

The weathering rates of silicates are variable and that seems to be true even for similar phases occurring in slags from different sites. For example, in the first study of Zn slags from Poland, willemite was indicated as being more susceptible to weathering than melilite,[13] whereas in the Zn slags studied by Vanaecker *et al.*[4] melilite (var. hardystonite, $Ca_2ZnSi_2O_7$) was shown to be more affected by secondary transformations than willemite. Incipient weathering of silicates is rarely observed and is limited to willemite from Zn slags, which was observed as reaction rims in Zn slags from Japan.[12] In fact, silicates, such as forsterite were often shown to be stable, even when the surrounding glass was affected by weathering (Figure 4.5).[7,12] Furthermore, in Cu–Pb slags from Tsumeb (Namibia), Pb feldspar remains unweathered in a matrix composed of secondary arsenates.[11] On the other hand, extensive dissolution of silicates occurred in Zn–Pb slags from Świętochłowice

(Poland), where remnants of silicates occur within porous HFO grains (Figure 4.4D).[14,34] This is consistent with silica having been removed during weathering and other components of dissolved silicates being either mobilized or precipitated as secondary phases. Therefore, the consensus is that conditions leading to silicate dissolution may occur but are not commonly observed as incipient reactions at slag heaps.

The main components of Fe and steel slags are Ca-rich silicates, such as larnite (Ca_2SiO_4), merwinite ($Ca_3Mg(SiO_4)_2$), and hatrurite (Ca_3SiO_5), Ca-rich pyroxenes, and melilites.[16] Some of these phases are considered to be rather reactive. For example, larnite weathers according to eqn (4.6), resulting in the consumption of acid and the generation of alkaline conditions as well as removing CO_2, a greenhouse gas, from the atmosphere.

$$Ca_2SiO_{4(s)} + 4H^+_{(aq)} + 2CO_{2(g)} + O_{2(g)} \rightarrow 2CaCO_{3(s)} + 2H_2O_{(l)} + SiO_{2(s)} \quad (4.6)$$

The dissociation of larnite also releases Ca ions, which may form calcite, the most reported secondary phase for ferrous slag (Table 4.1). An experimental study by Kasina *et al.*[45] revealed that rankinite and larnite are prone to dissolution in deionized water, with the latter being more soluble. This is consistent with a study of basic oxygen furnace (BOF) slag from steel production in Austria composed of larnite, brownmillerite ($Ca_2(Al,Fe)_2O_5$), wüstite (FeO), and lime, where, during leaching tests, larnite and lime were observed to be dissolved in preference of the other phases.[46] On the other hand, in BOF slags, of the silicates present, only hatrurite was identified as reactive, whereas larnite was considered as unweathered.[47] Potential initial weathering of silicates was also demonstrated using sequential extraction dissolution coupled with powder X-ray diffraction analyses of steel slags. The study by Spanka *et al.*[48] indicated that dissolution of Ca silicates (hatrurite and larnite) starts as a result of leaching the sample with distilled water (the first stage of the sequential extraction procedure). Furthermore, under acidic leaching conditions of pH equal to approximately 2.5, these silicates undergo complete dissolution.

4.4.4 Glass

Weathering of glass is often observed at the surface of slags and along cracks.[7,35,49] The reactions occurring at the glass boundary are complex and involve glass alteration by hydration, interdiffusion, and hydrolysis, as well as precipitation of phases, with the components being sourced both from the circulating fluids and from the dissolved glass itself.[49,50] Precipitated products differ depending on the weathering conditions and glass composition. For base metal slags and some ferrous slags with Fe-rich glass, weathering products are often Fe-rich, as evidenced by the brownish color of the alteration products (Figure 4.5A). This is consistent with the general high mobility of Fe during slag weathering. Under oxidizing conditions, Fe (hydro)oxides are a common product.[35] In a single crack in a Pb–Zn slag, Sobanska *et al.*[35]

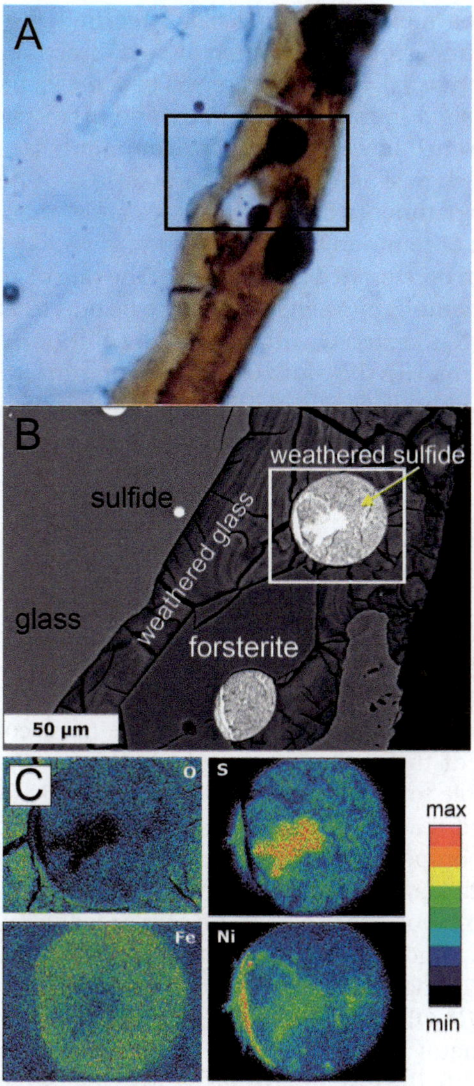

Figure 4.5 Weathering of glass and associated sulfides in Ni slag (Szklary site, Lower Silesia, Poland). (A) Weathered slag surface visible under a microscope with plane-polarized light as brownish crust, indicating precipitation of a complex mix of secondary phases within weathered glass. (B) Enlargement of zone from (A) in back-scattered electron image showing weathered sulfides and unweathered forsterite in weathered glass (please note that sulfides are not weathered in pristine glass). (C) Products of sulfide weathering from (B), as indicated by SEM elemental maps showing relative distribution of Ni, Fe, S, and O.

observed predominantly α-FeOOH and (Al,Fe) hydroxide, but also amorphous silica and $Fe(OH)_3$. On the other hand, in the anoxic conditions studied by Michelin *et al.*[49] for ancient ironworks slag, siderite ($FeCO_3$) was the dominant precipitate and ferrous silicate precipitated directly at the glass boundary. Glass alteration usually involves removal of some elements and the ones strongly impoverished in most cases are Ca and other modifier cations (K, Na, Mg, Fe if present); however, Si is also commonly removed.[5,32,33,40,49] In general, the altered glass is characterized by a loss of density.[35] However, glass alteration experiments using Si isotopes show that the reaction may go both ways with Si removed from the pristine glass, but also some Si condensed in and then precipitated from fluids.[50] The rate of glass alteration was estimated experimentally by Michelin *et al.*[49] to be 0.04 to 2.85 μm $year^{-1}$ with higher values for glass reacting with an Fe-rich solution. The rate decreases with time, consistent with the width of the reaction rims, which rarely exceed 50 μm.[7,49] This finding is consistent with similar widths observed in slags less than a century old[7] and, for example, anthropogenic glasses in archeological sites thousands of years old.[51]

4.4.5 Oxides

Oxides are common primary phases in slags with spinel-group (AB_2O_4, where A typically represents a divalent metal ion, such as Mg, Fe^{2+}, Ni, Mn, or Zn, and B typically represents a trivalent metal ion, such as Al, Fe^{3+}, or Cr) being the most common in base metal slags, whereas simple oxides of Ca, Fe, Mg, Mn (XO, XO_2, and X_2O_3), and spinel are the most common in ferrous slags. Regardless of the composition, spinels are generally considered as the least susceptible to weathering[52] and are commonly found intact in the strongly weathered matrix of sulfides and silicates.[35,53] However, some spinels of a specific composition may be weathered, as was the case for Cu-rich spinel from Cu–Pb slags in Namibia, which was transformed to Cu–Fe–Cr hydroxides.[11] In general, other oxides are more readily weathered than spinels. For example, zincite (ZnO) was identified as the phase least resistant to weathering in Zn slags from Poland and France.[4] Smithsonite ($ZnCO_3$) and goslarite ($ZnSO_4{\cdot}7H_2O$) were observed as products of zincite weathering.[4] However, base metal oxides can also be of secondary origin, making it difficult to identify the extent of oxide weathering. Secondary oxides may include zincite[54] or litharge.[6,11] Iron and steel slags contain different oxides compared to those observed in base metal slags with lime, periclase (MgO), wüstite, and brownmillerite ($Ca_2(Al,Fe^{3+})_2O_5$) being the most common. Of these, lime is extremely reactive and is easily hydrated to Ca hydroxide and then carbonated to Ca carbonate (see eqns (4.7) and (4.8)), whereas the other phases disintegrate much more slowly.[45,47,55]

$$CaO_{(s)} + H_2O_{(l)} \rightarrow Ca(OH)_{2(s)} \tag{4.7}$$

$$CaO_{(s)} + CO_{2(g)} \rightarrow CaCO_{3(s)} \tag{4.8}$$

For example, brownmillerite and wüstite in BOF slag from Austria were found to be generally unaltered during leaching tests, whereas lime and larnite preferentially dissolved.[46]

4.5 Advanced Weathering

The presence of secondary phases at slag disposal sites is an obvious manifestation of slag weathering. The signs of advanced weathering are important to study because they usually indicate mobility of metal(loid)s, which can constitute a potential threat to the environment and human health.[56,57] The phases are observed at different scales, from micrometer-sized crusts and aggregates in pores and cracks (Figure 4.6) to larger efflorescences occurring on surfaces of primary slag fragments, to accumulations at

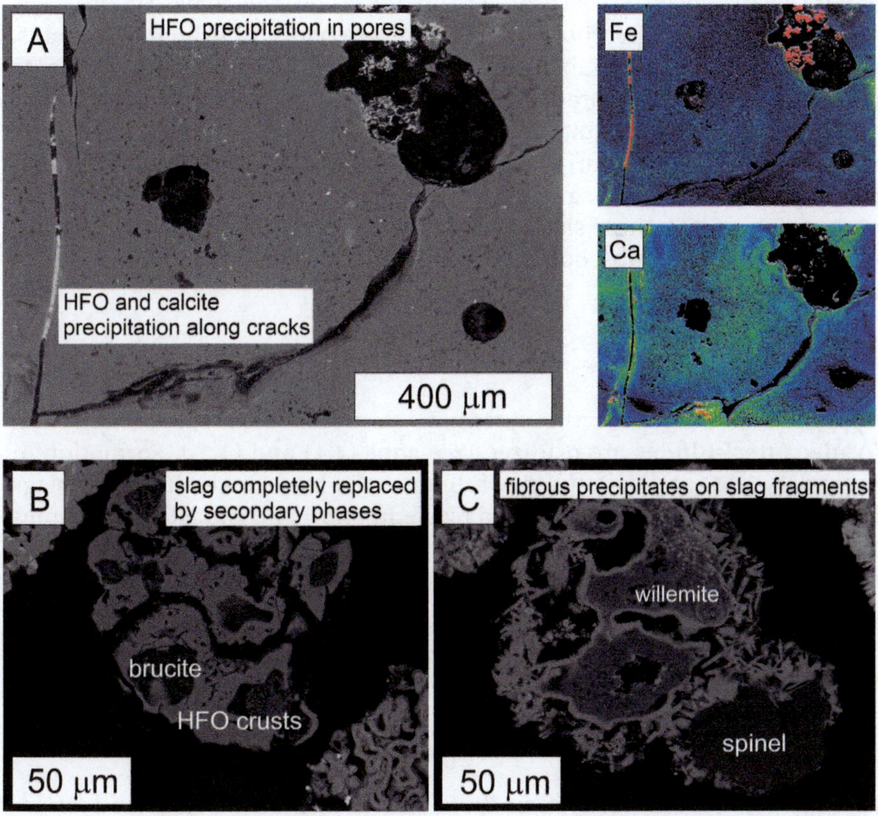

Figure 4.6 Mobility of elements in solution through slag pile (Zn–Pb slag, Świętochłowice site, Upper Silesia, Poland) as indicated by (A) secondary HFO and carbonates precipitated along cracks, (B) complete replacement of a slag fragment by secondary phases, and (C) formation of fibrous aggregates around slag surfaces.

seepage points at disposal sites (Figure 4.7). Fluids saturated with secondary phases can also move outside slag heaps where they react with basement and surrounding rocks. Such a situation was observed by Bayless and Schulz[15] who studied Fe slags used for coastal engineering on the shores of Lake Michigan (Indiana, USA). Numerous secondary phases were identified as a result of slag drainage entering and reacting with native sediments and groundwater. Furthermore, hyperalkaline solutions derived from slags caused further weathering of adjacent rocks, manifested by corrosion of quartz.[15]

A wide range of secondary phases may form as a result of slag weathering (Table 4.1). Their type depends primarily on the composition of the slag, which reflects that of the processed ore and the smelting conditions, and also on the accompanying rocks and climatic and hydrological conditions at the disposal site. In general, secondary phases are dominated by carbonates, oxyhydroxides (Ca, Fe), and gypsum for Fe and steel slags. On the other hand, a more diverse spectrum of secondary phases is observed in base metal slags, including a variety of carbonates, sulfates, arsenates, phosphates, silicates, and oxy(hydro)oxides (Table 4.1). Nevertheless, some secondary phases almost always occur as products of non-ferrous slag weathering, with HFO and gypsum being ubiquitous for most slag (Table 4.1). This is because numerous Fe- and Ca-bearing silicates and oxides as well as Fe-rich sulfides (*e.g.*, fayalite (Fe_2SiO_4), magnetite (Fe_3O_4), pyroxene, pyrrhotite, and bornite) occur commonly in these slags. Secondary HFO form as a result of oxidation of Fe^{2+} to Fe^{3+} with subsequent hydrolysis of Fe^{3+} and precipitation of HFO according to eqn (4.9):

$$Fe^{3+}{}_{(aq)} + 3H_2O_{(l)} \leftrightarrow Fe(OH)_{3(s)} + 3H^+ \tag{4.9}$$

Gypsum forms because the system is saturated with sulfate ions that formed after sulfides are oxidized (see eqns (4.1) and (4.2)). Formation of secondary HFO and sulfates may influence the further weathering of slags because the former generates acid (see eqn (4.9)) and the latter consumes acid. However, the acid consumption caused by the sulfate precipitation is rather temporary because these phases are often soluble and their redissolution will release the acidity (in a similar way this is observed for sulfide-rich mine waste,[39] see eqn (4.10) showing the example of coquimbite dissolution).

$$Fe_2(SO_4)_3{\cdot}9H_2O_{(s)} \rightarrow 2Fe(OH)_{3(s)} + 3SO_4{}^{2-}{}_{(aq)} + 6H^+ + 3H_2O \tag{4.10}$$

In contrast to acid generation from the formation of HFO and dissolution of sulfates, the precipitation of calcite and portlandite from the weathering of Ca silicates and Ca oxides in ferrous slags (eqns (4.7) and (4.8)) consume acid, leading to the alkaline conditions associated with these slags.

Figure 4.7 Advanced weathering as indicated by white secondary precipitates observed at slag-pile scale. (A, B) Zn–Pb slag pile (Katowice-Wełnowiec site, Upper Silesia, Poland) and slag sample. (C) Sample of Ni slag with green secondary phases (Szklary site, Lower Silesia, Poland). (D) Fragment of Cu slag with secondary brochantite (green; Leszczyna site, Lower Silesia, Poland). (E) Secondary white and green coatings on ferrous slag identified as aragonite and calcite on ferrous slag (Chicago area, Illinois, USA). (F) Ferrous slag sample with calcite coating (Chicago area, Illinois, USA). Photos: Jakub Kierczak and Nadine M. Piatak.

4.6 Preservation of Slags on Longer Timescales

The remaining question is why some slags are only slightly weathered and some are completely or extensively replaced by diverse secondary phases. Certainly, time is important, as observed in the natural rock cycle; once formed, all things will be destroyed by chemical and physical weathering after some time. The time limits on rock preservation in the geological record are less than the age of the Earth, as only scarce rocks are present with ages over 3 Ga and none with ages over 4 Ga; only the most resistant mineral, zircon, survived unweathered.[58] Slags are human-made and, as such, relatively young, with preservation potential discussed so far in the context of hundreds of years, not billions. For some slag occurrences, their interaction with the environment has lasted for centuries. In fact, Ettler *et al.*[11] showed more extensive weathering for Pb slags deposited at the beginning of the 14th century as compared to modern ones deposited in the same area and conditions (1980–2000). However, time dependency is not so straightforward, such that in Poland Cu slags deposited several centuries ago are only slightly weathered,[22] whereas Zn–Pb slags deposited in the last century are extensively altered.[34] This may be consistent with either the different stabilities of Cu and Zn–Pb slags or the different conditions of their disposal. Indeed, Zn–Pb slags in Poland were deposited in a strongly industrialized region subjected to acid rains in the second half of the 20th century.[59] External factors also seem to be the main driving force of extensive weathering in the historical base metal slags from Queensland (Australia), where additional acidity comes from acid mine drainage.[43]

Nevertheless, timescales of slag weathering are short compared to geological timescales. How long will slag be preserved if the conditions are not as adverse as those in Upper Silesia or Queensland? Comparing rates of slag weathering to natural weathering occurring over millions of years is tricky because it is difficult to find an ideal counterpart of slag in natural rocks. Natural volcanic glasses are a similar material to glassy slags and these glasses are usually prone to weathering.[60] As many slags are entirely to partly vitreous, weathering of natural glass can be a good approximation of timescales of slag preservation. The oldest preserved glass formed 300 million years ago is rhyolitic in composition[61] and therefore does not correspond well to the composition of slag glasses. Basaltic glasses contain more modifier cations, such as Fe and Ca, and are closer to slag glass composition. However, weathering of basaltic glasses is usually studied in the context of basalt–seafloor interaction[62,63] or hydrothermal alteration,[64] which is not representative of terrestrial weathering typical for slags. Nevertheless, the weathering of glass can be generalized and presented as four steps progressing with time.[60] After initial high rates of glass weathering and formation of altered rinds of a few micrometers to a few tens of micrometers wide (Step 1), the rate of weathering drops by several orders of magnitude (Step 2), making the glasses essentially stable (Step 3). However, if the local solution environment around the glass changes, weathering may restart and the rate of weathering may exceed any initial rates (Step 4).

For example, the formation of some secondary phases may be a trigger for the resumption of weathering.[60] Therefore, the best estimation is that glassy slags are probably preserved in close-to-surface terrestrial environments for millions of years, considering that conditions are favorable. In terms of silicates, many slag phases are considered to be non-resistant to chemical and mechanical weathering, *e.g.*, pyroxene is rarely preserved in mature sediments. However, its resistance to weathering is higher than that of the glass[65] and therefore it should be preserved for longer timescales.

4.7 Conclusions

Slags weather at different scales and rates depending on their chemical and phase composition as well as on the external conditions. Weathering of slags is a process that is not perfectly reflected by natural processes. Many slags contain both sulfide and metallic remnants surrounded by silicates, and a common silicate glass. As such, slag weathering resembles both weathering of ores and silicate rocks, but interactions between sulfidic, metallic, and silicate parts are unique for slags. The interactions are complex and dual in nature as the weathering of sulfides produces acidic fluids, whereas silicate and oxide weathering may induce acid consumption. This complexity is enhanced in slags by specific textural associations of sulfides, metals, and silicates/glass, intergrown with each other. Moreover, when the silicate portion of the slags is considered, it is often dominated by glass – the silicate phase usually prone to weathering, especially under alkaline conditions. Glass is not that common in rocks and this makes any comparisons of weathering more problematic. Currently, weathering of slags is of concern because of the environmental threats posed by the release of contaminants. On the other hand, diverse phase transformations caused by the weathering of ferrous slags are studied because of the reuse of these materials. Overall, the different perspectives motivating studies of base metal and ferrous slags lead to unique knowledge on how they behave environmentally. Furthermore, slags have been produced for several millennia and some are preserved relatively intact while some are strongly weathered after decades. This leads to an interesting dilemma: if slag is treated as a record of anthropogenic activity, one of the *Anthropocene* products, is it going to be preserved in the geological record and for how long? The answer is not simple and the preservation depends on many factors, but clearly only silicate-rich slags have the chance of becoming part of the rock record, whereas sulfide-rich ones will weather away.

Acknowledgements

The institutional funding from the University of Wrocław is acknowledged. Alexandra Courtin-Nomade (University of Paris Saclay), Robert Seal (US Geological Survey), and Vojtěch Ettler (Charles University, Prague) provided constructive comments that vastly improved the chapter.

References

1. M. J. Wilson, *Clay Miner.*, 2004, **39**, 233.
2. R. Warchulski, A. Gawęda, K. Kupczak, K. Banasik and T. Krzykawski, *Lithos*, 2020, **372–373**, 105666.
3. N. M. Piatak, in *Environmental Geochemistry: Site Characterization, Data Analysis and Case Histories*, Elsevier, 2nd edn, 2018.
4. M. Vanaecker, A. Courtin-Nomade, H. Bril, J. Laureyns and J. F. Lenain, *J. Geochem. Explor.*, 2014, **136**, 1.
5. A. Zdravković, V. Cvetković, K. Šarić, A. Pačevski, A. Rosić and S. Erić, *J. Geochem. Explor.*, 2020, **218**, 106629.
6. C. Gee, M. H. Ramsey, J. Maskall and I. Thornton, *J. Geochem. Explor.*, 1997, **58**, 249.
7. J. Kierczak, C. Néel, J. Puziewicz and H. Bril, *Can. Mineral.*, 2009, **47**, 557.
8. A. Schmukat, L. Duester, D. Ecker, H. Schmid, C. Heil, P. Heininger and T. A. Ternes, *J. Hazard. Mater.*, 2012, **227–228**, 257.
9. B. G. Lottermoser, *Neues Jahrb. Mineral., Abh.*, 2005, **181**, 183.
10. H. Kucha, A. Martens, R. Ottenburgs, W. De Vos and W. Viaene, *Environ. Geol.*, 1996, **27**, 1.
11. V. Ettler, Z. Johan, B. Kříbek, O. Šebek and M. Mihaljevič, *Appl. Geochem.*, 2009, **24**, 1.
12. Y. Sueoka and M. Sakakibara, *Minerals*, 2013, **3**, 412.
13. H. Bril, K. Zainoun, J. Puziewicz, A. Courtin-Nomade, M. Vanaecker and J. C. Bollinger, *Can. Mineral.*, 2008, **46**, 1235.
14. A. Pietranik, J. Kierczak, R. Tyszka and B. Schulz, *Minerals*, 2018, **8**, 513.
15. E. R. Bayless and M. S. Schulz, *Environ. Geol.*, 2003, **45**, 252.
16. N. M. Piatak, M. B. Parsons and R. R. Seal, *Appl. Geochem.*, 2015, **57**, 236.
17. A. Potysz, J. Kierczak, A. Pietranik and K. Kądziołka, *Appl. Geochem.*, 2018, **98**, 22.
18. F. Engström, D. Adolfsson, C. Samuelsson, Å. Sandström and B. Björkman, *Miner. Eng.*, 2013, **41**, 46.
19. A. Potysz, J. Kierczak, M. Grybos, A. Pędziwiatr and E. D. van Hullebusch, *J. Environ. Manage.*, 2018, **222**, 325.
20. A. Potysz, M. Grybos, J. Kierczak, G. Guibaud, P. N. L. Lens and E. D. van Hullebusch, *Appl. Geochem.*, 2016, **64**, 92.
21. A. M. Álvarez-Valero, R. Pérez-López, J. Matos, M. A. Capitán, J. M. Nieto, R. Sáez, J. Delgado and M. Caraballo, *Environ. Geol.*, 2008, **55**, 1797.
22. J. Kierczak, A. Potysz, A. Pietranik, R. Tyszka, M. Modelska, C. Néel, V. Ettler and M. Mihaljevič, *J. Geochem. Explor.*, 2013, **124**, 183.
23. M. Vítková, V. Ettler, Z. Johan, B. Kříbek, O. Šebek and M. Mihaljevič, *Mineral. Mag.*, 2010, **74**, 581.
24. V. Ettler, M. Mihaljevič, A. Jarošíková, A. Culka, B. Kříbek, V. Majer, A. Vaněk, V. Penížek, O. Sracek, B. Mapani and F. Kamona, *Appl. Geochem.*, 2020, **114**, 104473.
25. V. Cappuyns, V. Alian, E. Vassilieva and R. Swennen, *Waste Biomass Valorization*, 2014, **5**, 355.

26. A. Potysz, J. Kierczak, Y. Fuchs, M. Grybos, G. Guibaud, P. N. L. Lens and E. D. van Hullebusch, *J. Geochem. Explor.*, 2016, **160**, 1.
27. A. Jarošíková, V. Ettler, M. Mihaljevič, B. Kříbek and B. Mapani, *J. Environ. Manage.*, 2017, **187**, 178.
28. V. Ettler, J. Jehlička, V. Mašek and J. Hruška, *Mineral. Mag.*, 2005, **69**, 737.
29. V. Ettler, M. Komárková, J. Jehlička, P. Coufal, D. Hradil, V. Machovič and F. Delorme, *Chemosphere*, 2004, **57**, 567.
30. A. Potysz and J. Kierczak, *Minerals*, 2019, **9**, 542.
31. M. B. Parsons, D. K. Bird, M. T. Einaudi and C. N. Alpers, *Appl. Geochem.*, 2001, **16**, 1567.
32. V. Ettler, O. Legendre, F. Bodénan and J. C. Touray, *Can. Mineral.*, 2001, **39**, 873.
33. V. Ettler and Z. Johan, *Appl. Geochem.*, 2014, **40**, 97.
34. R. Tyszka, J. Kierczak, A. Pietranik, V. Ettler and M. Mihaljevič, *Appl. Geochem.*, 2014, **40**, 70.
35. S. Sobanska, D. Deneele, J. Barbillat and B. Ledésert, *Appl. Geochem.*, 2016, **64**, 107.
36. A. L. Morrison, Z. Swierczek and B. L. Gulson, *Environ. Pollut.*, 2016, **210**, 271.
37. M. Tuhý, T. Hrstka and V. Ettler, *Environ. Pollut.*, 2020, **266**, 115118.
38. G. S. Plumlee, M. J. Logsdon and L. F. Filipek, in *The Environmental Geochemistry of Mineral Deposits*, Society of Economic Geologists, 2020, p. 71.
39. B. G. Lottermoser, *Mine Wastes (Third Edition): Characterization, Treatment and Environmental Impacts*, Springer Berlin Heidelberg, 2010.
40. V. Ettler and Z. Johan, *C. R. Geosci.*, 2003, **335**, 1005.
41. N. M. Piatak, R. R. Seal and J. M. Hammarstrom, *Appl. Geochem.*, 2004, **19**, 1039.
42. C. Mahé-Le Carlier, C. Le Carlier De Veslud, A. Ploquin and J. J. Royer, *C. R. Acad. Sci., Ser. IIa*, 2000, **330**, 179.
43. B. G. Lottermoser, *Mineral. Mag.*, 2002, **66**, 475.
44. R. Tyszka, A. Pietranik, J. Kierczak, V. Ettler, M. Mihaljevič and A. Medyńska-Juraszek, *Environ. Sci. Pollut. Res.*, 2016, **23**, 24350.
45. M. Kasina, P. R. Kowalski and M. Michalik, *Mineralogia*, 2015, **45**, 27.
46. L. De Windt, P. Chaurand and J. Rose, *Waste Manage.*, 2011, **31**, 225.
47. P. Y. Mahieux, J. E. Aubert and G. Escadeillas, *Constr. Build. Mater.*, 2009, **23**, 742.
48. M. Spanka, T. Mansfeldt and R. Bialucha, *Environ. Sci. Pollut. Res.*, 2018, **25**, 23082.
49. A. Michelin, E. Leroy, D. Neff, J. J. Dynes, P. Dillmann and S. Gin, *Chem. Geol.*, 2015, **413**, 28.
50. A. Verney-Carron, L. Sessegolo, M. Saheb, N. Valle, P. Ausset, R. Losno, D. Mangin, T. Lombardo, A. Chabas and C. Loisel, *Geochim. Cosmochim. Acta*, 2017, **203**, 404.
51. J. L. Weaver, C. I. Pearce, R. Sjöblom, J. S. McCloy, M. Miller, T. Varga, B. W. Arey, M. A. Conroy, D. K. Peeler, R. J. Koestler, P. T. DePriest, E. P. Vicenzi, E. Hjärthner-Holdar, E. Ogenhall and A. A. Kruger, *Int. J. Appl. Glass Sci.*, 2018, **9**, 540.

52. V. Ettler, M. Mihaljevič, J.-C. Touray and P. Piantone, *Bull. Soc. Geol. Fr.*, 2002, **173**, 161.
53. A. Pietranik, J. Kierczak, R. Tyszka and B. Schulz, *Minerals*, 2018, **8**, 513.
54. N. V. Sidenko, R. Gieré, S. B. Bortnikova, F. Cottard and N. A. Pal'chik, *J. Geochem. Explor.*, 2001, **74**, 109.
55. N. M. Piatak and R. R. Seal, *Appl. Geochem.*, 2012, **27**, 623.
56. A. L. Morrison and B. L. Gulson, *Sci. Total Environ.*, 2007, **382**, 30.
57. V. Ettler, D. Štěpánek, M. Mihaljevič, P. Drahota, R. Jedlicka, B. Kříbek, A. Vaněk, V. Penížek, O. Sracek and I. Nyambe, *Chemosphere*, 2020, **260**, 127642.
58. S. A. Wilde, J. W. Valley, W. H. Peck and C. M. Graham, *Nature*, 2001, **409**, 175.
59. E. Pudlis, *Ambio*, 1983, **12**, 125–127.
60. J. S. McCloy, *J. Non-Cryst. Solids: X*, 2019, **4**, 100035.
61. A. Repstock, F. Heuer, J. Im, M. Hübner, B. Schulz, C. Breitkreuz, S. Gilbricht, F. Fischer and M. Lapp, *J. Volcanol. Geotherm. Res.*, 2019, **369**, 35.
62. H. Staudigel and S. R. Hart, *Geochim. Cosmochim. Acta*, 1983, **47**, 337.
63. H. Staudigel, H. Furnes, N. McLoughlin, N. R. Banerjee, L. B. Connell and A. Templeton, *Earth-Sci. Rev.*, 2008, **89**, 156.
64. A. W. Walton and P. Schiffman, *Geochem., Geophys., Geosyst.*, 2003, **4**, 8709.
65. S. Andò, E. Garzanti, M. Padoan and M. Limonta, *Sediment. Geol.*, 2012, **280**, 165.
66. L. Svirenko, J. Vergeles and O. Spirin, in *Approaches to Handling Environmental Problems in the Mining and Metallurgical Regions*, Springer Netherlands, 2003, p. 211.
67. C. Navarro, M. Díaz and M. A. Villa-García, *Environ. Sci. Technol.*, 2010, **44**, 5383.
68. K. Kądziołka, J. Kierczak and A. Pietranik, *Przegl. Geol.*, 2018, **67**, 164.
69. A. Navarro, E. Cardellach, J. L. Mendoza, M. Corbella and L. M. Domènech, *Appl. Geochem.*, 2008, **23**, 895.
70. A. M. Álvarez-Valero, R. Pérez-López and J. M. Nieto, *Am. Mineral.*, 2009, **94**, 1417.
71. N. M. Piatak and R. R. Seal, *Appl. Geochem.*, 2010, **25**, 302.
72. P. Costagliola, M. Benvenuti, L. Chiarantini, S. Bianchi, F. Di Benedetto, M. Paolieri and L. Rossato, *Appl. Geochem.*, 2008, **23**, 1241.
73. N. M. Piatak, R. R. Seal, D. A. Hoppe, C. J. Green and P. M. Buszka, *Minerals*, 2019, **9**, 468.
74. R. Warchulski, M. Mendecki, A. Gawęda, M. Sołtysiak and M. Gadowski, *Appl. Geochem.*, 2019, **109**, 104396.
75. C. A. Cravotta III, *Assessment of Characteristics and Remedial Alternatives for Abandoned Mine Drainage: Case Study at Staple Bend Tunnel Unit of Allegheny Portage Railroad National Historic Site, Cambria Country, Pennsylvania, 2004*, 2005.
76. M. Ashelford and D. B. Gore, *Miner. Eng.*, 2020, **154**, 106407.
77. R. Warchulski, A. Gawęda, M. Kądziołka-Gaweł and K. Szopa, *Mineral. Mag.*, 2015, **79**, 459.

CHAPTER 5

Slag Leaching Properties and Release of Contaminants

VOJTĚCH ETTLER*[a] AND MARTINA VÍTKOVÁ[b]

[a]Institute of Geochemistry, Mineralogy and Mineral Resources, Faculty of Science, Charles University, Albertov 6, 128 00 Prague 2, Czech Republic; [b]Department of Environmental Geosciences, Faculty of Environmental Sciences, Czech University of Life Sciences, Kamýcká 129, 165 00 Prague – Suchdol, Czech Republic
*E-mail: vojtech.ettler@natur.cuni.cz

5.1 Introduction

Metallurgical slags are exposed to weathering in various environments and the leaching of slag contaminants can impact surrounding soil, surface water, groundwater, and other environmental compartments. Laboratory leaching tests are used as proxy methods to simulate the release of contaminants under variable environmental conditions. The amount of contaminants leached from the slag is strictly dependent on the specific surface exposed to weathering at the slag–water interface. In real-life scenarios, the reactive surface area depends on the physical aspects of the slag material. Figure 5.1 shows typical slag disposal environments where leaching reactions may result in the release of contaminants. When slags are tapped into ladles at the furnace outlet and air-cooled, they generally occur as pebble- to boulder-size fragments, commonly greater than 5 cm in diameter, which have a relatively low specific surface area. When deposited on the dumps, they can be exposed to intermittent humidity and rain events (Figure 5.1A and B).[1–5]

Chemistry in the Environment Series No. 2
Metallurgical Slags: Environmental Geochemistry and Resource Potential
Edited by Nadine M. Piatak and Vojtěch Ettler

Published by the Royal Society of Chemistry, www.rsc.org

Nevertheless, highly weathered slags have been observed in the interiors of such slag dumps, presumably related to the high humidity and rainwater percolating through the dumps.[6] When slags are granulated (quenched by water) (Figure 5.1C) and subsequently deposited in heaps, the much finer grain size of the slag fragments (generally <5 mm in diameter) helps keep the water table a few centimeters below the surface of the heap and the slag may be exposed to water for longer periods of time (Figure 5.1D).[7–10] Piles of granulated slag are likely weathered under vadose zone conditions near the surface (*i.e.*, cm to m), and could be fully saturated for much of the year at greater depth, depending on the weather and topography of the dump. Another situation can occur at some sites, where slag dumps end up in water reservoirs and slag fragments are fully submerged in water (Figure 5.1E and F). Similarly, Schmukat *et al.*[11] and Duester *et al.*[12] reported that in Germany Cu slags had been used as armor stone in hydraulic engineering to protect coasts (in groynes) or to stabilize dams and shores of waterways (canals, rivers), which corresponds to the same scenario, where the slag is fully exposed to water. This also occurs along the Great Lakes in the USA, where slag from steelmaking centers is used to fill the wetlands and extend land into the lakes, providing the industry with solid ground to build/expand their steelmaking facilities.

This chapter describes how various leaching tests are used to understand the leaching of major elements and metal(loid) contaminants and how experimental results can be combined with detailed solid-phase characterization and/or thermodynamic modeling to predict the environmental behavior of slags under various conditions.

5.2 Leaching Tests

Leaching tests of various experimental protocols can be applied to slag materials and can be roughly divided into the four categories presented in Figure 5.2. Many parameters affect contaminant leaching from solid waste, especially the particle size (and related reactive surface area), liquid-to-solid (L/S) ratio, pH and chemical composition of the leaching solution, chemical and mineralogical composition of the slag, and time.[13,14] The following individual subsections focus on sample preparation and then describe the most commonly employed leaching tests, which have been used on metallurgical slags, and their results in terms of contaminant leaching.

5.2.1 Slag Sample Preparation

Most leaching tests require a particle-size reduction (see below) involving the crushing of samples manually using a rock hammer or mechanically using jaw crushers. The particle-size reduction is used (i) for sample homogenization and (ii) to prepare coarse-grained samples for extraction so the liquid–solid equilibrium is enhanced and mass transport through large particles is minimized.[15] On the other hand, particle-size reduction often leads to

Figure 5.1 Photos of representative environments where metallurgical slags can undergo leaching: (A) Dumps of air-cooled Pb smelting slags (Příbram, Czech Republic). (B) Large fragments of air-cooled Ni slag (Szklary, Poland). (C) Granulation of slag melt at a Ni smelter furnace outlet (Barro Alto, Brazil). (D) Eroded surface of a granulated Cu slag heap (Tsumeb, Namibia). (E) Copper slags submerged in lake water (near the Bingham Canyon Mine, Utah, USA). (F) Fragment of Cu slag that underwent weathering and leaching in lake water (near the Bingham Canyon Mine, Utah, USA) (Photos A–D: Vojtěch Ettler, photos E–F: Martina Vítková).

overestimating the release of contaminants and the obtained results may be considered a "worst-case" scenario.

For instance, for the normalized leaching tests (batch and column) adopted by the European Union (EU), the grain size of the tested material should be <4 mm (or <10 mm) in diameter (95% of the particles).[16,17] A grain size of <1 mm in diameter (95% of particles) is specified for materials tested by pH-dependent leaching tests.[18,19] Similarly, the majority of the leaching tests designed for the characterization of waste materials by the United States Environmental Protection Agency (USEPA) within the Leaching Environmental Assessment Framework (LEAF)[20] require a particle-size reduction to <0.3, 2, or 5 mm in diameter (85% of the particles) (see also Kosson *et al.*[13]). However, during the sample processing (crushing and sieving), many commercial laboratories discard some of the fine grain-size fractions, which can lead to an underestimation of the hazardous properties of the studied materials.[21] In contrast, this approach may also more closely resemble field conditions if very fine particles are not present at the slag disposal sites. A comparative investigation of different Cu slag grain-size fractions using the standardized EU protocol under a controlled pH indicated that the fine particles generated during crushing, which adhere to the surfaces of larger particles, can primarily be responsible for significant increases in contaminant leaching.[22] A similar observation (*i.e.*, element mobilization induced by the adhering dust fraction caused by the slag sample preparation) was reported by Pareuil *et al.*[23] for Mn-rich slags. On the contrary, the increase in the amount of fine dust particles with partial size reduction indirectly resulted in lower amounts of metals released from Cu slags as a response to a natural increase of the final pH due to higher dissolution of silicates.[22]

Despite the need for a clear definition of the sample processing prior to testing in standardized protocols, particle-size reduction is often unsuitable for predicting the environmental behavior of slags in real-life scenarios (see the large slag fragments in Figure 5.1A and B). For such cases, tank tests with monoliths (columns or rectangular blocks 5–10 cm in size, by coring or cutting a sample from a larger specimen) or compacted granular materials have been standardized to simulate continuous water-saturated conditions (*e.g.*, EN 15863[24] or USEPA Method 1315;[25] see also Kosson *et al.*[13] for details). Alternatively, polished thin sections placed into a leachant at a given liquid-to-surface area (L/A) ratio have been adopted to study the dissolution of the individual slag phases as a function of time and in various compositions of leaching solutions.[26,27]

5.2.2 Single-step Leaching Tests

Simple batch leaching tests (Figure 5.2A) have been adopted by many regulators for the purpose of determining the leachability of contaminants from waste materials and their subsequent classification. Thus, these tests are designated as "compliance tests" since they are used to determine whether the material complies with a specific behavior or with specific reference values

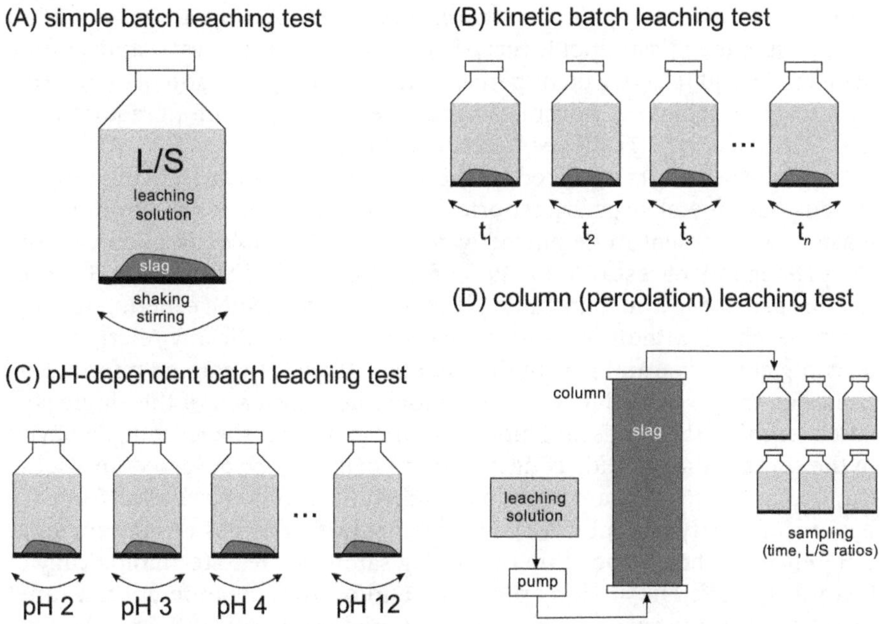

Figure 5.2 Experimental setups used for the leaching of slags. (A) Single-step batch leaching test. (B) Kinetic batch leaching test. (C) pH-dependent leaching test. (D) Column (percolation) leaching test.

(regulatory limits). For instance, the EU adopted four normalized batch tests, including EN 12457 (parts 1–4),[16] where a sample with a grain size of <4 mm (parts 1–3) or <10 mm (part 4) in diameter is leached in deionized water for 24 h in reactors that are continuously, but gently, shaken; part 1 (L/S = 2) and part 2 (L/S = 10) have most frequently been adopted for slags.[5,7,8,10,21–23,28–31] These two leaching tests are also used for waste classification before its acceptance into landfills according to EU legislation.[32,33]

The Toxicity Characteristic Leaching Procedure (TCLP) adopted by the USEPA has also been widely used[34] to assess the hazardous properties of slags.[2,7,35–38] This method is based on the extraction of a material with a grain size of <1 cm in diameter (surface area per gram of material must be ≥3.1 cm^2) at an L/S ratio of 20 for 18 h in an acetic acid solution at a pH of 4.93 (for a sample with a pH < 5) or 2.88 (for a sample with a pH > 5) with reactors shaken at 30 rpm. The TCLP is designed to simulate reactions in a weak organic acid solution, which could occur when the waste is co-disposed with an organic-rich municipal solid waste in a landfill site. Because this is generally not a plausible disposal scenario for slags, and also since the TCLP underestimates As leaching,[39] many researchers prefer to use the Synthetic Precipitation Leaching Procedure (SPLP),[40] which better simulates the potential interaction of slag and rainwater at disposal sites.[2,30,37,38,41,42] This method is based on the single-step extraction of material in a solution with a pH of 4.2 (for investigation of sites east of the Mississippi River) or 5.0 (for sites west of the Mississippi River) prepared from distilled water and sulfuric acid–nitric acid (60 : 40 weight percent mixture) at an L/S ratio of 20 for 18 h with reactors shaken at 30 rpm. Sample size reduction is often required and is specified as in the TCLP test described above.

Some of the results obtained by these single-step batch leaching tests on various base metal slags are reported in Table 5.1. The results from a 48 h leaching experiment in deionized water at an L/S ratio of 10 adopted from the pH-dependent test (EN 14429[18] or EN 14997;[19] see Subsection 5.2.3 for details) are also included in Table 5.1 for comparison. Such experiments represent leaching without any pH adjustments, being officially referred to as leaching at the "natural pH" of the material. Table 5.1 contains information compiled from selected studies to demonstrate how each of the single-step batch leaching standards and modifications to it affect the leaching behavior of the contaminants. Wide concentration ranges can be observed for As, Cd, Cu, Pb, and Zn, which are the most studied metal(loid)s released from the base metal slag types (Table 5.1). The data indicate that, besides the apparent differences in the composition of the slag samples due to technologically or historically different smelting processes and/or from the different raw input materials, the amount of metal(loid)s leached is strongly influenced by the various operational leaching parameters, such as the slag particle size or the experimental contact time. As previously mentioned, particle size represents a crucial variable that may dramatically affect the leachability of slags.[21,22] Interestingly, it is the main parameter that is modified compared to the standard protocol (Table 5.1) and, therefore, it directly influences the comparability of results obtained from the commonly used procedures.

Furthermore, many of these single-step leaching tests have been criticized due to the short time period of the leaching, which does not fulfill the criteria of attaining a leaching "pseudo-equilibrium".[13] For this reason, kinetic leaching tests with replicated leaching reactors for individual sampling times have been adopted for the slags to mimic much longer time scales and to investigate the time dependence of the leaching process (Figure 5.2B). It is also necessary to stress again that, due to particle-size reduction, regulatory tests may lead to overestimating the amount of contaminants leached, mostly due to rapid dissolution of fines formed by the slag crushing and adhering on the surfaces of larger slag grains. For the EU regulatory leaching tests, Vítková *et al.*[22] used only coarse slag fractions, which were ultrasonically cleaned in alcohol to remove the fines before leaching, and found a substantial decrease in contaminant leaching compared to slags containing smaller grain-size fractions.

Generally, the steady-state leaching of fine-grained materials is relatively quick, especially when highly soluble sulfides and metastable glass are the predominant carriers for the contaminant metal(loid)s. For instance, Yin *et al.*[43] used a 9-d leaching procedure and found that Pb–Zn slags with a grain size of 100 μm to 2 mm in diameter attained a steady-state after *ca.* 100 hours of leaching for most major and trace elements; the initial pulse of elements

Table 5.1 Leached concentrations of metals and metalloids (mg kg^{-1}) obtained during the single-step batch leaching tests in deionized water and TCLP and SPLP leachants. Data compiled from various sources.

Slag type (provenance)	Leaching test	Modification to standard procedure	As (mg kg^{-1})	Cd (mg kg^{-1})	Cu (mg kg^{-1})	Pb (mg kg^{-1})	Zn (mg kg^{-1})	Reference
Cu (India)	TCLP	<1 mm size fraction	0.6–1.4	0.2–0.4	40–60	Up to 2.8	0.6–2.4	Shanmuganathan *et al.*[36]
Cu (Namibia)	EN 12457-2	No particle-size reduction[a]	5.3–42	<DL	0.03–0.8	0.03–0.3	Up to 0.2	Jarošíková *et al.*[10]
Cu (Poland)	EN 12457-2	<1 mm size fraction	Up to 0.1	Up to 0.02	0.2–2.5	Up to 3.8	0.7–8.1	Mikoda *et al.*[30]
Cu (Poland)	SPLP	<1 mm size fraction	Up to 0.3	Up to 0.3	0.2–2.7	Up to 4.5	1.1–27	Mikoda *et al.*[30]
Cu (Singapore)	TCLP	None	1.6	0.6	672	10	370	Lim and Chu[35]
Cu (USA)	SPLP	<2 mm size fraction, 1 min shaking and 24 h settling	Up to 0.01	0.002–0.05	1.1–36	Up to 0.01	1.5–9.4	Piatak *et al.*[41]
Cu (USA)	SPLP	None	<DL	0.7–4	Up to 0.9	0.6–1.1	22–28	Parsons *et al.*[2]
Cu (USA)	TCLP	None	1.1–13	9.8–50	82–1000	400–600	1180–2200	Parsons *et al.*[2]
Cu (Zambia)	EN 12457-2	None or 4–0.5 mm size fraction	0.03–0.04	<DL	13–28	<DL	0.2	Vítková *et al.*[22]
Cu (Zambia)	EN 14997[b]	None or 5–0.5 mm or <5 mm size fractions	0.01–0.02	<DL	1.6–11	<DL	0.1–0.2	Vítková *et al.*[22]
Cu–Pb (Namibia)	EN 12457-2	<0.1 mm size fraction	4.0–37	0.02–0.26	0.4–0.8	0.2	0.05–0.10	Ettler *et al.*[7]

(*continued*)

Table 5.1 (*continued*)

Slag type (provenance)	Leaching test	Modification to standard procedure	As ($mg\ kg^{-1}$)	Cd ($mg\ kg^{-1}$)	Cu ($mg\ kg^{-1}$)	Pb ($mg\ kg^{-1}$)	Zn ($mg\ kg^{-1}$)	Reference
Cu–Pb (Namibia)	TCLP	<0.1 mm size fraction	105–150	5–151	361–3040	39–1020	104–258	Ettler *et al.*[7]
Ni (Brazil)	EN 12457-2	No particle-size reduction[a]	n.d.	<DL	<DL	n.d.	0.5	Ettler *et al.*[8]
Ni (Brazil)	EN 14997[b]	No particle-size reduction[a]	n.d.	0.004	0.3	n.d.	1.4	Ettler *et al.*[8]
Pb (Brazil)	SPLP	None	15	3.5	n.d.	96	n.d.	de Andrade Lima and Bernardez[37]
Pb (Brazil)	TCLP	None	<DL	0.2	n.d.	13	n.d.	de Andrade Lima and Bernardez[37]
Pb (Czech Rep.)	EN 12457-2	2–5 mm, 1 year	<DL	n.d.	0.3–1.1	1.1–53	0.3–400	Ettler *et al.*[45]
Pb (Czech Rep.)	EN 12457-2	<0.5 mm, 12 years	0.05–0.06	n.d.	0.3–1.7	0.2–15	0.4–8460	Ettler and Johan[29]
Pb (UK)	EN 12457-2	None or 4–0.5 mm size fraction	Up to 4.9	n.d.	n.d.	Up to 24	Up to 0.2	Zandi *et al.*[21]
Pb–Ag (USA)	SPLP	<2 mm size fraction, 1 min shaking and 24 h settling	1.0	0.01	17	220	7.4	Piatak *et al.*[41]
Zn (USA)	SPLP	<2 mm size fraction	Up to 0.04	5.0	Up to 15	Up to 1.8	5–528	Piatak and Seal[42]
Zn–Pb–V (Namibia)	EN 12457-2	None	0.2–0.5	<DL	0.03–0.4	Up to 0.6	0.01–0.3	Ettler *et al.*[31]

[a] Granulated slags.
[b] Leaching at the natural pH of the material (*i.e.*, without a pH adjustment). <DL – Values below the detection limit. n.d. – Values not determined.

released is kinetically controlled, followed by lower amounts of elements leached due to the precipitation of the newly formed phases. Another reason for this observation may be related to an initial release of the elements due to dissolution of fine fractions, as mentioned above.

Much longer tests are beneficial for determining the dissolution kinetics of primary phases and the role of secondary (newly formed) phases in attenuating the released contaminants (metal(loid)s) (see also Section 5.4). Ettler and Johan[29] presented extended results of the shorter-term tests published in Ettler[44] and Ettler *et al.*[45] (1 day to 1 year of leaching) and supplemented them with a 12 year leaching test to simulate the long-term interaction of a Pb–Zn slag with water under saturated conditions at an L/S ratio of 10 (Table 5.1). They found that much higher amounts of Pb and Zn were released from the primary slag (resulting from the processing of the Pb ores) when compared to the secondary slag (produced by the pyrometallurgical recycling of car batteries) (Figure 5.3). Whereas the leached concentrations of both Pb and Zn were below the EU leaching limit for non-hazardous waste after 1 day of leaching, these metals significantly exceeded this limit after more extended periods of leaching, but only in the case of the primary slag. In the long-term tests, the observed decrease in Pb leaching after 1 month is controlled by the precipitation of secondary sulfates (anglesite, $PbSO_4$), carbonates (cerussite, $PbCO_3$), and Pb phosphates and/or by Pb adsorption to newly formed hydrous ferric oxides (HFO) (Figure 5.3, see also Section 5.4). In contrast, Zn was continuously released as a function of time (Figure 5.3) and the leached concentration attained 8460 mg kg^{-1}, which exceeded the leaching limit for landfilling of hazardous waste by more than 40 times.[33] This observation indicates that Zn mostly remains in solution and its leaching is not efficiently controlled by the precipitation of or adsorption to newly formed phases. Moreover, compared to Pb, which is mostly bound in sulfides

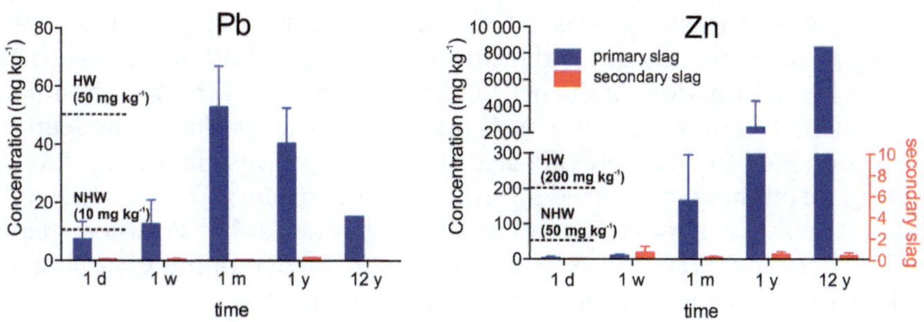

Figure 5.3 Example of the kinetic leaching of Pb and Zn from Pb smelting slags. The contaminant concentrations (in mg kg^{-1}) were leached at an L/S ratio of 10 as a function of time (1 day, 1 week, 1 month, 1 year, and 12 years). The EU limit values for non-hazardous waste (NHW) and hazardous waste (HW) used for classifying waste for potential landfilling are indicated.[32,33] (note the secondary *y*-axis for Zn leaching from the secondary Pb slag). Adapted from ref. 29 with permission from Elsevier, Copyright 2013.

and (inter)metallic compounds, Zn is partitioned among all the primary slag phases (oxides, silicates, glass, and sulfides) and the slow dissolution kinetics of the glass and primary silicates will lead to long-term Zn leaching from the Zn-bearing slags.[29]

The kinetic batch leaching test (L/S = 10 and 100, grain size <1 mm, over 30 days) was also used for basic oxygen furnace (BOF) steel slag to assess its leaching behavior in view of its possible reuse as an aggregate for road construction.[46] The combination of leach tests and geochemical modeling (see also Section 5.3) indicated that the progressive dissolution of the primary phases is kinetically controlled but the long-term release of major elements as well as metals (Cr, V) is mainly controlled by the precipitation of secondary phases, such as calcium–silicate–hydrates (C–S–H), oxides, and hydroxides (see also Table 4.1 in Chapter 4).

5.2.3 pH-dependent Leaching Tests

The pH value is one of the key parameters affecting the dissolution of primary slag phases, the subsequent release of major elements and contaminants, as well as their scavenging mechanisms (precipitation, sorption).[14,47] As a result, pH-dependent leaching tests provide information about leaching reactions in the relevant pH domains related to the different conditions and potential disposal scenarios and uses. For instance, highly acidic conditions (pH 1.5 to 2) simulate what can happen when slag dust is dissolved in a gastric fluid after incidental dust ingestion;[48,49] mildly acidic to circumneutral conditions can predict how metals are released from slag particulates when deposited into soils or leached by rainwater;[22] alkaline to highly alkaline conditions simulate the situation when the slag is cement-stabilized in a disposal site or is used as an aggregate during the preparation of cement mortars and concrete in civil engineering.[50,51] There are many advantages and limitations of pH-dependent leaching tests: (i) they help to identify the sensitivity of leaching to small pH changes and factors controlling metal(loid) release; (ii) they provide information on the pH conditions imposed by the surrounding environment; (iii) they serve as a basis for geochemical speciation modeling (see Section 5.4); (iv) they provide acid neutralization capacity (ANC) information; and (v) they can be used for mutual comparisons with other (waste) materials to assess similarities in the leaching behavior.[47,52] For these reasons, pH-dependent tests are considered essential for determining the short- and long-term leaching behavior under various conditions.

Two standardized protocols have been adopted as pH-dependent (or pH-static) leaching tests in Europe (the former technical standards CEN/TS): EN 14429[18] – Influence of pH on leaching with initial acid/base addition and EN 14997[19] – Influence of pH on leaching with continuous pH-control. A preliminary determination of the acid and base consumptions (ANC/BNC, acid/base neutralization capacity) is first needed to run these tests. The pH-static experiments are carried out at ambient temperature at an L/S ratio of 10 for 48 h (the suspensions are gently shaken on an end-over-end table shaker).

At least eight values (independent batch reactors) are selected, covering, at the minimum, the pH range 4–12 (Figure 5.2C). An acid (HNO_3) or base (NaOH or KOH) is added to adjust to the desired pH values. For EN 14429,[18] the pH is adjusted only at the beginning of the experiment and the pH can freely drift during the test (an equilibrium condition is established at different pH values as a result of the reaction between the pre-selected amounts of the acid or base and the test portions of the material).[53,54]

In contrast, in the case of EN 14997,[19] which has most frequently been used for slags during the last decade,[8,10,22,31,38,55] the pH is continuously controlled over the entire period and adjusted manually or automatically (*via* an automatic titrator) by the addition of an acid/base. Several researchers have slightly modified the original experimental protocol; for instance, Yin *et al.*[43] used an L/S ratio of 20 and the slag eluates were collected at different equilibration times (1 to 9 days). Similarly, to investigate the leaching rates, Ganne *et al.*[56] and Cappuyns *et al.*[38] studied the pH-dependent leaching behavior of slags in the time frame of 0.5 to 96 or 168 h.

Based on the European experience, and in the framework of the Leaching Environmental Assessment Framework[20,57] and the SW-846 package of methods,[58] the USEPA has recently adopted a similar pH-dependent leaching test entitled USEPA Method 1313 – Liquid–solid partitioning as a function of extract pH using a parallel batch extraction procedure.[15] There are several differences compared to the European pH-dependent leaching tests: (i) nine pH values in the range 2–12 are selected; (ii) apart from the test positions with the studied waste, three method blanks are used; and (iii) the reactors are tumbled in an end-over-end fashion for a specified contact time, which depends on the particle size of the sample (for instance, a grain size [85%] <0.3 mm, time = 24 h; a grain size <2 mm, time = 48 h; a grain size <5 mm, time = 72 h), (USEPA Method 1313;[15] see the rationale in Kosson *et al.*[13]). Many researchers prefer the USEPA method for fine-grained granulated slag samples owing to the shorter time equilibrations compared to the EU methods.[5,59]

The pH-dependent leaching trends for selected contaminants (As, Cu, Pb, and Zn) leached from the smelting slags compiled from the literature are reported in Figure 5.4 (the data for Cu, Pb, Zn, and Ni slags have been plotted). Interestingly, despite the significant variability in the slag chemistry and mineralogy and the differences in the solubility controls, the pH-dependent leaching curves (note the log scale on the *y*-axis) exhibit clear trends.

Moreover, the leaching trends for the key contaminants released from the slags are similar to the schematic leaching curves for the cationic (Cu, Zn), amphoteric (Pb), and oxyanionic (As) species (Figure 5.5), as shown by Kosson *et al.*,[13] and are used for the interpretation of the pH-dependent leaching data in the USEPA Method 1313.[15] This observation indicates that the pH-dependent leaching of the contaminants can follow similar release and controlling mechanisms for many industrial waste materials.[52] The leaching of As (which occurs in the leachate, mainly as oxyanionic species, such as $H_2AsO_4^-$, $HAsO_4^{2-}$, and AsO_4^{3-}) is relatively flat with slightly higher medians

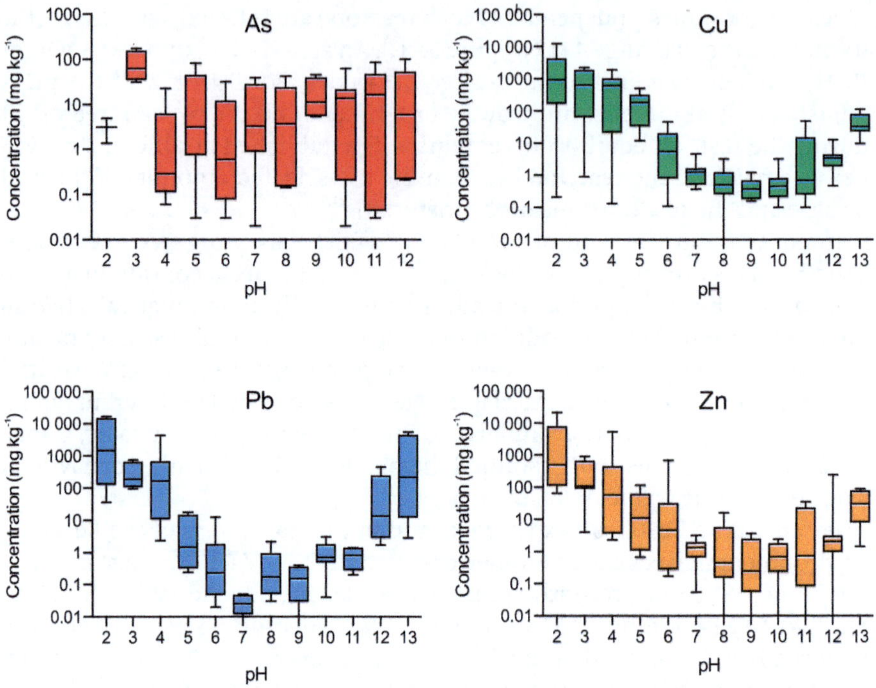

Figure 5.4 The pH-dependent leaching of the major contaminants (As, Cu, Pb, Zn) from non-ferrous smelting slags expressed in leached concentrations (mg kg^{-1}) at an L/S ratio of 10. Data compiled and redrawn from various sources.[5,8,10,22,38,59]

of leaching at a low pH and under alkaline/highly alkaline conditions (medians up to 62 mg kg^{-1} and up to 16 mg kg^{-1}, respectively; Figure 5.4). Copper and Zn exhibit mostly cationic behavior (L-shaped or moderately U-shaped curves) with the highest leaching under highly acidic conditions (Cu: median 921 mg kg^{-1}; Zn: median 494 mg kg^{-1} at pH 2), with the lowest leaching at slightly alkaline conditions (*ca.* 3–4 orders of magnitude lower than at pH 2) and with only a moderate increase at pH 12–13 (Figure 5.4). Lead shows a V-shaped leaching behavior, typical for amphoteric species, with the highest leaching at the pH extremes and the lowest leaching at pH 7 (up to 5 orders of magnitude difference between the extreme and neutral values of the pH) (Figure 5.4).

5.2.4 Column Leaching Tests

Column (percolation) leaching tests (Figure 5.2D) are designed to simulate what happens to a granular waste subjected to percolation with a solution as a function of the L/S ratio. The leachant (*e.g.*, deionized water) is percolated in an up-flow regime (*via* a peristaltic pump) through a column filled with waste

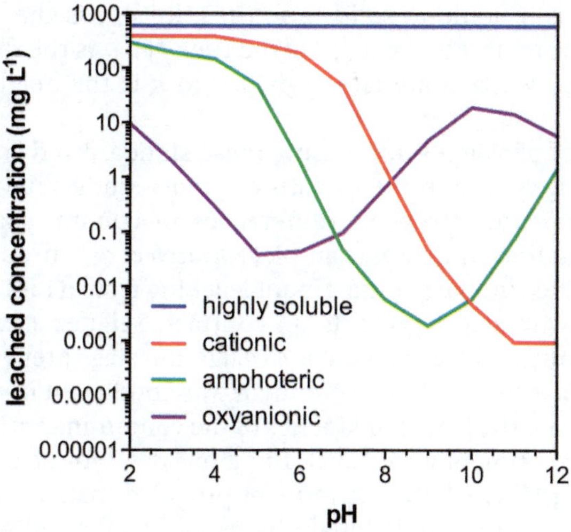

Figure 5.5 Schematic leaching curves for highly soluble, cationic, amphoteric, and oxyanionic species obtained during the pH-dependent leaching test.[15]

at a specified flow rate. The eluates are collected in several separate fractions corresponding to the fixed L/S ratios. Such a percolation test was adopted in the EU as technical standard CEN/TS 14405 in 2014 and was upgraded to European norm EN 14405 in 2017.[17] Granular waste (or waste after grain-size reduction; 95% must be <4 mm in diameter) is placed into a 5 cm or 10 cm diameter column (*ca.* 0.8 L of the sample is leached), the leachant is percolated at a flow rate of *ca.* 15 cm day^{-1}, and seven eluates at cumulative L/S ratios of 0.1, 0.2, 0.5, 1, 2, 5, and 10 L kg^{-1} are collected and analyzed. This method is also used for classifying waste before potential landfilling (inert, non-hazardous, hazardous waste categories).[33] In the footsteps of this leaching test implementation in the EU, the up-flow percolation leaching test was also adopted in 2013 as USEPA Method 1314;[60] it is based on a similar experimental protocol, but nine eluate fractions corresponding to 0.2, 0.5, 1, 1.5, 2, 4.5, 5, 9.5, and 10 L kg^{-1} are collected. Relative to batch leaching tests, there are numerous advantages of the percolation tests: (i) the solubility control *versus* the flushing of the non-interacting species can be identified; (ii) the low L/S data correspond well to the pore water concentrations measured in field leachates (the test allows a comparison with lysimeter data); and (iii) a projection towards the long-term waste behavior is possible.[47,61]

For the percolation-controlled scenario at waste disposal sites, there is a relationship between the L/S and the time, and the column test can, thus, serve as a good proxy method for the time-dependent predictions of the leaching behavior (eqn (5.1)):

$$\mathrm{L/S} = I \times t/(\rho \times h) \tag{5.1}$$

where the L/S is a liquid-to-solid ratio (in L kg^{-1}), I is the net infiltration of the precipitation in mm $year^{-1}$, t is the time (y), ρ is the dry bulk density of the deposited waste material (kg m^{-3}), and h is the height of the waste layer (m).[13,62]

Investigations of slag leaching using these standardized column test protocols are relatively rare in the literature;[63] most studies have modified the original experimental protocols (differences in column size, sample size and format, and flow rate). For example, Navarro *et al.*[64] used a crushed slag mixed with soil to simulate contaminant leaching from a highly polluted soil contaminated with slag fragments. In contrast, Seignez *et al.*[65,66] used column tests, in which both crushed Pb–Zn slags and slags prepared as polished sections were placed. The first experiment was conducted in a time frame of 2270 h to calculate the leaching kinetics of the contaminants (the mean rates for the entire experiment were 1.35×10^{-2} g m^{-2} day^{-1} for Pb and 6.81×10^{-3} g m^{-2} day^{-1} for Zn).[66] The latter experiment provided insight into the kinetics of the dissolution of the individual phases and confirmed that the Pb-rich droplets and glass were the most altered after the leaching test.[65]

5.2.5 Leaching Tests Simulating the Slag Exposure in Soils

Re-cultivation of slag dumps using soil covers followed by re-vegetation represents one possible scenario for managing slags at numerous smelting sites.[67] However, leaching of contaminants from slag in such soil environments may be accelerated owing to complexation with organic ligands derived from root exudates (*i.e.*, low-molecular-weight [LMW] organic acid ligands, such as citrate, oxalate, or acetate, or natural high-molecular-weight [HMW] organic acid ligands, such as fulvate or humate) commonly found in the soil pore waters. Ettler *et al.*[68] used citric acid solutions (8 mM and 20 mM) as simulators of root exudates to examine Pb slag dissolution in a soil solution (L/S = 10; time scale: 12–720 h). Higher contaminant releases were observed for stronger citrate solutions, but the secondary HFO, amorphous SiO_2, and microcrystalline calcite ($CaCO_3$), which formed on the slag surfaces, were partly responsible for attenuating the contaminants. Similarly, when Pb slags were leached in fulvic acid solutions and natural peat water (with *ca.* 50 mg L^{-1} of dissolved organic carbon, DOC), the organic species enhanced slag dissolution and the released Fe rapidly oxidized and re-precipitated as HFO.[69] Subsequent adsorption onto HFO surfaces appears to be an efficient sink mechanism for many contaminants (Table 5.2). This process is expected to be one of the key immobilization mechanisms during the long-term interactions of slags with real soil solutions in slag disposal sites covered with a soil layer;[69] however, it is problematic in highly organic environments, where Fe is complexed with the organic ligands, which prevent the HFO precipitation.

For similar simulations, Potysz *et al.*[5] used an artificial root exudate (ARE) solution containing glucose (9 mM), fructose (9.2 mM), sucrose (4.6 mM), citric acid (4.6 mM), lactic acid (9.2 mM), and succinic acid (6.9 mM). Their 30 day

Table 5.2 Comparisons of the normalized releases (mg m^{-2}) of selected metal(loid)s leached out of Pb–Zn slag in various media (deionized water and organic solutions). The maximum values and the values obtained after 112 days of leaching are shown.[69]

Leaching media	Normalized release (mg m^{-2})					
	As		Pb		Zn	
	Max	Final	Max	Final	Max	Final
Deionized water	5.19	**1.75**[a]	13.2	13.2	19.5	19.5
Fulvic acid solution[b]	18.1	**1.05**	59.4	**1.55**	14.7	**1.43**
Peat water[b]	23.2	**2.52**	51.0	**11.0**	20.1	20.1
Citrate (8 mM)[c]	56.8	19.0	80.1	**0.05**	70.5	**1.66**
Citrate (20 mM)[c,d]	149.7	149.7	304.6	304.6	211.4	196.5

[a]Values in **bold** indicate the strong contaminant scavenging mechanism by adsorption to newly formed hydrous ferric oxides (HFO).
[b]Dissolved organic carbon (DOC) concentrations are 50.8 mg L^{-1} and 47.1 mg L^{-1} for the fulvic acid solution and natural peat water, respectively.
[c]Data for the 30 day leaching procedure from ref. 68.
[d]High citrate concentration enables complexation of Fe in the solution and prevents the precipitation of HFO.

experiment was conducted with a Cu slag in a dynamic system with a solution renewal every 5 days. The slag was dissolved to a greater extent in ARE and the leached concentrations of the contaminants (Cu, Pb) were generally 1–2 orders of magnitude higher as compared to leaching in deionized water.[5]

However, in real soil environments, contaminant leaching from slags may be very complex and *in situ* leaching experiments are rare in the literature. Pareuil *et al.*[63] used a 1 year field experiment where plastic boxes filled with an Mn-rich slag (11 kg) covered with a soil layer (16 kg) were placed in a pasture area and equipped with a leachate collection system at the bottom. The presence of the slag in the soil induced changes in the pH (increased by a pH unit compared to the reference experiment with the soil only), which, in turn, led to the mobilization of the organic matter and the transport of metallic elements in colloidal forms. Comparisons of the field data with the results from the column (percolation) test (CEN/TS 14405) demonstrated that the laboratory leaching experiments strongly underestimate the releases of metallic elements compared to the real exposure of the slags in soil systems.[63]

5.3 Geochemical Modeling

Leaching experiments are often combined with geochemical modeling to determine the speciation of contaminants in the leachates and the degree of leachate saturation with respect to the solubility-controlling phases (this approach is called "speciation–solubility modeling"). To obtain reliable modeling results, none of the potentially important chemical constituents of the leachate should be neglected: the input data files should include not only the concentrations of the studied contaminants but also data on the major cations and anions and the physicochemical parameters (pH, redox potential, temperature). Over the last two decades, the PHREEQC geochemical code[70] was the most widely used for speciation-solubility calculations in slag leachates; however, the end-users are responsible for the correct selection of a suitable thermodynamic database, which contains all the relevant elements (*e.g.*, the default phreeqc.dat database does not contain species like As or Sb and cannot be used for speciation modeling of these elements). In many cases, the existing databases must be supplemented with additional thermodynamic data taken from the literature.[7,31] Numerous other geochemical codes have been adopted for the speciation-solubility modeling of slag leachates: MINTEQA2,[56,69,71] Visual MINTEQ,[23,43] EQ3NR,[2,45] and CHESS/HYTEC.[46] The majority of these codes have been applied as equilibrium thermodynamic models. Advanced modeling approaches, such as reaction path modeling (code EQ6[2]) and kinetic modeling (code KINDIS(P)[63]), require the implementation of the calculated/estimated dissolution rate constants, which are not available for the majority of the primary slag phases (*e.g.*, glass).

The pH-static leaching data can be directly fitted to and/or compared with the pH-dependent solubilities of the potential solubility-controlling phases,[10,71] which helps to determine the mobilities of elements and their mineralogical and geochemical controls under changing pH conditions. Figure 5.6A shows the key Pb-bearing phases that were predicted to control the release of Pb from slags as an example of geochemical modeling interpretations. A more sophisticated database/expert system (LeachXS™) has been recently developed to facilitate leaching data interpretations, scenario evaluations, and leaching test comparisons.[47,61] In combination with the pH-dependent leaching data, this software helps to identify the element speciation in the leachate as well as its partitioning in the solid phase.[53,54] The distribution of Cr species and prevailing binding mechanisms illustrated in Figure 5.6B represent a software output example. LeachXS™ Lite is freely downloadable from the Vanderbilt University website,[57] is available for applications related to the USEPA leaching tests within the LEAF framework, and can be used for any mineral waste, including slags.

5.4 Investigation of Newly Formed Phases

Geochemical modeling results must be interpreted with care and it is helpful to confirm the model predictions using independent methods. For instance, the precipitation of solubility-controlling phases predicted by the modeling should be verified by any conventional mineralogical techniques or combinations thereof. An X-ray diffraction analysis (XRD) can give rough information about the changes in the major phase compositions during the leaching tests (comparisons can be made between the original slag and the leached slag residue;[49] for details about mineralogical techniques see Chapter 3). Scanning electron microscopy (SEM) equipped with energy-dispersive spectroscopy

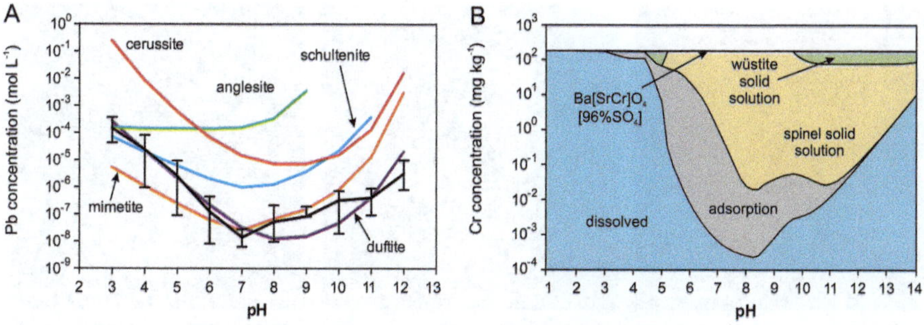

Figure 5.6 Modeling of the leaching data. (A) The pH-dependent dissolution curves for the potential Pb solubility-controlling phases calculated by PHREEQC and compared to the pH-static leaching data obtained from Cu smelting slags (in black). (B) The speciation and solubility controls for Cr leached from steel slag as calculated by LeachXS software. (A) Reproduced from ref. 10 with permission from Elsevier, Copyright 2016 and (B) Reproduced from ref. 54, https://doi.org/10.3390/min9090525, under the terms of the CC BY 4.0 license, https://creativecommons.org/licenses/by/4.0/. Phase formulae: anglesite ($PbSO_4$), duftite ($PbCu(AsO_4)(OH)$), cerussite ($PbCO_3$), mimetite ($Pb_5(AsO_4)_3Cl$), schultenite ($PbHAsO_4$), and wüstite (FeO).

(EDS) is an advantageous method for the visualization and chemical analysis of secondary products formed on the slag surfaces during leaching; it helps to understand the processes governing the dissolution of the primary phases (*e.g.*, visibly etched surfaces[2]) and phases contributing to the release and attenuation of contaminants (adsorption/precipitation).[5,10,29] Transmission electron microscopy (TEM) is also a useful mineralogical technique, especially when the visualization and analysis of smaller objects (less than a few μm in size) is needed.[4,43,45] Moreover, this method can be combined with EDS (chemical composition information), electron diffraction (SAED, selected area electron diffraction, which provides crystallographic information), and other analytical techniques.

Many investigations of leached base metal slags indicate that HFO are almost ubiquitous secondary alteration products and may serve as efficient sorbents for many contaminants (Figure 5.7B and D). Sulfates (anglesite, $PbSO_4$; barite, $BaSO_4$; and complex Cu sulfates), carbonates (cerussite, $PbCO_3$; smithsonite, $ZnCO_3$; and calcite, $CaCO_3$), and complex metal-bearing phosphates also commonly form as solubility-controlling phases during slag leaching (Figure 5.7A–C)[2,5,10,29,45,65,72] as well as during natural weathering of slags (see Table 4.1 in Chapter 4). In the case of As-rich slags (5280–9500 mg As kg^{-1}) originating from the processing of As-bearing Cu ores, high levels of leached As, Cu, and Pb may lead to precipitation of arsenates, such as duftite ($PbCu(AsO_4)(OH)$) or conichalcite ($CaCu(AsO_4)(OH)$), which can be important solubility-controlling products limiting the mobility of these contaminants (Figure 5.7E and F).[10]

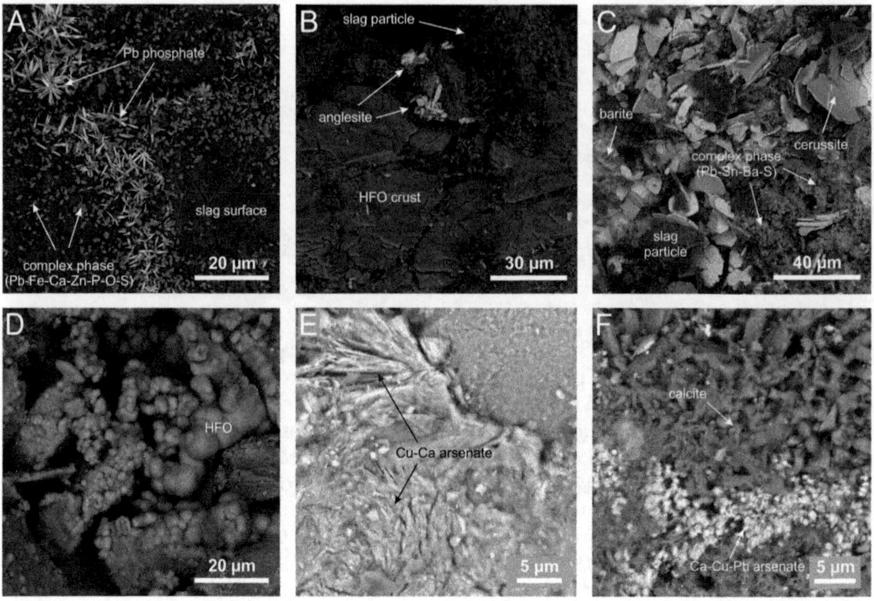

Figure 5.7 Secondary phases formed during base metal slag leaching tests. (A–D) Reproduced from ref. 29 with permission from Elsevier, Copyright 2013. (E–F) Reproduced from ref. 10 with permission from Elsevier, Copyright 2016. Phase formulae: HFO – hydrous ferric oxide ($Fe(OH)_3$), anglesite ($PbSO_4$), barite ($BaSO_4$), and cerussite ($PbCO_3$).

5.5 Discussion and Outlook

Many different leaching tests have been implemented by environmental regulators and standardization institutes (*e.g.*, CEN – the European Committee for Standardization – and the USEPA) over the last two decades to provide methods for the environmental assessment of solid waste materials. Owing to their experimental simplicity, single-step and pH-dependent tests have been the most applied to metallurgical slags to evaluate contaminant releases. Column leaching tests, in which experimental setups and subsequent interpretations of the leaching data are quite complicated, have not been widely reported in the slag-related literature. Indeed, more data for slag leaching using columns are needed because many non-ferrous metal smelters produce granulated slags, which, when deposited on the heaps, undergo a percolation regime (rainwater infiltration) (see Figure 5.1). Moreover, the first fractions obtained during the initial stages of the leaching in a column setup (L/S of 0.1–0.5 L kg^{-1}) can be directly compared with the low L/S conditions encountered in the field (*e.g.*, pore water collected by lysimeters).[61]

Field trials investigating the long-term behavior of slags in specific scenarios are also quite rare.[63,73] However, they are technically complicated (a lysimeter installation in slag heaps, the long-term monitoring of leachates and slag mineralogical transformations) and thus costly. Nevertheless, such *in situ* experiments would provide very useful datasets, which could be directly compared with the results of laboratory leaching tests conducted under controlled conditions. Souter and Watmough[9] studied water emanating from a Cu–Ni slag pile in the Sudbury area (Ontario, Canada). They found that concentrations of several trace metals in the slag runoff pond were relatively similar to those obtained by the laboratory column leaching experiments.[9] In contrast, Fällman and Hartlén[73] made comparisons of the results from laboratory column tests conducted on blast furnace and steel slags with those for the field lysimeter experiments and found strikingly different results. Significantly higher leaching of some metals (*e.g.*, Cu) in the field trials was attributed to atmospheric conditions influencing the physicochemical parameters of the leachates, which exhibited lower pH and higher redox potential compared to the leachates obtained by the laboratory experiments. Another pioneering study by Pareuil *et al.*[63] that exposed Mn-rich slag from recycling alkaline batteries in a soil layer for 1 year indicated a much slower dissolution under field conditions than during the laboratory percolation. Besides, natural weathering reactions in slag disposal sites and slag-containing environmental compartments can be extremely complex and are not fully understood. For example, the effect of microbes on the slag weathering/leaching process in the dumps/heaps has not yet been studied.

Geochemical modeling represents a useful tool for understanding the processes of contaminant leaching and immobilization. However, to date, most modeling has employed an equilibrium thermodynamic approach for data interpretation. In the future, kinetic approaches are essential to predict the long-term behavior of slags, even though the dissolution rates of individual slag phases are not well characterized at present.[2,28] In addition, geochemical modeling tools should not be used as "black boxes", where end-users blindly consider the modeling results as a matter of fact. Independent methods should verify the predictions obtained by geochemical modeling. Detailed mineralogical investigations of the slag transformations during the leaching process are of crucial importance for interpreting the leaching data, as demonstrated in many studies.[5,29,43,49,72]

To summarize, despite the usefulness of standardized leaching tests, more effort should be given to their combination with detailed mineralogical investigations using conventional techniques (XRD, SEM/EDS, TEM/EDS) and geochemical modeling to fully understand the processes affecting contaminant release from slags.

5.6 Conclusions

The leaching behavior of metallurgical slags is an important parameter that must be considered when decisions about slag disposal and/or reuse (*e.g.*, in civil engineering) are to be taken. In this chapter, we presented (i) the experimental protocols for many leaching tests (some of them implemented by environmental regulators), (ii) a comparison of experimental results, and (iii) implications for understanding the processes governing the release of contaminants (metals, metalloids) under different environmental conditions.

Each leaching test provides specific information and often combinations of various leaching are is needed. In addition, combining leaching tests with a detailed mineralogical investigation of the slag transformations (*e.g.*, precipitation of newly formed phases) and geochemical modeling can help to fully describe the contaminant release. There is an ongoing need for more long-term leaching tests and field experiments to help develop more reliable predictions of slag leaching in real-world environments.

Acknowledgements

Czech Science Foundation (GAČR) project no. 19-18513S and the institutional funding from the Centre for Geosphere Dynamics (UNCE/SCI/006) are acknowledged. Anna Potysz, Valérie Cappuyns, and Daniel Vollprecht are thanked for sending us the original data and/or figures. We are grateful to Michael B. Parsons (Natural Resources Canada) and Nadine M. Piatak (US Geological Survey) for their thorough reviews, which helped to substantially improve the original version of the chapter.

References

1. V. Ettler, O. Legendre, F. Bodénan and J. C. Touray, *Can. Mineral.*, 2001, **39**, 873.
2. M. Parsons, D. K. Bird, M. T. Einaudi and C. N. Alpers, *Appl. Geochem.*, 2001, **16**, 1567.
3. J. Puziewicz, K. Zainoun and H. Bril, *Can. Mineral.*, 2007, **45**, 1189.
4. J. Kierczak, C. Néel, J. Puziewicz and H. Bril, *Can. Mineral.*, 2009, **47**, 557.
5. A. Potysz, J. Kierczak, A. Pietranik and K. Kadziolka, *Appl. Geochem.*, 2018, **98**, 22.
6. R. Tyszka, J. Kierzcak, A. Pietranik, V. Ettler and M. Mihaljevič, *Appl. Geochem.*, 2014, **40**, 70.
7. V. Ettler, Z. Johan, B. Kříbek, O. Šebek and M. Mihaljevič, *Appl. Geochem.*, 2009, **24**, 1.
8. V. Ettler, J. Kvapil, O. Šebek, Z. Johan, M. Mihaljevič, G. Ratié, J. Garnier and C. Quantin, *Appl. Geochem.*, 2016, **64**, 118.
9. L. Souter and S. A. Watmough, *Sci. Total Environ.*, 2017, **605–606**, 461.
10. A. Jarošíková, V. Ettler, M. Mihaljevič, B. Kříbek and B. Mapani, *J. Environ. Manage.*, 2017, **187**, 178.
11. A. Schmukat, L. Duester, D. Ecker, H. Schmid, C. Heil, P. Heininger and T. A. Ternes, *J. Hazard. Mater.*, 2012, **227–228**, 257.
12. L. Duester, C. Brinkmann, T. A. Ternes and P. Heininger, *Crit. Rev. Environ. Sci. Technol.*, 2016, **46**, 434.
13. D. S. Kosson, H. A. van der Sloot, F. Sanchez and A. C. Garrabrants, *Environ. Eng. Sci.*, 2002, **19**, 159.
14. M. Vítková, *Environmental Characteristics of Mineral Waste from Metallurgy*, PhD thesis, Charles University, 2012.

15. USEPA Method 1313, *Liquid–Solid Partitioning as a Function of Extract pH Using a Parallel Batch Extraction Procedure*, USEPA, Washington, 2017, https://www.epa.gov/hw-sw846/sw-846-test-method-1313-liquid-solid-partitioning-function-extract-ph-using-parallel-batch.
16. EN 12457, *Characterization of Waste – Leaching – Compliance Test for Leaching of Granular Waste Materials and Sludges, Parts 1–4*, CEN, Brussels, 1999.
17. EN 14405, *Characterization of Waste – Leaching Behaviour Test – Upflow Percolation Test (Under Specified Conditions)*, CEN, Brussels, 2017.
18. EN 14429, *Characterization of Waste – Leaching Behaviour Tests – Influence of pH on Leaching with Initial Acid/Base Addition*, CEN, Brussels, 2015.
19. EN 14997, *Characterization of Waste – Leaching Behaviour Tests – Influence of pH on Leaching with Continuous pH-control*, CEN, Brussels, 2015.
20. LEAF: Leaching Environmental Assessment Framework, USEPA, Washington, https://www.epa.gov/hw-sw846/leaching-environmental-assessment-framework-leaf-methods-and-guidance#LEAF%20 Methods.
21. M. Zandi, N. V. Russel, R. G. J. Edyvean, R. J. Hand and P. Ward, *J. Environ. Monit.*, 2007, **9**, 1429.
22. M. Vítková, V. Ettler, M. Mihaljevič and O. Šebek, *J. Hazard. Mater.*, 2011, **197**, 417.
23. P. Pareuil, F. Bordas, E. Joussein and J. C. Bollinger, *Environ. Technol.*, 2010, **31**, 1565.
24. EN 15863, *Characterization of Waste – Leaching Behaviour Tests – Dynamic Monolithic Leaching Test with Periodic Leachant Renewal*, CEN, Brussels, 2015.
25. USEPA Method 1315, *Mass Transfer Rates of Constituents in Monolithic or Compacted Granular Materials Using a Semi-dynamic Tank Leaching Procedure*, USEPA, Washington, 2017, https://www.epa.gov/hw-sw846/sw-846-test-method-1315-mass-transfer-rates-constituents-monolithic-or-compacted-granular.
26. V. Ettler, M. Mihaljevič, J. C. Touray and P. Piantone, *Bull. Soc. Geol. Fr.*, 2002, **173**, 161.
27. A. Potysz and J. Kierczak, *Minerals*, 2019, **9**, 542.
28. P. Pareuil, F. Bordas, E. Joussein, P. Vieillard and J. C. Bollinger, *Appl. Geochem.*, 2010, **25**, 1187.
29. V. Ettler and Z. Johan, *Appl. Geochem.*, 2014, **40**, 97.
30. B. Mikoda, H. Kucha, A. Potysz and E. Kmiecik, *Appl. Geochem.*, 2018, **98**, 459.
31. V. Ettler, M. Mihaljevič, A. Jarošíková, A. Culka, B. Kříbek, V. Majer, A. Vaněk, V. Penížek, O. Sracek, B. Mapani and F. Kamona, *Appl. Geochem.*, 2020, **114**, 104473.
32. EU, Council Directive 99/31/EC of 26 April 1999 on the landfill of waste, *Off. J. Eur. Communities*, 1999, **L182**, 1.
33. EU, Council Decision of 19 December 2002 establishing criteria and procedures for the acceptance of waste at landfills pursuant to Article 16 of and Annex II to Directive 1999/31/EC, *Off. J. Eur. Communities*, 2003, **L11**, 27.

34. USEPA Method 1311, *Toxicity Characteristic Leaching Procedure (TCLP)*, USEPA, Washington, 1992, https://www.epa.gov/hw-sw846/sw-846-test-method-1311-toxicity-characteristic-leaching-procedure.
35. T. T. Lim and J. Chu, *Waste Manage. Res.*, 2006, **24**, 67.
36. P. Shanmuganathan, P. Lakshmipathiraj, S. Srikanth, A. L. Nachiappan and A. Sumathy, *Resour., Conserv. Recycl.*, 2008, **52**, 601.
37. L. R. P. de Andrade Lima and L. A. Bernardez, *J. Hazard. Mater.*, 2011, **189**, 692.
38. V. Cappuyns, V. Alian, E. Vassilieva and R. Swennen, *Waste Biomass Valorization*, 2014, **5**, 355.
39. A. Ghosh, M. Mukiibi and W. Ela, *Environ. Sci. Technol.*, 2004, **38**, 4677.
40. USEPA Method 1312, *Synthetic Precipitation Leaching Procedure (SPLP)*, USEPA, Washington, 1994, https://www.epa.gov/hw-sw846/sw-846-test-method-1312-synthetic-precipitation-leaching-procedure.
41. N. M. Piatak, R. R. Seal II and J. M. Hammarstrom, *Appl. Geochem.*, 2004, **19**, 1039.
42. N. M. Piatak and R. R. Seal II, *Appl. Geochem.*, 2010, **25**, 302.
43. N. H. Yin, Y. Sivry, F. Guyot, P. N. L. Lens and E. D. van Hullebusch, *J. Environ. Manage.*, 2016, **180**, 310.
44. V. Ettler, *Etude du potential pollutant de rejets anciens et actuels de la métallurgie du plomb dans le district de Příbram (République tchèque)*, PhD thesis, Université d'Orléans and Charles University, 2000.
45. V. Ettler, P. Piantone and J. C. Touray, *Mineral. Mag.*, 2003, **67**, 1269.
46. L. De Windt, P. Chaurand and J. Rose, *Waste Manage.*, 2011, **31**, 225.
47. leachxs.com, last accessed February 2021.
48. V. Ettler, M. Vítková, M. Mihaljevič, O. Šebek, M. Klementová, F. Veselovský, P. Vybíral and B. Kříbek, *Environ. Geochem. Health*, 2014, **36**, 919.
49. V. Ettler, D. Štěpánek, M. Mihaljevič, P. Drahota, R. Jedlicka, B. Kříbek, A. Vaněk, V. Penížek, O. Sracek and I. Nyambe, *Chemosphere*, 2020, **260**, 127642.
50. N. Saikia, G. Cornelis, G. Mertens, J. Elsen, K. Van Balen, T. Van Gerven and C. Vandecasteele, *J. Hazard. Mater.*, 2008, **154**, 766.
51. N. Saikia, G. Cornelis, O. Cizer, C. Vandecasteele, D. Van Gemert, K. Van Balen and T. Van Gerven, *J. Mater. Cycles Waste Manage.*, 2012, **14**, 102.
52. H. A. van der Sloot, R. N. J. Comans and O. Hjelmar, *Sci. Total Environ.*, 1996, **178**, 111.
53. M. Loncnar, H. A. van der Sloot, A. Mladenovič, M. Župančič, L. Kobal and P. Bukovec, *J. Hazard. Mater.*, 2016, **317**, 147.
54. S. Neuhold, A. van Zomeren, J. J. Dijsktra, H. A. van der Sloot, P. Drissen, D. Algermissen, D. Mudersbach, S. Schüler, T. Griessacher, J. G. Raith, R. Pomberger and D. Vollprecht, *Minerals*, 2019, **9**, 525.
55. A. Król, K. Mizerna and M. Bozym, *J. Hazard. Mater.*, 2020, **384**, 121502.
56. P. Ganne, V. Cappuyns, A. Vervoort, L. Buvé and R. Swennen, *Sci. Total Environ.*, 2006, **356**, 69.
57. vanderbilt.edu/leaching, last accessed February 2021.

58. *Test Methods for Evaluating Solid Waste: Physical/Chemical Methods, Hazardous Waste Test Methods/SW-846*: https://www.epa.gov/hw-sw846.
59. A. Potysz, J. Kierczak, Y. Fuchs, M. Grybos, G. Guibaud, P. N. L. Lens and E. D. van Hullebusch, *J. Geochem. Explor.*, 2016, **160**, 1.
60. USEPA Method 1314, *Liquid–Solid Partitioning as a Function of Liquid–Solid Ratio for Constituents in Solid Materials Using An Up-flow Percolation Column Procedure*, USEPA, Washington, 2017, https://www.epa.gov/hw-sw846/sw-846-test-method-1314-liquid-solid-partitioning-function-liquid-solid-ratio-constituents.
61. H. A. van der Sloot, J. C. L. Meeussen, A. van Zomeren and D. S. Kosson, *J. Geochem. Explor.*, 2006, **88**, 72.
62. H. A. van der Sloot, D. S. Kosson and O. Hjelmar, *Waste Manage.*, 2001, **21**, 753.
63. P. Pareuil, F. Bordas, E. Joussein and J. C. Bollinger, *Environ. Pollut.*, 2010, **158**, 1311.
64. A. Navarro, E. Cardellach, J. L. Mendoza, M. Corbella and L. M. Domènech, *Appl. Geochem.*, 2008, **23**, 895.
65. N. Seignez, A. Gauthier, D. Bulteel, M. Buatier, P. Recourt, D. Damidot and J. L. Potdevin, *J. Hazard. Mater.*, 2007, **149**, 418.
66. N. Seignez, A. Gauthier, D. Bulteel, D. Damidot and J. L. Potdevin, *Appl. Geochem.*, 2008, **23**, 3699.
67. S. Sobanska, B. Ledésert, D. Deneele and A. Laboudigne, *C. R. Acad. Sci.*, 2000, **331**, 271.
68. V. Ettler, M. Komárková, J. Jehlička, P. Coufal, D. Hradil, V. Machovič and J. Jehlička, *Chemosphere*, 2004, **57**, 567.
69. V. Ettler, J. Jehlička, V. Mašek and J. Hruška, *Mineral. Mag.*, 2005, **69**, 737.
70. D. L. Parkhurst and C. A. J. Appelo, *Description of input and examples for PHREEQC version 3–A computer program for speciation, batch-reaction, one-dimensional transport, and inverse geochemical calculations*, US Geological Survey, Techniques and Methods, book 6, ch. A43, 2013.
71. R. Barna, H. R. Bae, J. Méhu, H. van der Sloot, P. Moszkowicz and C. Desnoyers, *Waste Manage.*, 2000, **20**, 115.
72. N. Seignez, D. Bulteel, D. Damidot, A. Gauthier and J. L. Potdevin, *WIT Trans. Ecol. Environ.*, 2006, **92**, 31.
73. A. M. Fällman and J. Hartlén, *Stud. Environ. Sci.*, 1994, **60**, 39.

CHAPTER 6

Environmental Impact of Slag Particulates

VOJTĚCH ETTLER*[a] AND JAKUB KIERCZAK[b]

[a]Institute of Geochemistry, Mineralogy and Mineral Resources, Faculty of Science, Charles University, Albertov 6, 128 00 Prague 2, Czech Republic; [b]Institute of Geological Sciences, University of Wrocław, Cybulskiego 30, 50-205 Wrocław, Poland
*E-mail: vojtech.ettler@natur.cuni.cz

6.1 Introduction

Mining operations represent potential sources of airborne metal(loid) contaminants.[1] Apart from the wind erosion of legacy mine tailings and direct dust emissions from active mines, non-ferrous metal smelters are especially responsible, through many technological operations, for the generation of small particles, mostly composed of slag, which can become atmospheric particulates that are transported in the air and can be deposited into various environmental compartments.[2] For instance, granulation, that is quenching of the slag melt by water, produces glassy slags with a grain size generally of <5 mm in diameter, which are then deposited on the dumps (Figure 6.1A; see also other chapters in this book). For the recovery of sulfides and other dense metal-bearing phases, granulated slags are often reprocessed by milling and flotation[3] with the resulting waste (slag tailings) being deposited on disposal dumps. Figure 6.1B shows that slag tailings may be used as a cover on mine tailings generated after flotation of concentrated ore. Owing to slag tailings' substantially higher density, they are thought to limit dust dispersion from

Chemistry in the Environment Series No. 2
Metallurgical Slags: Environmental Geochemistry and Resource Potential
Edited by Nadine M. Piatak and Vojtěch Ettler

Published by the Royal Society of Chemistry, www.rsc.org

the mine tailings. Small slag particles eroded and transported from the dumps can end up in the nearby soil, where they undergo weathering processes, as has been commonly observed at numerous former smelting sites (Figure 6.1C and D). Furthermore, legacy slags enriched in valuable metals may be crushed for re-smelting, which generates dust. As shown in Figure 6.1E, up to 80 year-old deposits of Co-rich (0.3–2.6 wt%) Cu slags are crushed to <15 mm in diameter, generating slag dust, which can be transported by the wind into the environment. The crushed slag is then mixed with fluxes (*e.g.*, lime, coal, rutile), and re-smelted in an electric direct current (DC) arc furnace to recover Co and also Ni and Cu.[4] Another source of slag particulates is flue dust emitted from smelter stacks, which contains slag-like particles of variable sizes and shapes.[5–11]

Fine-grained slag deposits are vulnerable to wind erosion, especially in arid and semi-arid areas (Figure 6.1A and B). Soils represent a typical target reservoir for these slag particulates and smelter-derived emissions (see Figure 6.1C, D, and F and Ettler[12] and references therein). Slag particulates have also been documented in highly weathered rock surfaces in smelter areas with high emission loads[13,14] and snow collected near active smelters.[15,16] Windblown slag particles can also be attached to the surfaces of plants (*e.g.*, leaves, tubers, and fruits) and, in the case of agriculturally important crops, might represent a human health risk.[17,18] Slag-like particles are an integral part of aerosols (particulate matter <10 μm in size, PM_{10}) as well as coarser "nuisance" dust in smelter production sites[19] and smelter-affected industrial areas.[20] Direct contact with contaminant-rich slag dust particles and their incidental ingestion/inhalation (Figure 6.1G) is another pathway for contaminant transport to target reservoirs.[10,21]

6.2 Slag Particulates in Soils

6.2.1 Methods of Investigation

Smelter-impacted soils used for the investigation of metal(loid)-bearing particles (including slags) are commonly prepared by standard sieving procedures to <2 mm in diameter to obtain the geochemically "reactive" part of the soil.[22] Since metal(loid)s are concentrated in the fine-grained soil fractions, sieving to <0.5 mm has been adopted in numerous studies.[23–25] Lanteigne *et al.*[26,27] used magnetic separation to obtain smelter-derived magnetic particulates from soils near historical Ni–Cu smelting operations at Sudbury, Ontario, Canada, where over 10 000 tons of particulate matter has been emitted over the last 100 years. Lastly, heavy mineral fractions (after separation in heavy liquids such as 1,1,2,2-tetrabromethane with a density of 2.96 g cm^{-3} or sodium tungstate, Na_2WO_4, with a density of 4.18 g cm^{-3}) have been studied because they concentrate the majority of metal(loid)-bearing heavy particulates.[12,22,28,29]

Further sample processing is strictly dependent on the objective of the study and the methods subsequently used. Given that scanning electron microscopy (SEM) equipped with energy dispersion spectrometry (EDS) is

Figure 6.1 Photos of the sources and environmental fate of slag particulates. (A) Wind- and water-eroded granulated Cu slag deposit (Tsumeb, Namibia). (B) Mine tailing impoundments covered with denser Cu slag tailings to limit dust dispersion (Tsumeb, Namibia). (C) Cultivated field near a Ni slag dump (Szklary, Poland). (D) Small fragment of weathered Ni slag collected in the agricultural soil from (C) (scale of the grid is 1 mm). (E) Crushing of air-cooled Cu slag near the Nkana Cu smelter (Kitwe, Zambia) for subsequent slag reprocessing in the nearby Chambishi smelter to recover Co. (F) Metal(loid)-rich soil profile (Spolic Technosol Calcaric Arenic) located near the Tsumeb Cu smelter with the 0–30 cm layer containing a mixture of mine tailings and slags, partly deposited after the historical tailings pond dam failures as well as corresponding to the current emissions from the smelter stack and/or windblown particles from the waste disposal sites. (G) Children playing on Pb–Zn slag dumps in Kabwe (Zambia); note the fine slag dust adhering to their hands and legs. Photos: Vojtěch Ettler, Jakub Kierczak, and Martin Mihaljevič.

frequently used for the visualization and chemical characterization of slag particulates, the samples are prepared either (i) by sprinkling the sample material or attaching individual grains onto a carbon-impregnated tape or (ii) by fabrication of a polished section. Both types of sample are then coated with a layer of conductive material (*e.g.*, carbon, gold, platinum), a process called sputtering, before the SEM observation and qualitative or semi-quantitative EDS analysis. Specimens can be observed under secondary electrons (SE) for visualization of the surface morphology or back-scattered electrons (BSE) for visualization of the distribution of elements, where brightness is related to the mean sample mass (the higher the average atomic number, the brighter it appears in the image); see Figures 6.2 and 6.3. The spatial distribution of the elements in the specimen can be acquired using X-ray mapping (Figure 6.3). Modern instruments with a field emission gun (FEG) electron source have a much higher image resolution than SEM instruments with a conventional tungsten wire cathode. An electron probe microanalyzer (EPMA) equipped with one or several wave-dispersion spectrometers (WDS) may be used for the quantitative determination of the chemical compositions of the slag phases (larger than 1 μm in size) with detection limits at trace levels (routinely around 0.01 wt% or down to around 0.001 wt% with optimized settings for modern instruments).[30] Many trace elements of importance occur in slag particulates in concentrations undetectable by EPMA and, for this reason, laser ablation-inductively coupled plasma-mass spectrometry (LA-ICP-MS) has been used;[27] the disadvantage of this analysis is related to the relatively low spatial resolution (generally tens of μm). Transmission electron microscopy (TEM), often combined with a focused ion beam (FIB) preparation of electron-transparent foils, offers higher spatial resolution (nanoscale), which is commonly necessary to visualize and analyze fine-grained smelter-derived particulates.[14,26,31,32]

For the recognition of modal compositions (the percentage abundance of phases or minerals in a sample), SEM-based automated mineralogy (initially developed by QEMSCAN® for the mining industry[33]) has also been adopted to understand the distributions and mineralogical compositions of slag particulates in smelter-emitted dusts,[19] smelter-polluted soils,[29] and soils directly developed on slag heaps.[34] This method is particularly useful for quantitative calculation of the contaminants' partitioning (deportment) within the individual phases in the studied material. Such information is of crucial importance for assessing the environmental fate of the contaminants in soils polluted by slag particulates.

6.2.2 Origins of Slags in Soils

Soils in the vicinity of non-ferrous smelting operations may exhibit extremely high concentrations of metal(loid) contaminants (thousands to tens of thousands of mg kg^{-1}).[12,22] These contaminants are often bound in metal(loid)-bearing particles, which originate from mining activities (mining and mechanical ore processing) as well as from smelters and smelter waste

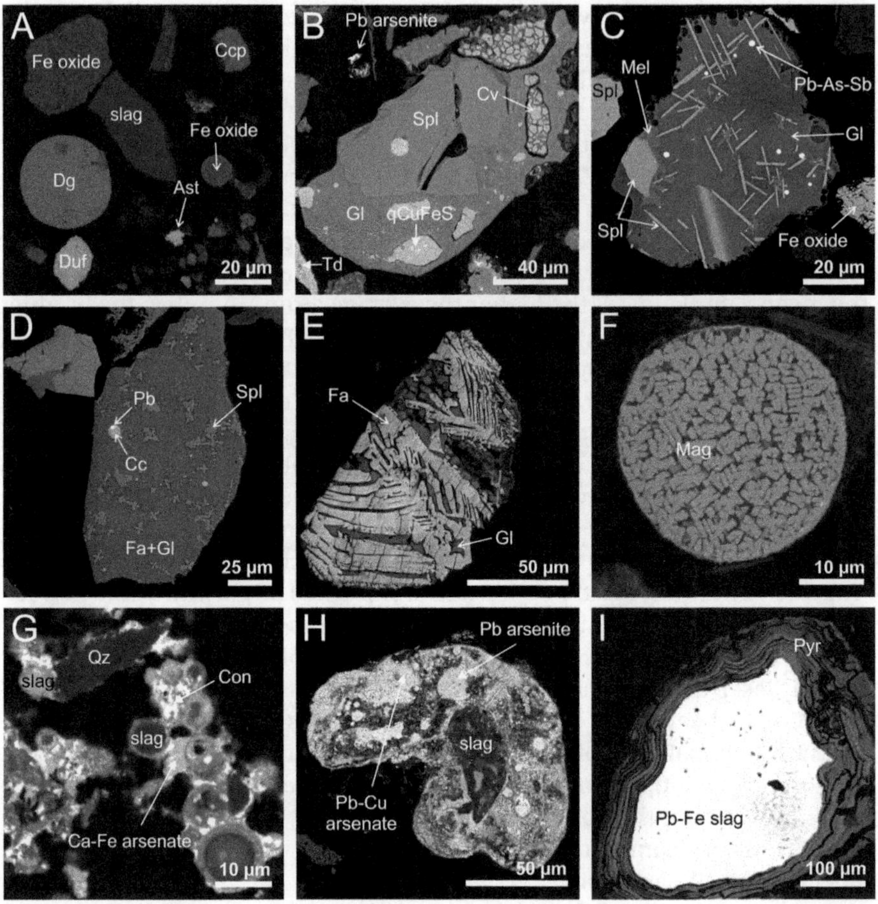

Figure 6.2 Scanning electron photomicrographs of slag particulates from smelter-polluted soils (images in back-scattered electrons, BSE). (A) Spherical digenite (Dg; Cu_9S_5) particle and glassy slag fragment (origin: smelter) associated with chalcopyrite (Ccp; $CuFeS_2$), duftite (Duf; $PbCu(AsO_4)(OH)$), and arsentsumebite (Ast; $Pb_2Cu(AsO_4)(SO_4)(OH)$) grains (origin: mine tailing). (B) Smelter slag particle composed of spinel (Spl; $(Zn,Mg,Fe)(Fe,Al,Cr)_2O_4$) and glass (Gl) with inclusions of quenched Cu–Fe–S phase (qCuFeS) and weathered covellite (Cv; CuS) (origin: smelter), associated with Pb arsenite (origin: mine tailing) and soil-derived Pb-bearing todorokite (Td; $Pb_{1-x}Mn_6O_{12}$). (C) Slag particle composed of melilite (Mel; $Ca_2(Mg,Fe,Zn,Al)Si_2O_7$), spinels (Spl), and intermetallic compounds occurring in the glass (Gl) matrix associated with pedogenic Fe oxide. (D) Slag particle composed of fayalite (Fa; Fe_2SiO_4) and glass (Gl) matrix, dendrites of spinel (Spl), and inclusion formed by metallic Pb and chalcocite (Cc; Cu_2S). (E) Slag composed of skeletal fayalite (Fa) embedded in glass (Gl). (F) Spherical slag-like particle composed of magnetite (Mag; Fe_3O_4) and glass matrix. (G) Spherical smelter slag particles and angular grain of quartz (Qz; SiO_2) cemented with conichalcite (Con; $CaCu(AsO_4)(OH)$) and Ca–Fe arsenate phase. (H) Pb arsenite and Pb–Cu arsenate phases associated with a weathered slag grain (origin: smelter, weathering in soil).

disposal sites. Many smelting operations are located in heavily industrialized sites near active and/or abandoned mining areas, and the resulting contamination corresponds to the mixture of particles from more than one operation.[35] The pioneering study by Davis *et al.*[36] devoted to waste and contaminated soils in the Pb mining site of Butte, Montana, USA, demonstrated that, apart from primary ore minerals (*e.g.*, galena, PbS) and their weathering products (*e.g.*, anglesite, $PbSO_4$; cerussite, $PbCO_3$; Pb oxides and phosphates) and soil-derived Pb-bearing Fe and Mn oxides, slag particles have also been frequently detected.

Slag particles deposited in soils have variable morphologies and sizes. Particulates windblown from slag deposits are typically from a few to hundreds of micrometers in size and are of irregular (angular) shapes (Figure 6.2A–E).[25,28,29,35,37] Many of them are predominantly composed of glass because fine-grained granulated slags, which are more susceptible to wind transport than air-cooled slags, are often glass-rich owing to the rapid slag melt cooling.[3,38] Moreover, they typically contain dendritic crystals of silicates and spinel oxides, as well as small metallic/sulfide inclusions (Figure 6.2A–E). In contrast, slag-like particles originating from smelter emissions generally exhibit spherical shapes; they correspond to droplets of the slag melt emitted from the furnace and quenched during the flue-gas cleaning process.[23–27,35] They are commonly smaller than 60 μm in diameter, which is generally finer than slag particles windblown from the slag heaps. However, Lanteigne *et al.*[26] also observed much larger spherical slag particulates (up to 200 μm in diameter) in the polluted soils near a Cu–Ni smelter in Sudbury, Canada. These emitted particles are often composed of a silicate glassy matrix with dendrites of spinels (magnetite, Fe_3O_4; trevorite, $NiFe_2O_4$; cuprospinel, $CuFe_2O_4$) or other oxides (hematite/maghemite, Fe_2O_3; delafossite, $CuFeO_2$; tenorite, CuO).[24,26,29,35] Many of the spherical slag particles exhibit a rim, which is thought to be formed by the rapid cooling process (*e.g.*, a collision of finer dust particles with slag melt droplets of lower viscosity) and/or the subsequent weathering and phase transformation.[26,27] The rims are thus often enriched in contaminants such as Cu, Ni, Zn, Pb, or As (Figure 6.3).[26,27] The soil contamination, as well as the amount of spherical slag particles in surface soils, decreases as a function of the distance from the smelter.[2,12,23] Very fine-grained smelter-emitted particulates (aerosols <1 μm) can travel a greater

---

(I) Pb- and Fe-rich glassy slag with a pyromorphite (Pyr; $Pb_5(PO_4)_3Cl$) weathering rim. (A)–(C) and (H) Topsoils near the Tsumeb Cu–Pb smelters (semi-arid Namibia). (D) and (E) Soils near the Legnica Cu smelter (Poland). (F) Topsoil near the Mufulira Cu smelter (Zambia). (I) Garden soil from the Pb–Zn mining/smelting village of Stratoni (Greece). (A, B, F, G, H) Reproduced from ref. 35 with permission from Elsevier, Copyright 2016. (C) Reproduced from ref. 29 with permission from Elsevier, Copyright 2020. (D, E) Reproduced from ref. 28, https://doi.org/10.1007/s11356-016-7655-4, under the terms of the CC BY 4.0 license, http://creativecommons.org/licenses/by/4.0/. (I) Reproduced from ref. 39 with permission from Springer Nature, Copyright 2013.

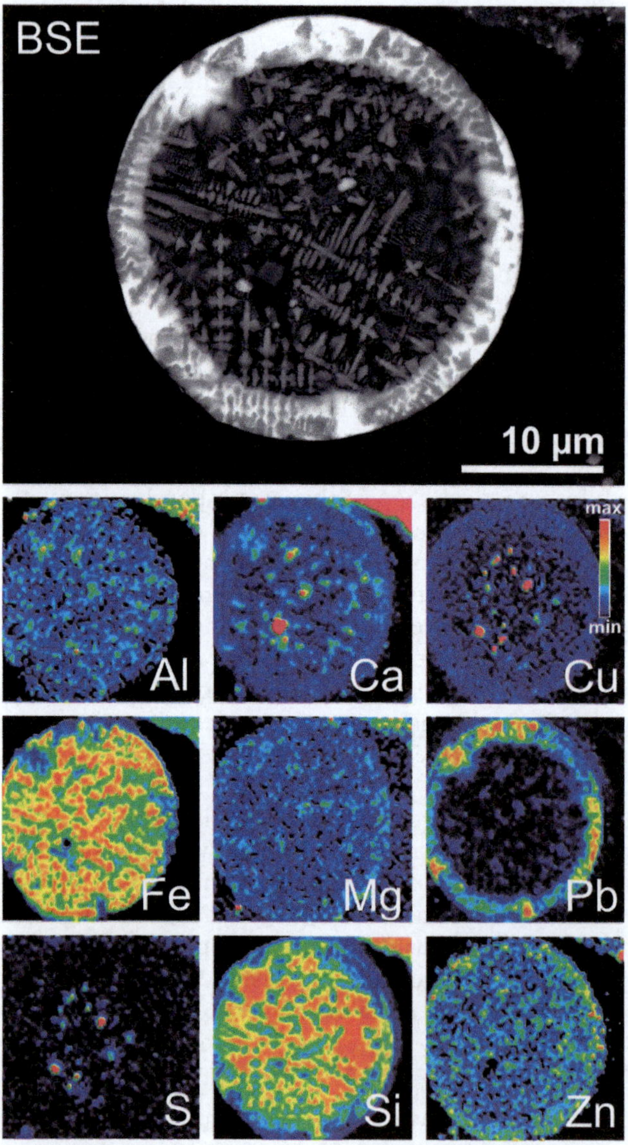

Figure 6.3 A spherical slag-like particle emitted by Cu–Pb smelters in Tsumeb (Namibia) separated from the biomass-rich topsoil in the vicinity of the smelting operation. Back-scattered electron image and electron probe microanalyzer elemental maps showing the spatial distribution of the elements.

distance from the point source of the pollution and are often more enriched in metal(loid)s than coarser dust particles.[1,2] Examination of such ultrafine aerosol particles sampled directly from the air[1] or soils/rock coatings[14] represents a challenge and advanced techniques with nano-scale resolutions are needed (FEG-SEM, FIB-TEM) for their detailed mineralogical investigation.

6.2.3 Slag Weathering in Soils: Descriptive and Experimental Approaches

The weathering features observed on slag particulates are dependent on the chemical/mineralogical compositions of the particulates and surrounding soil substrate, the residence time after being deposited into the soil, and climatic factors (*e.g.*, temperature, precipitation). Comparisons of smelter-polluted soils in hot semi-arid areas and humid subtropical areas indicate that smelter-derived particles are significantly more weathered in the latter environments, as reflected by SEM observations and a higher proportion of "labile" forms of metal(loid)s obtained by chemical extractions.[35] Many slag particles observed in smelter-affected soils do not exhibit any weathering features, probably related to the relatively short residence time after their deposition in soil (Figure 6.2A–F). They are considered more resistant than other smelter-derived dusts because they are composed of a relatively stable silicate glass, silicates, and spinel-type oxides.[24] The occurrence of non-weathered slag particles in soils often reflects their continuous deposition from active smelting areas[29,35] or uncovered slag disposal sites.[37] However, the dissolution of slag glass and precipitation of secondary phases (*e.g.*, metal arsenates and phosphates) forming rims around the slag particles have also been observed at several places (*e.g.*, in Tsumeb, Namibia, where As-rich polymetallic ores were smelted,[35] Figure 6.2G and H, or in Stratoni, Greece, where Pb–Zn ores were processed,[39] Figure 6.2I). For example, the study by Kierczak *et al.*[40] evidenced that although fragments of the Ni slag present at the waste dump are almost entirely unweathered, small slag particles present in the cultivated soil located near the disposal site display more apparent traces of weathering in the form of a brownish crust (Figure 6.1C and D). Lanteigne *et al.*[26] compared the shapes and weathering features of slag particulates separated from polluted soils near Cu–Ni smelting sites in Sudbury, Canada, and suggested weathering sequences that caused the breakdown of the matrix glass, metal-rich sulfide inclusions, and Fe oxide transformations. The "collapsed skeleton" morphologies of Cu–Fe oxides (*e.g.*, delafossite, $CuFeO_4$) indicate that they underwent intensive weathering, as observed at numerous sites near base metal smelting operations,[26,35] and are substantially less resistant to alteration than Ni–Fe spinel oxides.

Even though many slag particles seem to be mostly unaffected by weathering (Figure 6.2A–F), the rate of weathering in the soils can be accelerated by the interaction with organic ligands (*e.g.*, humic substances, root exudates). Experimental leaching tests indicate that both low-molecular-weight (LMW) and high-molecular-weight (HMW) organic acids increase metal leaching from base metal slags.[41–43] Mobilized Fe, released from slag matrix silicates and glass, may be oxidized and then precipitate as hydrous ferric oxides, and these act as an efficient contaminant trap.[41,42]

The examples cited above as well as those presented in Chapter 4 clearly show that many factors influence the weathering of slag particulates. Slag weathering also influences the properties of the soil in which they are located. Principally, this is related to the leachate chemistry generated as a result of the slag phase's dissolution reaction (*i.e.*, acidic to neutral for base metal slags and/or alkaline in the case of ferrous slags). More details about the leaching tests in general, including in simulated soil solutions, are given in Chapter 5 of this book. At this point, however, it is worth mentioning that secondary slag transformations in soil may accelerate their further weathering. A perfect example of such an influence of slag particle dissolution on soil properties is the study by Gee *et al.*[44] They pointed out that the breakdown of calcium silicates, being major components of a Pb slag, led to an increase in the soil pH and resulted in preferential precipitation of carbonates. Experimental investigations depicting the kinetics of the slag particles weathering in the soils and subsequent release of the contaminants into the soil pore water/matrix are scarce in the literature. Birkefeld *et al.*[45] exposed Cu slag particles attached to a plastic polymer support in acidic soils for 18 months and found that, in comparison with the PbO and Cu sulfide concentrate, they were resistant to weathering. Similarly, Pareuil *et al.*[46] carried out a field trial, where they exposed a layer of Mn-rich slag in a soil layer and collected the leachates throughout a year. Both studies compared these *in situ* experiments with laboratory leaching tests and concluded that the faster slag dissolution in the field was caused by the more intensive biological activity (*e.g.*, the effect of the root exudates from the plant roots and fungi hyphae).[45,46]

6.3 Slag Particulates and Human Health Implications

The mining and processing of metal ores generates vast amounts of waste materials, and dust particles mobilized from the waste disposal sites can have direct human health impacts on communities living in the former/active mining areas (Figure 6.1G).[47,48] The fraction of contaminants in the dust, which, *via* ingestion, inhalation, or dermal contact, reaches the systemic circulation (blood compartment) and is absorbed by the receptor (human or animal) is called the "bioavailable fraction". In contrast, the "bioaccessible fraction" is the contaminant fraction, which is soluble in a gastrointestinal or lung environment and is potentially available for absorption.[49] Many *in vitro* bioaccessibility methods have been developed and validated by *in vivo* bioavailability studies conducted with pig or mouse models.[50–52] A large number of studies have been devoted to determining the bioaccessibility of metal(loid) contaminants from mining-polluted soil, urban dust, and mining waste, but relatively limited information on the bioaccessibility of contaminants from slag dust is available in the literature.[10,21,38,53–57] This section presents the principal methodological approaches used for the bioaccessibility testing and the results obtained for slag dusts from several mining areas.

6.3.1 Bioaccessibility Tests

Incidental dust ingestion and inhalation are major pathways for the human population's exposure to dust particles. Many experimental protocols have been suggested to simulate what happens to dust particles in the airways after inhalation (*e.g.*, extraction in a simulated lung fluid, SLF) and in the gastrointestinal tract after ingestion (*e.g.*, in a simulated gastric fluid, SGF). Several critical review papers by Kastury *et al.*[58] (inhalation), Ruby *et al.* (ingestion),[59] and Deshommes *et al.*[60] (ingestion) summarize the advantages and limitations of these bioaccessibility testing protocols.

For the bioaccessibility testing of slag dusts, ingestion has mainly been considered as the major exposure pathway;[21,38,53,56] for this reason, extraction in highly acidic SGF, which promotes the dust dissolution, represents the "worst-case" scenario in terms of the contaminant release. The oral bioaccessibility test according to the USEPA's[50] protocol (also known as SBRC-G; gastric phase extraction using the Solubility Bioaccessibility Research Consortium assay) is often used due to its simplicity and comparability with *in vivo* tests for Pb[50,60] and As.[61,62] The dust sample is extracted for 1 h, which is the estimated residence time in the stomach, in a 0.4 M solution of glycine adjusted to a pH of 1.5 ± 0.05 by reagent-grade HCl, at a liquid-to-solid (L/S) ratio of 100 and a temperature of 37 °C. The obtained extract is analyzed, and the bioaccessible concentrations of the metal(loid)s are expressed in mg kg^{-1} and converted to their percent amount of the total contents (*i.e.*, the bioaccessible fraction, BAF). The standard protocol prescribes testing a <250 μm fraction of the dust, suitable for ingestion *via* hand-to-mouth behavior. However, recent experimental investigations demonstrated that the size fractions adhering to the hands are much smaller. Siciliano *et al.*[63] suggested that a 45 μm sieve should be used as the cut-off mesh size for the sample preparation before the oral bioaccessibility testing.

In contrast to the highly acidic SGF, SLF is near-neutral and many experimental solutions with variable compositions have been suggested for the simulation of pulmonary surfactants. One example solution is Gamble's solution, a mixture of inorganic salts with a pH of 7.4 (see Ettler *et al.*[10] and Kastury *et al.*[58] and references therein). Variable L/S ratios have been tested, but an L/S ratio of 100 is the most frequently used. The test is performed at a temperature of 37 °C and for over 24 h, but the kinetics of dissolution for the fine-grained smelter dusts are generally much faster and steady-state is attained within *ca.* 6 h.[10] Fractions that are <10 μm in diameter, presumably the size entering the airways, are used for these *in vitro* inhalation bioaccessibility tests; however, the larger sized inhaled particles (approximately 4–10 μm) are cleared from the airways by the mucociliary escalator, swallowed, and transported into the gastrointestinal tract, and only particles approximately <4 μm in size can access the deep lung.[58] For this reason, <10 μm slag dust fractions have also been recently investigated by oral bioaccessibility tests.[21,38]

6.3.2 Bioaccessible Metal(loid)s in Slags

A compilation of bioaccessible concentrations and BAF values obtained from slag materials from smelting sites around the world is reported in Table 6.1. Large variability in the BAFs of individual contaminants has been observed: from very low values for Ni slags from Brazil (<2% of the total)[57] to 100% for Pb–Zn slags from Zambia.[21] The bioaccessible concentrations obtained by extraction in SGF and SLF are strikingly different; BAFs are one to two orders of magnitude lower for SLF due to the significantly higher pH of the SLF solution (Table 6.1).[10] The bioaccessible concentrations depend on numerous factors, particularly the bulk concentrations, mineralogy, phase textures, and sample granulometry. Morrison *et al.*[56] used an automated mineralogy approach carried out by the QEMSCAN® instrument to visualize Pb-bearing phases in the slag particle-size fractions and quantify Pb bioaccessibility and its relationship to the grain morphology. This approach helped to determine the distribution and proportions of the main contaminant-hosting phases for the individual slag particles, as shown for a slag particle in Figure 6.4A, and to classify them according to their accessibility: "liberated" phases are fully exposed to the leaching solution; "accessible" phases are partly subjected to the leaching, and "locked" phases are embedded in other, more resistant, phases (*e.g.*, silicate glass) and are isolated from the leaching solution (Figure 6.4B). Significantly higher proportions of "liberated" and "accessible" contaminant-containing phases were found in finer slag fractions.

Indeed, the grain size, proportional to the reactive surface of the sample, also affects contaminant leaching in the simulated body fluids. Morrison and Gulson[53] compared the total Pb and bioaccessible Pb concentrations in sieved fractions of granulated Pb slags from Australia and found a significant increase in both parameters with decreasing particle size. Generally, higher bulk concentrations and bioaccessibilities of the contaminants were also found in finer slag fractions from other base metal smelting sites.[21,38] However, exceptions can also be found: for instance, Co occurred in slightly lower concentrations in the <10 μm fraction from a site in Zambia than in the original slag dust. This phenomenon was attributed to its partial binding in the silicate phases (*e.g.*, olivine, glass), which predominantly form larger grains.[4,10]

The bioaccessible concentrations obtained from the extraction tests in the simulated body fluids can be used for the prediction of exposure of

Table 6.1 Examples of the bioaccessible concentration (mg kg^{-1}) and bioaccessible fractions (BAFs, % of the total) of the contaminants in the slag dusts as obtained by extractions in simulated gastric fluid (G) and simulated lung fluid (L).

Site	Sample	Size fraction	Test	Contaminant	Bioaccessible concentration (mg kg^{-1})	BAF (%)	Reference
Niquelândia and Barro Alto, Brazil	Ni slag	Granulated (as is) (mostly 0.5–2 mm)	G	Ni	9.02–19.1	0.6–2	Ettler *et al.*[57]
				Cr	11.8–106	0.2–1.6	
				Co	0.3–1.6	0.3–1.7	
Mitrovica, Kosovo	Pb slag	<250 μm	G	As	NA[a]	1–34	Boisa *et al.*[55]
				Cd	NA	41–53	
				Cu	NA	1–73	
				Pb	NA	1–71	
				Zn	NA	5–67	
Chambishi, Zambia	Cu slag	<200 μm	G	As	21	32	Ettler *et al.*[10]
				Cd	0.3	29	
				Co	2340	26	
				Cu	16 200	50	
				Pb	79.8	55	
				Zn	72	26	
		<10 μm	L	As	5.8	2.26	
				Cd	0.28	2.59	
				Co	18.39	0.21	
				Cu	946	1.82	
				Pb	0.05	0.01	
				Zn	2.31	0.42	

(*continued*)

Table 6.1 (*continued*)

Site	Sample	Size fraction	Test	Contaminant	Bioaccessible concentration (mg kg^{-1})	BAF (%)	Reference
Tsumeb, Namibia	Cu slag	Granulated (as is) (mostly <300 μm)	G	As	678–4380	13–58	Ettler *et al.*[38]
				Cd	13.6–48.1	52–97	
				Cu	1570–20 400	11–39	
				Pb	2120–5700	22–27	
				Sb	86.9–1060	12–33	
				Zn	4440–13 800	28–43	
		<10 μm	G	As	1050–5230	50–57	
				Cd	27.6–110	54–67	
				Cu	3370–22 000	25–59	
				Pb	5630–20 100	60–76	
				Sb	110–1100	12–20	
				Zn	5970–14 300	40–49	
Berg Aukas, Namibia	Zn slag	<600 μm	G	Pb	3670	49	Ettler *et al.*[38]
				Zn	28 400	46	
		<10 μm	G	Pb	16 900	93	
				Zn	35 700	62	
Southeast Brazil	Pb slag	<180 μm	G	Pb	NA	80	Bosso and Enzweiler[54]
Lake Macquarie, Australia	Pb slag	<250 μm	G	Pb	2012–25 893	31–63	Morrison and Gulson[53]
		32–53 μm	G	Pb	16 951–48 818	66–86	
Kabwe, Zambia	Pb–Zn slag	<48 μm	G	Cd	10.4–51.2	42–79	Ettler *et al.*[21]
				Pb	9762–47 294	24–92	
				V	424–2070	21–92	
				Zn	14 042–102 112	54–79	
		<10 μm	G	Cd	16.1–72.6	54–82	
				Pb	13 549–60 439	29–96	
				V	523–2195	21–100	
				Zn	16 467–154 862	58–81	

[a]NA – not available.

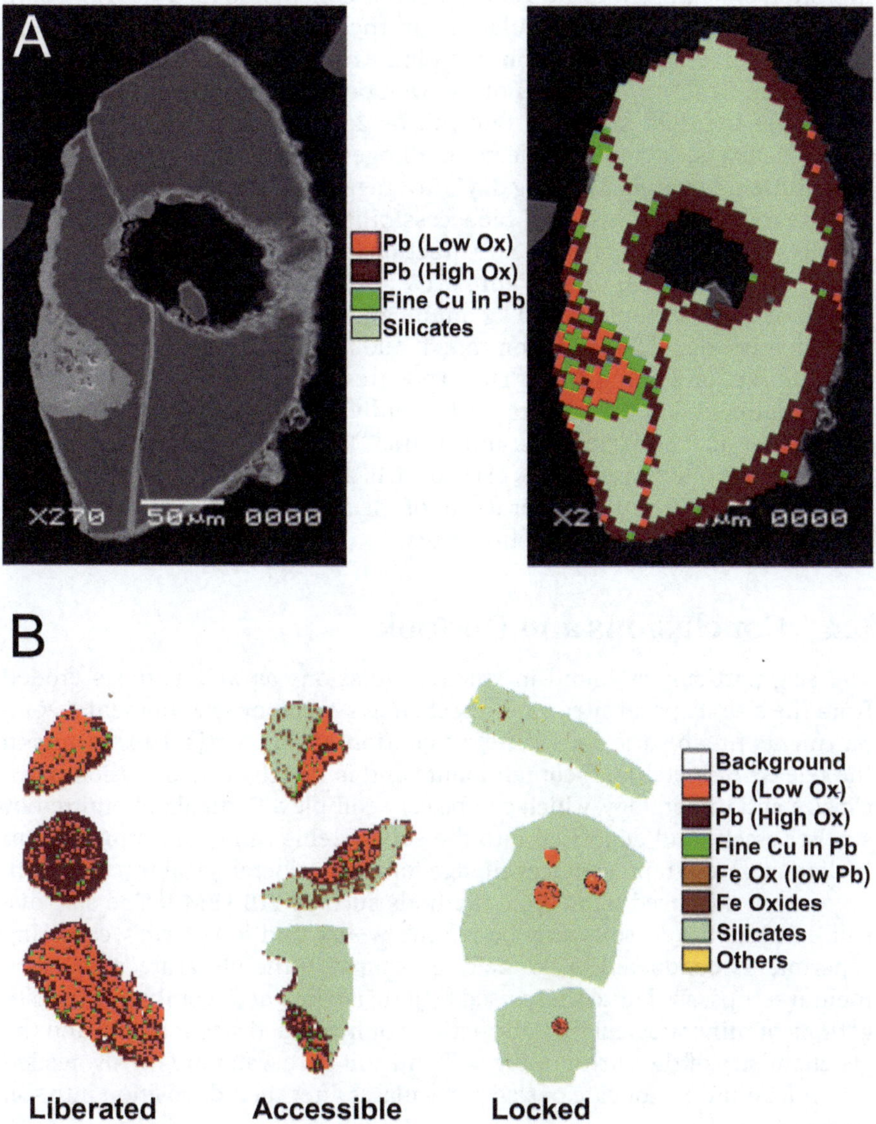

Figure 6.4 Examples of the automated mineralogy performed on slag particles used for the assessment of the contaminant bioaccessibility. (A) Back-scattered electron image and superimposed color-coded phase map of a complex slag particle with the color key for the individual phases. (B) Smelter slag particles showing the reaction surface accessibility categorization of the Pb and phase mapping as obtained by the automated mineralogy (Ox: oxygen, oxide). Reproduced from ref. 56 with permission from Elsevier, Copyright 2015.

humans to metal(loid) contaminants occurring in slag dust. Whereas dust ingestion of 50 mg day^{-1} and 100 mg day^{-1} is considered for adults and children, respectively, in the calculations of the oral exposure route,[64] recent studies from less developed countries indicate much higher dust ingestion rates. Data from mining areas of the Democratic Republic of Congo indicated that the dust ingestion rate can be 280 mg day^{-1} (geometric mean for all datasets, including objects of all ages), but is substantially higher for children, reaching 1700 mg day^{-1} (average value for children, age group 4–15 years).[48] Using the metal bioaccessibility data obtained on slag dusts for the <10 µm and <48 µm size fractions collected near an old Pb–Zn smelter in Kabwe (Zambia) (Figure 6.1G), Ettler *et al.*[21] calculated the daily intake of contaminants for 70 kg adults and 10 kg children considering a conservative slag dust ingestion rate of 100 mg day^{-1} and a probably more realistic rate of 280 mg day^{-1}. The exposures were compared with the tolerable daily intake (TDI) limits for the individual contaminants compiled by Baars *et al.*[65] and Tiesjema and Baars.[66] The data shown in Figure 6.5 demonstrate that these metals exceed TDIs and, in particular, the Pb daily intakes are significantly higher for all of the slag dust samples and age categories, even at low dust ingestion rates.[21]

6.4 Conclusions and Outlook

The slag particulates found in smelter emissions as well as dusts eroded from the slag disposal sites have direct effects on many environmental compartments (mainly the soils). The dissolution of slag particulates results in the release of metal(loid) contaminants and is strictly related to their particle size and mineralogy, which can be very complex. Particulates emitted by smelter stacks and deposited into the soil systems can be ultrafine (<1 µm in diameter) and represent a challenge for their mineralogical investigation, even using advanced nano-scale methods such as FIB-TEM.[14] The slag dissolution kinetics in soils may be relatively slow and long-term weathering experiments conducted in the field are sparse in the literature.[45,46] Experimental setups, similar to those used by Jarošíková *et al.*,[67] combining investigations of mineralogical transformations of incubated slag material and the geochemistry of the surrounding soil and soil pore water are highly needed for understanding the fate of slag particulates after their deposition into soil systems.

It is obvious that slag dust particles represent a risk for human health, particularly in situations where former smelter areas with slag heaps are unfenced and accessible to local populations.[21] Screening bioaccessibility studies investigating mineralogical and geochemical forms of contaminants and their leaching from slag dusts should be conducted to find problematic smelting areas, where urgent environmental remediation measures are needed and where incidental slag dust ingestion/inhalation potentially causing adverse health effects should be prevented.

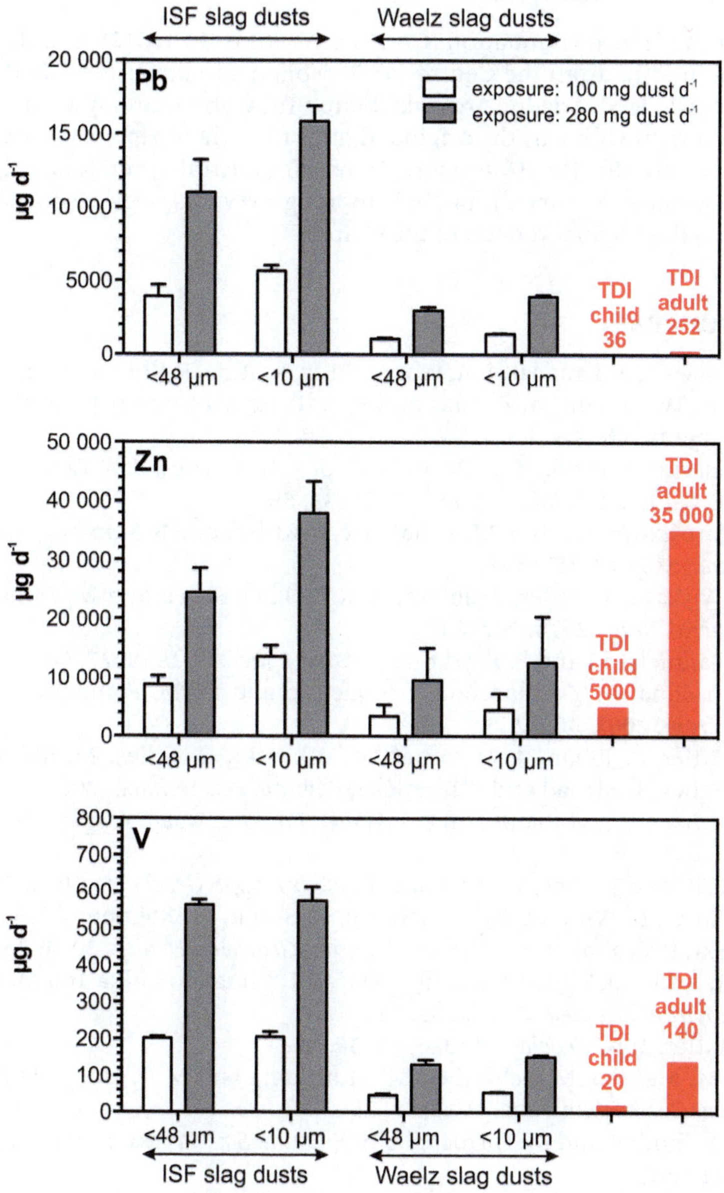

Figure 6.5 Daily intakes of Pb, V, and Zn (in μg day^{-1}) obtained by the bioaccessibility testing of slag dusts (fractions <48 μm and <10 μm) from Kabwe (Zambia), where the Imperial Smelting Furnace (ISF) and Waelz slags are deposited. Two slag dust intake scenarios were assumed: 100 mg day^{-1} and 280 mg day^{-1}. The values are compared with the tolerable daily intakes (TDI) calculated for a 10 kg child and a 70 kg adult. Adapted from ref. 21 with permission from Elsevier, Copyright 2020.

Acknowledgements

The Czech Science Foundation (GAČR) project no. 19-18513S and the institutional funding from the Centre for Geosphere Dynamics (UNCE/SCI/006) are acknowledged. Ariadne Argyraki, Marek Tuhý, and Anthony Morrison are thanked for providing us the original figures from their papers. We are grateful to Anthony Morrison (Macquarie University, Australia) and Nadine M. Piatak (US Geological Survey) for their thorough reviews, which substantially improved the original version of the chapter.

References

1. J. Csavina, A. Landázuri, A. Wonaschütz, K. Rine, P. Rheinheimer, B. Barbaris, W. Conant, A. E. Sáez and E. A. Betterton, *Water, Air, Soil Pollut.*, 2011, **221**, 145.
2. J. Csavina, J. Field, M. P. Taylor, S. Gao, A. Landázuri, E. A. Betterton and A. E. Sáez, *Sci. Total Environ.*, 2012, **433**, 58.
3. A. Jarošíková, V. Ettler, M. Mihaljevič, B. Kříbek and B. Mapani, *J. Environ. Manage.*, 2017, **187**, 178.
4. M. Vítková, V. Ettler, Z. Johan, B. Kříbek, O. Šebek and M. Mihaljevič, *Mineral. Mag.*, 2010, **74**, 581.
5. C. Samuelsson and B. Björkman, *Scand. J. Metall.*, 1998, **27**, 54.
6. A. A. Piña, G. T. Villaseñor, P. S. Jacinto and M. M. Fernández, *Atmos. Environ.*, 2002, **36**, 5235.
7. V. Ettler, Z. Johan, A. Barronet, F. Jankovský, C. Gilles, M. Mihaljevič, O. Šebek, L. Strnad and P. Bezdička, *Environ. Sci. Technol.*, 2005, **39**, 9309.
8. J. M. Skeaff, Y. Thibault and D. J. Hardy, *Environ. Monit. Assess.*, 2011, **177**, 165.
9. M. Vítková, J. Hyks, V. Ettler and T. Astrup, *Appl. Geochem.*, 2013, **29**, 117.
10. V. Ettler, M. Vítková, M. Mihaljevič, O. Šebek, M. Klementová, F. Veselovský, P. Vybíral and B. Kříbek, *Environ. Geochem. Health*, 2014, **36**, 919.
11. A. Jarošíková, V. Ettler, M. Mihaljevič, P. Drahota, A. Culka and M. Racek, *J. Environ. Manage.*, 2018, **209**, 71.
12. V. Ettler, *Appl. Geochem.*, 2016, **64**, 56.
13. N. M. Mantha, M. Schindler, M. Murayama and M. F. Hochella Jr, *Geochim. Cosmochim. Acta*, 2012, **85**, 254.
14. M. Schindler and M. F. Hochella Jr, *Environ. Sci.: Processes Impacts*, 2017, **19**, 1016.
15. D. Gregurek, C. Reimann and E. F. Stumpfl, *Sci. Total Environ.*, 1998, **221**, 189.
16. D. Gregurek, F. Melcher, V. A. Pavlov, C. Reimann and E. F. Stumpfl, *Mineral. Petrol.*, 1999, **65**, 85.
17. B. Kříbek, V. Majer, I. Knésl, I. Nyambe, M. Mihaljevič, V. Ettler and O. Sracek, *J. Afr. Earth Sci.*, 2014, **99**, 713.
18. B. Kříbek, V. Majer, I. Knésl, J. Keder, B. Mapani, F. Kamona, M. Mihaljevič, V. Ettler, V. Penížek, A. Vaněk and O. Sracek, *Appl. Geochem.*, 2016, **64**, 75.

19. J. A. Small, C. J. G. van Hoek, K. Schollbach, E. Moosavi-Khoonsari, F. J. L. van der Does, S. Melzer, T. P. R. de Jong, T. de Bie, R. A. Versfeld, M. de Roo and S. R. van der Laan, *Environ. Sci. Eur.*, 2020, **32**, 135.
20. A. Francová, V. Chrastný, H. Šillerová, M. Vítková, J. Kocourková and M. Komárek, *Environ. Pollut.*, 2017, **220**, 286.
21. V. Ettler, D. Štěpánek, M. Mihaljevič, P. Drahota, R. Jedlicka, B. Kříbek, A. Vaněk, V. Penížek, O. Sracek and I. Nyambe, *Chemosphere*, 2020, **260**, 127642.
22. V. Ettler, M. Mihaljevič, B. Kříbek, V. Majer and O. Šebek, *Geoderma*, 2011, **164**, 73.
23. P. J. Henderson, I. McMartin, G. E. Hall, J. B. Percival and D. A. Walker, *Environ. Geol.*, 1998, **34**, 39.
24. R. D. Knight and P. J. Henderson, *Geochem.: Explor., Environ., Anal.*, 2006, **6**, 203.
25. P. J. Harvey, M. Rouillon, C. Dong, V. Ettler, H. K. Handley, M. P. Taylor, E. Tyson, P. Tennant, V. Telfer and R. Trinh, *Sci. Total Environ.*, 2017, **584–585**, 505.
26. S. Lanteigne, M. Schindler, A. M. McDonald, K. Skeries, Y. Abdu, N. M. Mantha, M. Murayama, F. C. Hawthorne and M. F. Hochella Jr, *Water, Air, Soil Pollut.*, 2012, **223**, 3619.
27. S. Lanteigne, M. Schindler and A. McDonald, *Can. Mineral.*, 2014, **52**, 453.
28. R. Tyszka, A. Pietranik, J. Kierczak, V. Ettler, M. Mihaljevič and A. Medynska-Juraszek, *Environ. Sci. Pollut. Res.*, 2016, **23**, 24350.
29. M. Tuhý, T. Hrstka and V. Ettler, *Environ. Pollut.*, 2020, **266**, 115118.
30. *Electron Probe Microanalysis, 2nd Edition, Essential Knowledge Briefings*, John Wiley and Co., Chichester, 2018.
31. V. Ettler, Z. Johan, M. Vítková, R. Skála, M. Kotrlý, G. Habler and M. Klementová, *J. Hazard. Mater.*, 2012, **221–222**, 298.
32. H. Mantha, M. Schindler and M. F. Hochella, *Environ. Sci.: Nano*, 2019, **6**, 163.
33. D. Pirrie and G. K. Rollinson, *Geol. Today*, 2011, **27**, 226.
34. A. Pietranik, J. Kierczak, R. Tyszka and B. Schulz, *Minerals*, 2018, **8**, 513.
35. V. Ettler, Z. Johan, B. Kříbek, F. Veselovský, M. Mihaljevič, A. Vaněk, V. Penížek, V. Majer, O. Sracek, B. Mapani, F. Kamona and I. Nyambe, *Sci. Total Environ.*, 2016, **563–564**, 329.
36. A. Davis, J. W. Drexler, M. V. Ruby and A. Nicholson, *Environ. Sci. Technol.*, 1993, **27**, 1415.
37. E. I. B. Chopin and B. J. Alloway, *Sci. Total Environ.*, 2007, **373**, 488.
38. V. Ettler, M. Cihlová, A. Jarošíková, M. Mihaljevič, P. Drahota, B. Kříbek, A. Vaněk, V. Penížek, O. Sracek, M. Klementová, Z. Engel, F. Kamona and B. Mapani, *Environ. Int.*, 2019, **124**, 205.
39. A. Argyraki, *Environ. Geochem. Health*, 2014, **36**, 677.
40. J. Kierczak, C. Néel, J. Puziewicz and H. Bril, *Can. Mineral.*, 2009, **47**, 557.
41. V. Ettler, M. Komárková, J. Jehlička, P. Coufal, D. Hradil, V. Machovič and J. Jehlička, *Chemosphere*, 2004, **57**, 567.

42. V. Ettler, J. Jehlička, V. Mašek and J. Hruška, *Mineral. Mag.*, 2005, **69**, 737.
43. A. Potysz, J. Kierczak, M. Grybos, A. Pędziwiatr and E. D. van Hullebusch, *J. Environ. Manage.*, 2018, **222**, 325.
44. C. Gee, M. H. Ramsey, J. Maskall and I. Thornton, *J. Geochem. Explor.*, 1997, **58**, 249.
45. A. Birkefeld, R. Schulin and B. Nowack, *Geochim. Cosmochim. Acta*, 2006, **70**, 2726.
46. P. Pareuil, F. Bordas, E. Joussein and J. C. Bollinger, *Environ. Pollut.*, 2010, **158**, 1311.
47. J. A. Entwistle, A. S. Hursthouse, P. A. M. Reis and A. G. Stewart, *Curr. Pollut. Rep.*, 2019, **5**, 67.
48. E. Smolders, L. Roels, T. C. Kuhangana, K. Coorevits, E. Vassilieva, B. Nemery and C. B. L. Nkulu, *Environ. Sci. Technol.*, 2019, **53**, 7851.
49. M. R. Cave, J. Wragg, S. Denys, C. Jondreville and C. Feidt, Oral Bioavailability, in *Dealing with Contaminated Sites*, ed. F. A. Swartjes, Springer Science+Business Media B.V., Berlin, 2011, pp. 287–324.
50. USEPA, Estimation of relative bioavailability of lead in soil and soil-like materials using *in vivo* and *in vitro* methods. *Office of Solid Waste and Emergency Response, OSWER 9285.7-77*, USEPA, Washington, 2007.
51. J. W. Drexler and W. J. Brattin, *Hum. Ecol. Risk Assess.*, 2007, **13**, 383.
52. S. Denys, J. Caboche, K. Tack, G. Rychen, J. Wragg, M. Cave, C. Jondreville and C. Feidt, *Environ. Sci. Technol.*, 2012, **46**, 6252.
53. A. L. Morrison and B. L. Gulson, *Sci. Total Environ.*, 2007, **382**, 30.
54. S. T. Bosso and J. Enzweiler, *Environ. Geochem. Health*, 2008, **30**, 219.
55. N. Boisa, G. Bird, P. A. Brewer, J. R. Dean, J. A. Entwistle, S. J. Kemp and M. G. Macklin, *Environ. Int.*, 2013, **60**, 56.
56. A. L. Morrison, Z. Swierczek and B. L. Gulson, *Environ. Pollut.*, 2016, **210**, 271.
57. V. Ettler, L. Polák, M. Mihaljevič, G. Ratié, J. Garnier and C. Quantin, *Environ. Geochem. Health*, 2018, **40**, 1699.
58. F. Kastury, E. Smith and A. L. Juhasz, *Sci. Total Environ.*, 2017, **574**, 1054.
59. M. V. Ruby, R. Schoof, W. Brattin, M. Goldade, G. Post, M. Harnois, D. E. Mosby, S. W. Casteel, W. Berti, M. Carpenter, D. Edwards, D. Cragin and W. Chappell, *Environ. Sci. Technol.*, 1999, **33**, 3697.
60. E. Deshommes, R. Tardif, R. Edwards, S. Sauvé and M. Prévost, *Chem. Cent. J.*, 2012, **6**, 138.
61. A. L. Juhasz, E. Smith, C. Nelson, D. J. Thomas and K. Bradham, *Environ. Sci. Technol.*, 2014, **48**, 11646.
62. A. L. Juhasz, P. Herde, C. Herde, J. Boland and E. Smith, *Environ. Sci. Technol.*, 2014, **48**, 12962.
63. S. D. Siciliano, K. James, G. Zhang, A. N. Schafer and J. D. Peak, *Environ. Sci. Technol.*, 2009, **43**, 6385.
64. J. Bierkens, M. Van Holderbeke, C. Cornelis and R. Torfs, Exposure Through Soil and Dust Ingestion, in *Dealing with Contaminated Sites*, ed. F. A. Swartjes, Springer Science+Business Media B.V., Berlin, 2011, pp. 261–286.

65. A. J. Baars, R. M. C. Theelen, P. J. C. M. Janssen, J. M. Hesse, M. E. van Apeldoorn, M. C. M. Meijerink, L. Verdam and M. J. Zeilmaker, Re-evaluation of Human-toxicological Maximum Permissible Risk Levels, *RIVM Report 711701025*, National Institute of Public Health and the Environment, Bilthoven, 2001.
66. B. Tiesjema and A. J. Baars, Re-evaluation of some Human-toxicological Maximum Permissible Risk Levels Earlier Evaluated in the Period 1991-2001, *RIVM Report 711701092*, National Institute of Public Health and the Environment, Bilthoven, 2009.
67. A. Jarošíková, V. Ettler, M. Mihaljevič, V. Penížek, T. Matoušek, A. Culka and P. Drahota, *Environ. Pollut.*, 2018, **237**, 83.

CHAPTER 7

Diverse Applications of Slags in the Construction Industry

IVANKA NETINGER GRUBEŠA*[a] AND IVANA BARIŠIĆ[a]

[a]Faculty of Civil Engineering and Architecture Osijek, Josip Juraj Strossmayer University of Osijek, Vladimir Prelog Street 3, 31000 Osijek, Croatia
*E-mail: nivanka@gfos.hr

7.1 Introduction

In the past few decades, advances in industry and technology have led to a significant increase in the amounts and types of waste generated. Industrial wastes include slags, red mud, fly ash, sludge, and a number of other types of waste listed on the European List of Waste.[1] These wastes can present a problem to the environment because of air pollution and possible leaching of toxic chemicals when they are dumped in landfills and quarries. Lately, environmental sustainability has become a major concern from the point of view of natural resources and their preservation. The construction industry is the largest user of natural materials and is therefore a domain of interest for using such waste. Some types of waste have been used in the construction industry for a long time. For example, slags have been used for various purposes through the centuries. In the USA, some of the first known uses of blast furnace (BF) slag are in road construction in 1830, as a railroad ballast in 1875, as concrete *aggregate* in 1880, and in bituminous surfaces in 1900.[2] The first BF slag cement was produced in Germany in 1800 by mixing slag and lime.[3] A major milestone in the use of BF slag was in 1862 with Emil Langen's

Chemistry in the Environment Series No. 2
Metallurgical Slags: Environmental Geochemistry and Resource Potential
Edited by Nadine M. Piatak and Vojtěch Ettler

Published by the Royal Society of Chemistry, www.rsc.org

discovery of the latent hydraulic properties of slag.[4] Since then, it has been used as a cement additive. Half a century later, in 1909, the first German standard that permitted the use of slags in cement production was published; this standard legally confirmed the use of BF slag in civil engineering.[5] In the USA, slag cement production began in 1896.[6] In 1940, Purdon[7] proposed dry mixing of slag and solid sodium hydroxide (NaOH) and subsequently adding water in order to prepare a mortar mixture.

From the early 20th century onward, slag production grew along with industrial development, which led to an increased need for research on the possible applications of slag in the construction sector. Consequently, numerous scientific and technical papers have been published on the results of laboratory and field research. In the following sections, some of the most common applications of slag based on studies by researchers worldwide, which highlight the wide range of possible uses of this valuable material, will be presented. Figure 7.1 shows civil engineering structures and materials that may contain slag as one of the components. Slag can be used as a partial or complete replacement of *binder* and as fine aggregate in *mortar* or *concrete* and also as coarse aggregate in concrete. Within road construction, slag can be used as partial aggregate or a binder replacement in all pavement courses (asphalt, hydraulically bound base courses, unbound base courses (UBC)) and also in road subbases and embankments.

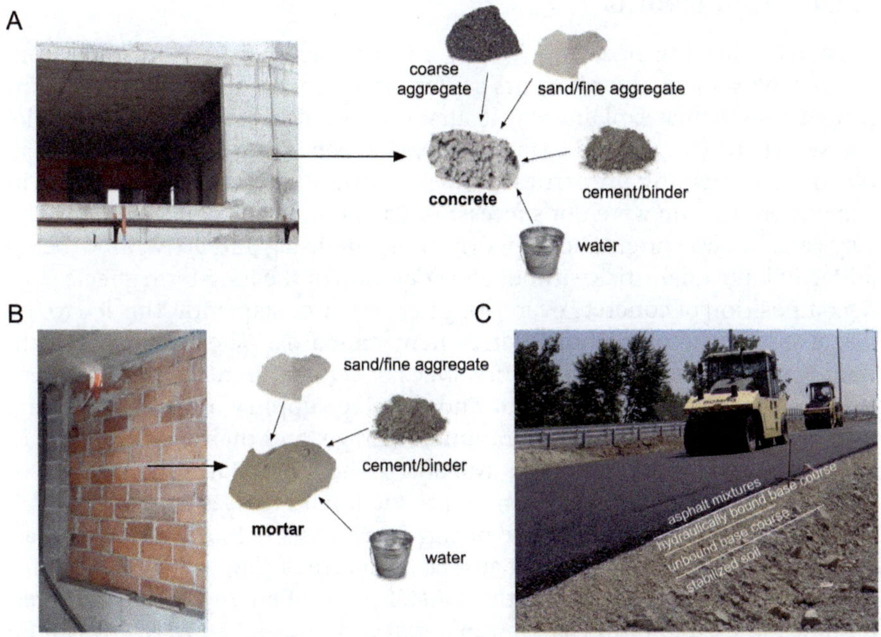

Figure 7.1 Civil engineering structures and materials containing slag. (A) Reinforced concrete wall with concrete components. (B) Masonry with wall mortar components. (C) Road structure with marked pavement layers (photos: Ivanka Netinger Grubeša, Ivana Barišić).

There are several associations in the world dealing with the reuse of ferrous slag: the Australasian (Iron and Steel) Slag Association (https://www.asa-inc.org.au) operating in the area of Australia, the European Slag Association (EUROSLAG; https://www.euroslag.com) operating in Europe, the Cementitious Slag Makers Association (https://ukcsma.co.uk) and the Quarry Products Association (https://www.mineralproducts.org) operating in the UK, the National Slag Association (http://nationalslag.org) and the Slag Cement Association (https://www.slagcement.org) operating in the USA, and the Nippon Slag Association (https://www.slg.jp/e/index.htm) operating in Japan. The main focus of most of these organizations is ferrous slags used in technical applications; however, the European Slag Association and the Nippon Slag Association provide detailed statistical data on the use of each type of slag in civil engineering.[8,9] Production and use of BF and steel slags in Europe and Japan, according to the latest available data, are given in Figures 7.2 and 7.3, respectively.

An explanation of the technical terms (glossary) and a list of abbreviations used in this chapter are provided at the end of this chapter. Technical terms are marked by italic letters when they are first used in the text.

7.2 Slag Application in the Cement Industry

7.2.1 Applications

Granulated blast furnace slag (GBFS), a by-product of iron production, consists of the same major elements as Portland cement, although in different proportions, which explains why its first use was in the cement industry. The glassy structure of ground granulated blast furnace slag (GGBS) is responsible for its cementing properties. However, early attempts at using BF slag in cement production were not successful. The poor performance of the initial slag cements was often due to poorly granulated slag and *accelerators* being added in large quantities without consideration of the long-term effects (*e.g.*, decomposition of concrete over a long period).[6] Consequently, this led to the use of any slag being prohibited in cement standards.[6] The reasons for abandoning the production of these first types of cement in most countries were the inability of manufacturers to find suitable supplies of granulated slag, the sensitivity of slag to deterioration in storage, and their low strengths in comparison to Portland cements available at the time.[6] Currently, a wealth of research exists on slag use in cement production. Processes have been developed for production and the use of slag in cements is more advanced than its initial use. Probably the most attractive feature of slag use in cement production is related to savings in the consumption of energy usually required for limestone exploitation as a raw material and, consequently, savings in the cement production costs. According to the Slag Cement Association,[10] 4.57 MBTU (MBTU = 1000 British thermal units) is required to produce 1 tonne of Portland cement while 0.62 MBTU is required to produce 1 tonne of slag cement.

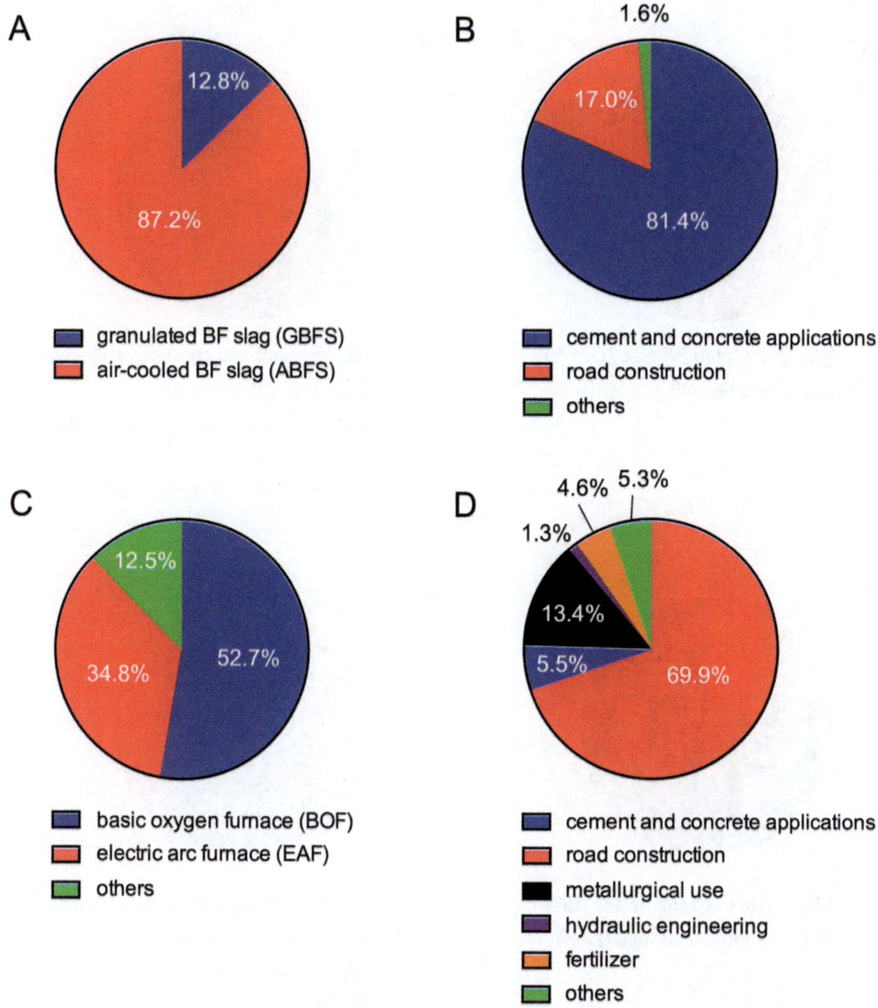

Figure 7.2 Production and use of blast furnace (BF) and steel slags in Europe in 2018. (A) Distribution of BF slags by slag type. (B) Distribution of BF slags by specific areas of their application. (C) Distribution of steel slags by slag type. (D) Distribution of steel slags by specific areas of their application.[8]

Currently, cement produced from BF slag is known as "slag cement." This term usually refers to the use of either combinations of Portland cement and GGBS or GGBS alone in cement. All cements containing GGBS covered by current European cement standards along with their properties and applications are given in Table 7.1.

The use of steel slag as a raw material for the production of *cement clinker* has also been recorded.[19] The presence of tricalcium silicate (C_3S), dicalcium silicate (C_2S), tetracalcium aluminate ferrite (C_4AF), and dicalcium ferrite

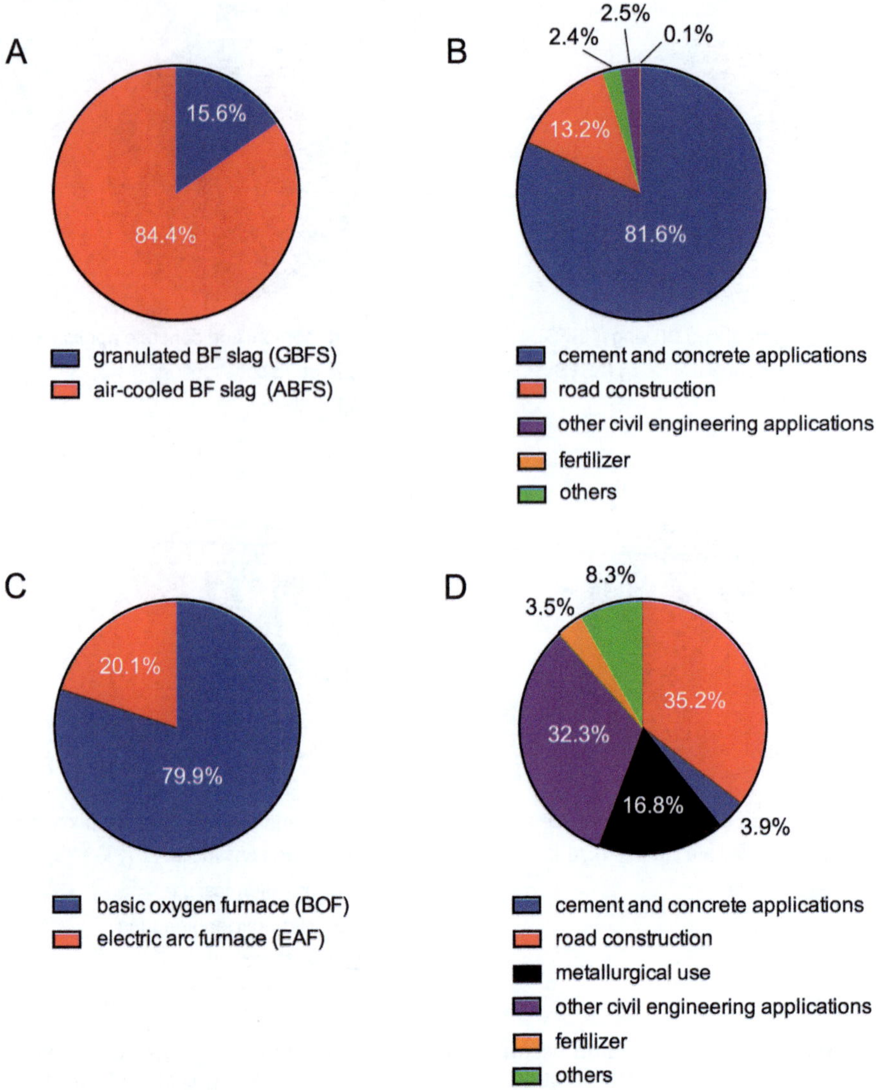

Figure 7.3 Production and use of blast furnace (BF) and steel slags in Japan in 2017. (A) Distribution of BF slags by slag type. (B) Distribution of BF slags by specific areas of their application. (C) Distribution of steel slags by slag type. (D) Distribution of steel slags by specific areas of their application.[9]

(C_2F) provides steel slag with cementing properties. However, the tricalcium silicate content of steel slag is much lower than that in Portland cement, which explains the prevailing opinion that steel slag used for raw mix preparation leads to inferior-quality clinker. Comparative chemical analysis of Portland cement clinker obtained by conventional charging furnace and

Table 7.1 Blast furnace (BF) cements covered by European cement standards and their properties and applications.

Relevant standard	Designation	Notation	BF slag content (%)	Cement properties	Cement uses
EN 197-1 : 2011 [11]	Portland-slag cement	CEM II/A-S	6–20	• High resistance to freeze–thaw cycles, chemicals, and seawater[14,15] • Sets more slowly than ordinary Portland cement[16] • Lower heat of hydration than ordinary Portland cement[16]	• Concrete structures that require high durability[15] • Concrete structures where quick setting is not required[16] • Very low heat cements are typically used for foundations[17] and grouts for encapsulating nuclear waste[18]
		CEM II/B-S	21–35		
	BF cement	CEM III/A	36–65		
		CEM III/B	66–80		
		CEM III/C	81–95		
	Composite cement	CEM V/A	18–30		
		CEM V/B	31–49		
EN 14216 : 2015 [12]	BF cement of very low heat	VLH III/B	66–80		
		VLH III/C	81–95		
	Composite cement of very low heat	VLH V/A	18–30		
		VLH V/B	31–50		
EN 15743 : 2015 [13]	Supersulfated cement	SSC	≥75	• Excellent resistance to aggressive agents (*e.g.*, seawater, sulfates, chlorides, alkali hydroxides, and weak acids)[6] • More susceptible to carbonation during storage and requires extra care during the initial curing period to keep the surface moist compared with ordinary Portland cement[6]	• Seawater works, concrete pipes exposed to aggressive groundwater, or in chemical plants[6]

clinker obtained by replacing the raw mix with 10.5 wt% electric arc furnace (EAF) slag, from scrap smelting in electric arc furnaces, showed that the presence of steel slag in the furnace did not significantly influence the mineralogical composition of cement clinker (C_3S, C_2S, tricalcium aluminate (C_3A), and C_4AF were formed in both clinker types) or its *microstructure*.[19] Furthermore, the *grindability*, *setting time*, *compressive strength*, and *soundness of cement* containing EAF slag are similar to those of the reference Portland cement.[19] Chen *et al.*[20] concluded that with low addition of desulfurization slag (*i.e.*, slag from the desulfurization process in steelmaking, EAF slag), the resulting clinker had a mineralogical composition similar to that of the reference clinker; the addition of desulfurization slag can lower the clinkerization temperature and increase the amount of Ca_3SiO_5 at 1300 °C, which may be beneficial to energy conservation in clinker burning. According to Bernardo *et al.*,[21] replacing commonly used materials (*e.g.*, limestone and clay) with 20 wt% EAF slag gives better raw mix burnability in clinker production. With the use of 6 wt% and 14 wt% fine fraction (<160 μm in diameter) ladle slag (*i.e.*, from refining stainless steel) in the production of cement clinker, cement with a slightly lower water demand for standard *consistency* and delayed early and late cement binding was obtained.[22] Despite the high content of calcium oxide (CaO) and magnesium oxide (MgO) in steel slag presenting potential volume instability problems, cement clinker with ladle slag seems to have acceptable soundness test results[22] and will not cause *volume changes* in cement composites. The compressive strength development (increase) with time (*e.g.*, 2, 7, and 28 day duration) for steel slag clinker cement is similar to that for ordinarily produced cement clinker.[19,22]

Although researchers are more focused on the use of steel slag in clinker production than on the use of BF slag, a few papers examine the use of both slag types for this purpose. Konovalov *et al.*[23] confirmed that both slag types can be used in cement clinker production, whereby the combination of BF slag and open-hearth furnace (OHF) slag is able to completely replace the use of clay in cement clinker production.

7.2.2 Environmental Implications

The utilization of slag in cement production can result in significant environmental footprint reduction. At the same time, the quality of the final products would not be compromised. A reduction in the amount of slag disposed of at a site can be attained. However, the most significant environmental impact is in the energy consumption and gas emission reduction – the slag does not need to be mined but only requires grinding, contributing to energy savings and gas emission reductions. According to the National Slag Association,[24] the reduction in the energy required to produce cement is 28% when 40 wt% of cement clinker is replaced with BF slag. Furthermore, the reduction can reach 42% when 60 wt% of cement clinker is replaced with BF slag. Iacobescu *et al.*[22] calculated the CO_2 emission reduction based on the chemical composition of the raw materials to be approximately 5% when 6 wt% ladle slag is used and 11% for 14 wt% ladle slag, compared to the reference sample (ordinary Portland cement). Osmanović *et al.*[25] studied the effect of replacing cement clinker with GBFS on CO_2, NO_x, SO_x, and CO emissions; they concluded that increasing the proportion of GBFS leads to a sharp reduction in the emissions of these gases.

7.3 Slag Application as a Binder in Cement Composites

7.3.1 Applications

Slag has been investigated as an independent binder, where its binding properties are enhanced by *alkali activator*s. Slag in this case is called alkali-activated slag or alkali-activated slag cement. In general, alkali-activated cement is a system in which an alkaline activator promotes one or more reactions in a natural or artificial inorganic solid, generating a material with cementitious characteristics. One such material is alkali-activated slag cement, which is produced by mixing GBFS and alkaline substances. Among all the potential activators, NaOH, Na_2CO_3, $Na_2O{\cdot}nSiO_2$, and Na_2SO_4 are the most widely available and economical.[15,26]

The *workability*, *durability*, and strength of alkali-activated slag are affected by factors such as the type and concentration of the alkaline activator, slag type and *fineness*, *curing* conditions, activating solution/slag ratio, and the use of admixtures and fibers.[27] Increasing the amount of activator accelerates the strength development and increases the 28 day strength of the paste.[28] According to Bakharev *et al.*,[29] heat treatment has a significant accelerating effect on the strength development of GGBS pastes: a higher curing temperature will positively influence the strength development. Burciaga-Díaz *et al.*[30] concluded that the highest strengths of GBFS mortars can be achieved either with a higher concentration of alkali solution at a lower temperature or with a lower concentration of alkali solution at a higher temperature. These conditions resulted in strengths of at least 6 and 20 MPa at 1 and 28 days, respectively, which are relatively high values for mortars.[30]

Research on alkali-activated slag has intensified remarkably in recent decades because of its numerous advantages compared with Portland cement, such as rapid strength development, high resistance to chemical attacks, and low heat release.[28] In addition, alkali-activated binders are generally believed to provide good *fire resistance* owing to their ceramic-like properties. The results of Abdel-Ghani *et al.*[31] showed the high stability and fire resistance of an alkali-activated binder formed by activating air-cooled blast furnace slag (ABFS) with 3% NaOH and 3% Na_2SiO_3 at up to 500 °C, while an alkali-activated binder formed using 3% NaOH (different activator) behaved similarly to Portland cement. Karahan *et al.*[26] studied the fire performance of a Portland cement mortar and alkali-activated GGBS mortars at 1000 °C. The reduction in compressive strength of the Portland cement mortar was 94%, whereas that of the alkali-activated GGBS using NaOH, Na_2CO_3, and Na silicates was in the range of 63–88%. According to Palacios *et al.*,[32] alkali-activated GGBS concretes and mortars have a lower total *porosity* and a larger fine pore structure than the respective Portland cements, which suggest lower permeability and subsequently higher durability of alkali-activated mortars/concretes. The high durability of alkali-activated GGBS binder was confirmed by Bernal *et al.*[33] Despite the promising advantages of alkali-activated slag over other conventional binders, its production and use are being hindered by the following: quick setting time and poor workability,[31] high *shrinkage*,[32,34] complicated chemistry, varying slag composition, and lack of a standard, which is probably the most significant reason.[35] Publishing a European standard for alkali-activated binders will have a positive effect on the wider use of this binder in the European Union (EU). A significant step forward was made at the global level when the British Standards Institute released a Publicly Available Specification, PAS 8820.[36] PAS 8820 defines a performance-based testing approach, allowing the use of innovative materials without imposing a set of restrictions on cement selection or mix design, as is the case in the current standards BS 8500[37] or EN 206.[38] PAS 8820[36] specifies that a candidate alkali-activated concrete should be tested in parallel with a reference concrete designed to meet the requirements of BS 8500[37] for the intended exposure class. To be acceptable for use, the candidate alkali-activated material must meet or exceed the performance of the reference concrete.

7.3.2 Environmental Implications

Since slag is an industrial by-product, the use of alkali-activated slag has many environmental advantages, such as low energy cost, use of secondary raw material, and low pollutant gas emissions.[28] These advantages make it preferable to conventional binders in practice. One of the major newer applications of alkali-activated slag is in waste management, including radioactive waste management and immobilization of toxic metals.[39]

7.4 Slag Application as Partial Cement Replacement in Cement Composites

7.4.1 Blast Furnace Slag Application

Because of its latent hydraulic properties, the use of GGBS as partial cement replacement in cement composites is beneficial owing to its fresh and hardened properties. Because of their higher fines content, low-slag-content concretes are easier to finish; however, setting times can be extended with high levels of cement replacement by slag and at low ambient temperatures,[40,41] but extra care may be required to guard against moisture loss.[40] GGBS causes slower strength development under standard curing conditions because of its low initial rate of hydration.[42] Low *early strength* development of GGBS can be beneficial owing to the low *heat of hydration*, which can decrease the thermal cracking of mass concrete.[40,43] If thermal cracks form, the crack would allow the ingression of aggressive agents, such as chlorides, carbonates, and sulfates, and can lead to corrosion of reinforcement bars and degradation of concrete. This slower strength development (the exception here is GGBS with high alumina content) explains why GGBS is not recommended for use in applications where high early-age strength is required, as previously mentioned. However, the early-age strength development can be influenced by higher temperatures, that is strength gain is much more rapid, even with an increase in curing temperature of only 10 °C.[41] The early and *ultimate strengths* of concrete with GGBS can also be positively influenced by curing in a high moisture environment.[40,43] For the same water–cement ratio, concrete containing GGBS as a partial cement replacement achieves a higher ultimate strength than concrete made of Portland cement.[41,44] According to the results presented by Yeaua *et al.*,[45] up to 55 wt% replacement of cement with GGBS results not only in improved *mechanical properties* but also in higher durability (expressed through chloride *diffusion coefficients*) of concrete. The positive effect on concrete durability results from the slag's latent hydraulic properties; secondary hydration reduces concrete porosity by blocking the capillary pores, resulting in lower permeability and diffusion coefficients than in ordinary Portland cement concrete.[42] This also makes the concrete microstructure denser (less porous), decreasing the penetration and diffusion of chloride ions and consequently achieving much longer transition periods and much lower chloride ion concentrations.[45] Pores become finer and the change in the mineralogy of cement as it hydrates leads to a reduction in the mobility of chloride ions.[46] In cement pastes with a high percentage of GGBS, self-healing of cracks can be achieved more efficiently than in ordinary Portland cement concrete, again contributing to durability.[47] This enhanced durability of concrete with BF slag favors its use in special fields of application, such as in constructing waterproof basements, dams, marine structures, bridges, telecommunication towers, and containment structures. A considerable improvement in durability against acid attack, water permeability, and chloride attack when cement is partially replaced by GGBS was reported by Bhat *et al.*[48] Lubloy *et al.*[49] heated conventional concrete and concrete with GBFS in different amounts to up to 800 °C and the residual compressive strengths were compared. According to the results, the residual compressive strength of concrete containing GBFS was significantly higher than that of the reference concrete. The authors attributed such results to the lower calcium content present in slag cement, leading to a lower amount of the hydration product portlandite ($Ca(OH)_2$). It is known that portlandite usually decomposes at high temperatures, causing lower fire resistance of concrete. In contrast to the study by Lubloy *et al.*,[49] the concrete with GGBS studied by Li *et al.*[50] experienced a slightly higher mass loss, lower residual compressive strength, and reduced *modulus of elasticity* when heated to 700 °C, which means that it has lower fire resistance than conventional concrete.

GGBS can replace up to 70 wt% of Portland cement in cement composites.[51] At higher percentages of cement replacement with GGBS (greater than 70 wt%), lower resistance to *carbonation* was recorded, possibly owing to the effect of the reduced amount of portlandite.[52] In the presence of carbonation, scaling can occur under the combined stressors of freezing–thawing and de-icing salt. Subsequently, other attacks on the structure are more likely to dramatically decrease the durability of concrete.[43] Owing to the reduction in alkalinity arising from carbonation, corrosion is initiated in steel rebar. According to Song *et al.*,[46] addition of GGBS causes a reduction in the pH value of the binder but does not adversely affect the concrete corrosion resistance. The results presented by Song *et al.*[46] showed that replacement of cement by 40 wt% GGBS had no significant influence on the corrosion rate of rebar in concrete; however, replacement by 60 wt% significantly reduced the corrosion rate. In contrast, according to the data presented by Yeaua *et al.*,[45] replacement of cement by 40 wt% GGBS or higher can increase the corrosion resistance of a steel bar by more than two times. When using BF slag in place of cement, providing a concrete cover of sufficient thickness to delay chloride ion attacks in an aggressive environment and protect against corrosion is advised.

7.4.2 Other Types of Slag Application

A sparsity of research exists dealing with the application of other types of slag besides BF slag as a partial replacement for cement in cement composites. According to Liu *et al.*,[53] cement replacement by steel slag powder results in a more workable concrete that is generally much easier to compact by vibration. Liu *et al.*[54] noticed that partial replacement of cement by steel slag decreases the strength of the concrete but increases the ratio of flexural to compressive strength and thus improves the ductility of cement-based materials. Cho *et al.*[55] used ferronickel slag powder to partially replace cement and detected a decrease in the compressive strength of the concrete; however, when ferronickel slag powder was mixed with fly ash in a 50:50 ratio, the compressive strength of such concrete was higher than that of the reference concrete.

7.4.3 Environmental Implications

The embodied carbon footprint of different mortar mixtures was evaluated by Onn *et al.*[51] through the global warming potential (kg CO_2 eq.) per volume (m^3) of each mixture. It was concluded that a 10 wt% increase in GGBS content in a mortar mixture decreases global warming potential by 10%. Namely, the use of waste materials in general reduces greenhouse gases emission and greenhouse gases are responsible for the absorption of infrared radiation, which would further lead to trapping heat and holding it in the atmosphere, thus contributing to global warming. It was also indicated that a higher cement replacement level with GGBS of up to 70 wt% will provide a better compressive strength/carbon footprint ratio, whereas cement replacement with up to 50 wt% BF slag is recommended when cost efficiency is considered. The calculated environmental impact per tonne of concrete, relative to a reference Portland cement concrete for concretes of the same 28 day strength shown by the Cementitious Slag Makers Association,[56] indicated that replacing 50 wt% of Portland cement with GGBS results in a 40% reduction in the carbon dioxide emissions, a 29% reduction in the primary energy use, and an 8% reduction in mineral extraction.

7.5 Slag Application as Aggregate in Cement Composites

If slag is used as an aggregate, special attention should be paid to its mineralogy owing to the potential for some slag phases to expand. Potential volume changes in steel slags are attributed to the content of free CaO and MgO.[2,57] Calcium oxide and MgO under the influence of moisture, leading to volume changes. In particular, CaO hydrates rapidly, causing volume changes in the first few weeks. Magnesium oxide hydrates slowly, contributing to long-term expansion, which can extend to several years, even when using aged slag.[57] Before using slag as an aggregate, verification of its volume stability is required in accordance with the current standard. Slag aging by exposure to atmospheric conditions for long periods (weathering) and occasional mixing of disposed slag are the simplest methods of controlling volume stability, but they do not necessarily ensure stability.[58] Aging by steam was developed in Japan, which consists of covering the steel slag with a sheet, injecting steam for a period of 48 h, uncovering the slag, and cooling by natural ventilation; the entire procedure lasts 6 days.[58] Another method of accelerated aging involves exposing the steel slag to steam at a pressure of 0.5 MPa for a period of 3 h.[58] As described by Pang *et al.*,[59] the expansion of BOF slag can be reduced by carbonation for 4 h, which will effectively reduce the free CaO content. The carbonation process described by Pang *et al.*[59] is conducted in a carbonation reactor that is sealed at 70 °C and vacuumed to −0.3 MPa before carbonation. Then, CO_2 is introduced into the reactor until the pressure reaches 0.3 MPa. After the carbonation treatment, the amount of portlandite decreased while the ratio of calcite to portlandite increased, indicating the reaction of $Ca(OH)_2$ with CO_2. When the slag volume is stable, it can be used as an aggregate in cement composites.

7.5.1 Blast Furnace Slag Applications

There are many studies using BF slag as a replacement for commonly used aggregates in cement composites. Nataraja *et al.*[60] and Nadeem *et al.*[61] observed that increasing the amount of GBFS used as a fine aggregate replacement (natural sand) results in a decreased mortar workability. In contrast, Pacheco *et al.*[62] reported that using BF slag as a coarse aggregate in concrete reduces concrete workability slightly. The difference in the size of aggregates studied by Nataraja *et al.*[60] (sand was much coarser than fine slag aggregate) was greater than the difference in size of aggregates studied by Pacheco *et al.*[62] (coarse ordinary aggregate and coarse slag of similar grain size), explaining the difference in the results of these two studies. Additionally, finer material requires more water for wetting the surface. Concerning mechanical properties, Nadeem *et al.*[61] noted that the replacement of up to 75 wt% fine aggregates with GBFS increases the compressive and split *tensile strengths* of the mortar. The improvement in compressive strength with the increase in the quantity of sand being replaced by BOF slag was recorded by Ouda *et al.*[63] as well: the strength increased proportionally with the percentage of slag in the mortar mixture. BF slag, used as a coarse aggregate in concrete, increases the compressive strength, whereas the splitting tensile strength and modulus of elasticity are similar to those of conventional concrete.[62] Self-compacting concrete with BF slag used as a fine aggregate shows an increase in compressive strength at the age of 365 days.[64] Higher drying and chemical shrinkage of self-compacting concrete were recorded by Valcuende *et al.*,[64] and the rates increased with the percentage of aggregate replaced by GGBS; this was attributed to the higher porosity of slag and slag's reactivity. Regarding durability, Humam *et al.*[65] concluded that the partial replacement of fine aggregates with BF slag can significantly enhance the sulfate resistance and reduce rapid chloride permeability as a result of the improvement in the microstructure of the mortar mixes. A reduction in chloride ion penetration in mortar using BF slag as a fine aggregate was confirmed by Takanashi *et al.*,[66] who also noted an improvement in freezing–thawing durability under a salty environment. Concrete containing BF slag sand shows higher resistance to sulfate attack[67] because BF slag fine aggregate prevents accumulation of calcium hydroxide around the fine aggregate particles and the formation of ettringite ($Ca_6Al_2(SO_4)_3(OH)_{12}{\cdot}26H_2O$). According to the same authors,[67] the BF slag concrete achieved higher resistance to freeze–thaw cycles and slower progression of carbonation compared to the reference concrete. A slightly lower degree of damage was recorded in concrete mixtures with GBFS as a fine aggregate when exposing concrete to high temperatures in comparison to the reference mixture.[68]

7.5.2 Steel Slag Applications

The high hardness of steel slag motivated its possible application as a concrete aggregate and many studies have been conducted with this purpose. In the study by Qasrawi *et al.*,[69] steel slag was used as a partial or complete replacement for the fine aggregate fraction. As expected, because steel slag has a higher density than natural aggregates, an increase in slag content caused a density increase in the fresh concrete mix. The resulting density with steel slag aggregate application in concrete can be 5–20% higher than that of conventional concrete, which is considered as one of the most important characteristics of steel slag aggregate concrete.[70] Furthermore, increasing the proportion of the fine fraction of steel slag in concrete decreases its consistency.[69,71,72] In terms of mechanical properties, the replacement of the fine aggregate fraction with up to 50 wt% steel slag achieves approximately the same or even higher values of compressive and tensile strengths, while concrete made of only steel slag aggregate attains lower compressive and tensile strengths than the reference concrete.[69] The positive effect of steel slag on the mechanical properties of concrete is more pronounced in the lower *concrete strength class*. The decline in compressive strength with more than 50 wt% fine steel slag aggregate is due to the greater amount of fine material (particles/grains smaller than 0.15 mm in diameter) in slag compared to sand. A higher amount of slag in concrete increases the total amount of fine material, requiring additional cement to coat the grains. Eliminating excessive amounts of fine particles in slag and replacing the fine aggregates with steel slag has a positive effect on the compressive strength at all maturing stages of concrete. The strength of concrete containing steel slag is not as high compared to that of conventional concrete but the target strength is achieved at the end of 28 days, showing reliability.[71] In a study by Maslehuddin *et al.*,[73] the effect of partial replacement of the coarse aggregates in concrete by steel slag was examined. All tested mixtures were made with the same amount of cement and the same water:cement ratio. Using the same coarse and fine aggregate ratio, the steel slag concrete attained a slightly higher compressive strength and a slightly lower tensile strength compared to the reference concrete. Increasing the coarse steel slag portion in the total mass of aggregate increases the compressive and tensile strengths of concrete.[73,74] The *ultrasound pulse velocity* increases with the increase in slag proportion, suggesting a greater modulus of elasticity of concrete with slag as an aggregate compared with the reference concrete.[73]

When analyzing the effect of steel slag aggregate on the mechanical properties of concrete (*i.e.*, the compressive and tensile strengths) the *interfacial transition zone (ITZ)* between the aggregate particles and the cementitious matrix needs to be considered. This ITZ is regarded as a weak zone of concrete. In this zone, the appearance of *microbleeding* around the aggregate particles, porosity, and some microstructural features depends on several factors, such as aggregate quality and size, water:cement ratio, binder, and the age of the mix.[70] In addition, the morphology (quality) of ITZ plays an important role in concrete permeability and durability. The presence of steel slag in concrete creates different ITZ morphologies. In particular, the smaller and less hollow ITZ in steel slag aggregate concrete compared to that of natural rock aggregate concrete is due to slow migration of CaO from the core of the steel slag grains to its surface, resulting in chemical evolution of CaO to calcium carbonate ($CaCO_3$).[70] Owing to the stronger ITZ in steel slag concrete, better concrete mechanical properties will be achieved. The method of enhancing slag aggregate quality in terms of improved ITZ is by accelerated carbonation process treatment of steel slag aggregate.[59] This process creates changes in the aggregate pore structure. It reduces pores with diameters larger than 1 μm and increases pores with diameters less than 1 nm.[59] After carbonation, the steel slag aggregate water absorption rate decreases while its impermeability increases. The strengths of carbonated steel slag and natural aggregate as well as the strengths of concrete prepared with these aggregate types were compared by Pang *et al.*[59] Although the crushing strength of carbonated slag aggregate was lower than that of natural aggregate, the compressive strength of concrete made with carbonated steel slag aggregate was improved. Sezer *et al.*[75] reported that the replacement of natural aggregate with coarse steel slag will impact the compressive strength positively, whereas the replacement of natural aggregate with fine steel slag will have a negative effect. The mechanical properties (compressive strength, *flexural strength*, and modulus of elasticity) of concrete with steel slag as a coarse aggregate are slightly worse than those of concrete with natural aggregate.[76]

The *absorption* and porosity of a mixture decrease with the increase in coarse slag portion, suggesting possible better durability of such concrete.[73] In Maslehuddin *et al.*,[73] a substantial increase in the corrosion resistance of steel embedded in concrete containing slag compared to the reference concrete was found and steel corrosion initiation was more delayed with increasing slag aggregate proportion. Concrete containing EAF slag accelerates corrosion of reinforcing steel embedded in it but does not show potential for *alkali–silica reaction.*[77] Moreover, concrete with one of two different steel slags as coarse aggregates produces a negative effect on corrosion, which potentially indicates an increased risk that corrosion of reinforcement will occur if that slag is used to prepare concrete that will be exposed to aggressive conditions, such as a marine environment.[76] The water penetration depth is greater and the *freeze–thaw cycle resistance* is better in concrete containing steel slag as a coarse aggregate than in concrete containing steel slag as a fine aggregate.[75] Better durability, in terms of chloride penetration resistance, of concrete containing steel slag aggregate as the coarse fraction was confirmed by Sezer *et al.*[75] Slag as an aggregate may be expected to improve the fire resistance of concrete because it is generated at high temperatures. However, concrete with EAF slag as a coarse aggregate experienced a higher damage degree than natural aggregate concrete when exposed to temperatures of up to 1000 °C.[78] Specifically, the most commonly used aggregates experience expansion during heating, while cement paste experiences shrinkage at the same time[75] and differences in the thermal behavior of the aggregate and cement paste result in the build-up of stresses, which can lead to weakening and disruption of concrete at high temperatures.[79,80] The difference between the thermal behavior of steel slag and cement paste is even higher than that between natural aggregate and cement paste, which causes a higher damage degree in concrete containing steel slag as an aggregate during exposure to high temperatures. When steel slag was heated to a target temperature of 1000 °C before being incorporated in concrete, the damage was diminished. This indicates higher fire resistance of concrete with thermally treated slags (EAF slag) in the lower temperature range (up to 400 °C) than that of concrete with natural aggregate.[78]

7.5.3 Base Metal Slag Applications

Copper (Cu), lead (Pb), zinc (Zn), tin (Sn), nickel (Ni), and ferronickel slags can also be used as aggregates in cement composites, bearing in mind their possible limitations. Namely, the major problem in using non-ferrous slags as a secondary material in road construction and concrete preparation is the potential presence of high amounts of toxic elements, such as Pb, Zn, and Cu, which may influence cement hydration and may leach when in contact with water.[81]

When sand in concrete was replaced with Cu slag, at the same workability, there was a general increase in the compressive strength as the Cu slag proportion increased.[82] Saikia *et al.*[81] concluded that the addition of Pb slag as a partial substitute for natural fine aggregate increases the workability of fresh mortar mixtures as well as the mortar setting time. Partial replacement of sand by Pb slag decreased the compressive strength of concrete in the study by Knežević *et al.*,[83] while there was no unambiguous conclusion on the effect of using Pb slag as a substitute for sand (up to 35 wt%) on the compressive and flexural strength as well as on the modulus of elasticity of mortar according to Saikia *et al.*[81] However, when the sand in concrete was replaced with granulated Pb–Zn slag, the compressive strength of concrete was higher than that of ordinary concrete.[84] According to the results shown by Mosavinezhad *et al.*,[85] replacing sand in the amount of 30 wt% by Zn slag negatively influenced the compressive and flexural strengths of concrete, whereas replacing sand by Pb slag kept the mechanical properties at the same level as in the reference concrete. Total replacement of sand by Sn slag did not influence the compressive strength of concrete, whereas the replacement of sand with Sn slag by 20 wt% did increase the compressive strength.[86] The use of Ni slag at up to 40 wt% as a sand replacement in concrete produced a positive effect on the compressive strength.[87] Additionally, replacement of sand with ferronickel slag in mortar at up to 50 vol% also led to an increase in the compressive strength.[88] The chloride ion penetrability of concrete samples with a mixture of ferronickel and BF slag as the fine fraction aggregate was lower than that of the reference sample and the carbonation depth of the samples decreased with the increase in the slag replacement ratio.[72]

7.5.4 Environmental Implications

There are many environmental effects of replacing commonly used aggregates with slag. According to Ouda *et al.*,[63] full sand replacement by BOF slag has significant effects on the efficiency of thick shields as it reduces the captured gamma rays better than normal mortar incorporating sand. Furthermore, the typical energy consumption for slag aggregate production is lower than that of natural rock aggregate: the production of aggregates from natural rock requires 50 MJ per tonne, whereas that of aggregates from steel slag consumes 40 MJ per tonne.[89]

Leaching tests of Pb slag and cement mortars with and without Pb slag were conducted by Saikia *et al.*[81,90] to evaluate the environmental suitability of this material for application as a constituent in cement mortar and concrete. Leaching tests were conducted for slag according to EN 12457-2[91] standard and for mortar with slag as fine aggregate according to EN 12457-2[91] standard and NEN 7345[92] standard. Lead slag leaches Pb, selenium (Se), and antimony (Sb) in significant amounts while the concentrations of the majority of toxic elements (except Cu, Pb, and Zn) in the leachates generated from mortar containing Pb slag were significantly lower than those present in the leachate generated from slag alone.[81,90] Saikia *et al.*[81] explained that the lower leaching of some toxic elements from cement mortars containing slag compared to slag alone observed during EN 12457-2 testing is due to the various physicochemical processes that occur in cement hydration while the observed higher leaching of Pb, Zn, and Cu is due to dissolution of their metal hydroxides, formed during cement hydration. The cumulative leaching amounts of elements from cement mortars with Pb slag after immersion in leachate for 64 days obtained according to NEN 7345 showed that all toxic elements in the leachates were significantly lower than standard requirements and that there is no environmental restriction on the use of slag in the investigated mixes of cement mortars and concrete. Toxic metals present in slag become stable if slag is bound in cement composite.

The greatest environmental effect of using slag in cement composites is in the preservation of natural resources. In particular, aggregates occupy 70–80% of the total volume of concrete;[89] replacing natural aggregates with slag aggregates will significantly contribute to the preservation of natural resources and reduce the disposal of this type of waste.

7.6 Slag Utilization in Road Construction

Table 7.2 summarizes slag applications in pavement structure. Research results for slag application in road construction will be presented within the next few subsections.

7.6.1 Soil Stabilization

In addition to their application in cement and cement composites, the *latent hydraulic properties* of BF slag and alkaline activation are also exploited for *soil* stabilization in earthworks in road and railway construction. Sulfate in soil is known to cause serious swelling problems in stabilized clay, which is associated with the formation of ettringite. In South Africa and Australia, GGBS cements and a combination of lime and GGBS are commonly used owing to their resistance to expansion and swelling.[93] Using combinations of lime and GGBS for soil stabilization resulted in a slower rate of early strength development, prolonged time for construction works (particularly compaction and levelling after slag addition), ability to self-heal in the case of early-life

Table 7.2 General evaluation of slag application in road construction.

Slag use	Advantages	Disadvantages	Literature
Soil stabilization (embankment, subgrade, subbase)	Increasing the shear strength, stiffness, compressive strength, angle of internal friction, decreased settlement	Potential source of toxic substances released into surface and groundwater	Higgins,[93] Mäkikyrö,[94] Corrêa-Silva *et al.*,[95] Salimi *et al.*,[96] Mahedi *et al.*[102]
Unbound base course	Increased strength, resilient modulus, early bearing capacity, freeze–thaw resistance, resistance to deflection	Expansive nature	Tikkakoski *et al.*,[105] Shiha *et al.*,[108] Behiry *et al.*[109]
Cement-treated aggregate base course	Increased OMC,[a] MDD,[a] compressive strength, stiffness, freeze–thaw resistance	Compaction difficulties, expansive nature	Singh *et al.*,[112] Autelitano *et al.*,[114] Liu *et al.*[115]
Asphalt	Increased durability and friction coefficient	High affinity for asphalt cement, expansive nature	Chen *et al.*,[122] Emery,[124] Kehagia,[128] Tao *et al.*[129]

[a]OMC – optimal moisture content; MDD – maximal dry density.

overloading damage, long-term increased strength, and inhibited deleterious swelling caused by sulfates or sulfides.[93] GGBS activated by cement has the potential for clay stabilization in *deep stabilization* when the aim is to achieve high *pillar strengths*.[94] GGBS can also be activated with NaOH, resulting in soft soil stabilization with the effect of significantly increasing the *shear strength* and *stiffness* and a stress–strain behavior typical of artificially cemented soils with ordinary Portland cement.[95] Combinations of GBFS and BOF slag activated with CaO and MgO are used for chemical stabilization of soft clays. The produced slag–clay-based geopolymers, particularly using BOF slag, showed improved behavior in terms of decreased settlement and enhanced compressive strength of the soft soil.[96]

Steel slag is also used for soil stabilization in two approaches. First, it is used as a binder, although its *pozzolanic activity* (ability to react with calcium hydroxide) is significantly less than that of GGBS. Stabilization of compacted soils, typically composed of quartz, feldspar, and volcanic glass in different proportions, with steel slag showed increases in soil resistance and stiffness with curing time due to pozzolanic reactions.[97] The second approach is using steel slag as an aggregate in clay soil stabilization, influencing its basic characteristics as a building material: *dry density*, *plasticity*, *swelling potential* as well as *cohesion* intercept decreased, and the *California bearing ratio (CBR)* and *angle of internal friction* increased.[98]

Adding ladle slag to clayey soil results in similar behavior to that achieved through stabilization with lime.[99] Improvements in *plasticity index*, expansiveness, *bearing capacity*, *free swelling*, and durability can be achieved by ladle slag stabilization.[99,100]

During the Pidgeon process to produce Mg, 5–6 tonnes of Mg slag is produced for 1 tonne of Mg, with the estimation of 4 million tonnes of Mg slag production worldwide annually.[101] Due to it having the same physical, chemical and mineral properties as cement,[101] Mg slag has potential for use in the construction industry. The use of Mg slag and cement in clayey sand soil stabilization increases pH, *maximum dry density (MDD)*, *optimum moisture content (OMC)*, uniaxial compressive strength, and modulus of elasticity as well as granulation of soil particles, which represents a material improvement; the optimum contents of cement and Mg slag were recorded as 6% and 20%, respectively, in dry soil mass.[101]

Construction of stabilized soil for embankments or road bases requires addressing its potential environmental risks since these layers are in direct contact with the surrounding terrain and potential contaminants may leach into the groundwater. Generally, slag has been identified as a potential source of toxic substances and may pose environmental risks by leaching metal(loid)s into soil and surface and groundwater.[102] Therefore, in addition to studies on material mechanical characterization, detailed research on potential environmental pollution is necessary when analyzing the potential use of slag in road construction. Usually, for materials intended for road construction purposes, simulated leaching tests, such as the toxicity characteristic leaching procedure (TCLP), are performed,[95,101,102] depending on applicable and available standards. The results of leaching tests conducted on soft soil with the addition of GGBS–NaOH indicate that this material does not pose a risk for soil contamination.[95] Additionally, the results presented by Mahedi *et al.*[102] indicated that an increase in GGBS content in GGBS-cement-stabilized soil may not necessarily increase the effluent metal concentrations. Despite the fact that deposition of Mg slag can lead to the entry of metals into surface water and subterranean water, a TCLP test performed on Mg slag and Mg slag-cement-stabilized soft soil resulted in lower concentrations of metals (cadmium (Cd), chromium (Cr), and Pb) leached from Mg slag-cement-stabilized soil compared to Mg slag, which can minimize environmental damage by Mg slag.[101]

7.6.2 Pavement Unbound Base Course

Based on design consideration, pavements can be categorized into flexible pavements (surfaced with asphalt wearing course) and rigid pavements (surfaced with concrete wearing course) (Figure 7.4). Common to both systems is the construction of an unbound base course and a subbase course (Figure 7.4), which are mostly made of natural materials. Since the thickness of these layers is substantially greater than that of other layers (20 to 45 cm compared to 2 to 8 cm for the asphalt wearing course), the use of alternative materials significantly contributes to environmental conservation. An overview of slag utilization in unbound pavement base courses (UBC) is presented in Table 7.2.

Figure 7.4 Basic pavement structure types.

One of the first steel slag applications in road construction was for the building of these layers, initially without any wearing course. The first slag road was built in 1830 in the USA, in 1866 slag was used as a base for a macadam-type pavement in Chicago, and the first utilization of slag in concrete pavement was in 1912 in Ohio.[103]

The basic advantage of using GBFS in UBC is its rough surface, high stability, and high friction angle (40° to 45°), which result in a highly stable layer.[104] The angularity and high friction angle of GBFS contribute to high bearing capacity with a CBR greater than 100%, which allows good load transfer for pavements placed on soft, weak subgrades. Combining BF slag and ladle slag results in an enhanced road construction material with increased strength (uniaxial strength and CBR value) and freeze–thaw resistance.[105]

The construction of UBC using steel slag as a replacement for natural aggregate has multiple advantages. Similar to the utilization of GBFS, angular steel slag grains are prone to "fixing" (rearranging grains in a more compact form) after installation and compaction, forming a very solid, compact, and durable surface that can accommodate heavy traffic loads.[106] Similarly, UBC containing steel slag can be built on weak soils since good grain interconnectivity provides good load transfer to the formed soil and the occurrence of heavy rain during construction does not affect the quality of the built layers.[107] The increase in CBR, shear strength, and *resilient modulus* for UBC with EAF slag is a result of its rough surface compared to limestone aggregate.[108] Another advantage is carbonate hardening due to the free CaO in steel slag. This results in a higher bearing capacity immediately after compaction compared to crushed gravel,[106] which meets the requirements for pavement sensitivity to external influences during construction due to moving heavy machinery, which could cause damage to the freshly installed course before the pavement is finished. Steel slag in UBC leads to increases in the MDD, CBR, resilient modulus, resistance to deflection, and *vertical strain* with optimal replacement of limestone aggregate by 70 wt%.[109] By comparing different types of slags and natural aggregates, Gökalp *et al.*[110] concluded that slags (EAF and ferrochromium slag) displayed higher water absorption rates and dry unit weights. In addition, the slags have superior resistance to abrasion, fragmentation, weathering, and polishing compared with natural aggregates.[110] A potential limitation of slag use in UBC is its expansive nature due to free CaO and MgO. The test results presented by Wang *et al.*[111] showed that the allowable free lime content for samples to be safely used as a granular base or subbase under a 10 cm thick upper layer of pavement structure is in the range of 3.7–3.9 wt%.

7.6.3 Pavement Cement-treated Aggregate Base Course

Heavy traffic pavements are usually designed with a bound aggregate layer, also known as a hydraulically bound or stabilized base course, and cement is commonly used as a binder. An overview of slag utilization in cement-treated aggregate base courses is presented in Table 7.2.

Optimal values of MDD and OMC are very important during construction of the cement-stabilized base course (CBC) to obtain a well-compacted and thus good load-bearing layer with less compaction energy. The latent hydraulic properties of GBFS can be used for the construction of these base courses, where the addition of a fly ash–cement mixture results in a well-graded mass, with a higher MDD and lower OMC of the compacted mixture.[112] The reaction of free CaO, reactive silica, and alumina in slag during latent hydraulic reaction and additional formation of calcium–silicate–hydrates (C–S–H) gel results in an increase in the CBR and compressive strength.

Because of its low pozzolanic activity, steel slag cannot be applied as the main constituent of hydraulic binders in road construction, but its free oxides can be used to activate other materials, such as BF slag.[113] Most of the research on slag application in CBC involves slag as an aggregate because of its grain characteristics (*i.e.*, rough surface, angularity, and hardness). The result is high interparticle friction and interlocking in the cementitious matrix, which improves the mechanical properties but creates compaction difficulties.[114]

As an aggregate in CBC, steel slag increases the OMC and MDD because of its higher water absorption compared to limestone aggregate.[115] According to the same research, the stiffness and strength both increase for mixtures with up to 50 wt% slag content in aggregate, creating more hydration products and making the structure tighter and of lower porosity. A further slag content increase results in strength reduction. A similar trend is observed for cement content increase.[116] In mixes with steel slag and gravel aggregate, compressive and indirect tensile strengths increase at 2 wt% cement content. However, the values generally decrease in mixes with higher cement contents (4 wt% and 6 wt%) with increasing slag content.[116] Generally, CBC is characterized by a low cement content and filling the surface voids of porous slag grains with cement paste causes an inconsistent improvement in strength. Similar trends are observed for the measured modulus of elasticity, where the beneficial effect of slag is more pronounced in mixtures with 2 wt% cement; however, there is generally an increase in the static and dynamic elasticity moduli with increasing steel slag and cement content.[117] The addition of steel slag to a gravel aggregate can improve the performance of CBC during freeze–thaw cycles in the presence of NaCl owing to its porous and nonhomogeneous grain surfaces, causing difficulty in sodium chloride (de-icing agent) crystallization.[118]

The potentially expansive nature of steel slag could be a limiting factor for its use in CBC owing to the development of *reflective cracks* on the asphalt wearing course, causing a shorter design life. Comparing the dry shrinkage and temperature shrinkage characteristics of CBC with steel slag to those of macadam aggregate, Li *et al.*[119] concluded that using coarse grade steel slag can effectively reduce dry shrinkage cracks, whereas intermediate grade steel slag is not suitable for use as a CBC aggregate. In Belgium, during the construction of storage plateaus, steel slag from stainless steel production was used as an aggregate for CBC.[120] The volume change was measured and recorded as 2.3% on average, which did not present a major problem in the construction and use of the plateaus.

7.6.4 Asphalt Mixtures

Blast furnace slag is mainly used in concrete for various purposes, including concrete slabs for pavement surface courses. All advantages and disadvantages of this utilization have been analyzed earlier in the text and Table 7.2 summarizes slag applications in pavement structures.

Blast furnace slag can also be used as a coarse or fine aggregate in asphalt mixtures, where its use yields numerous benefits. Although the high affinity for asphalt cement (need for higher binder content during asphalt mix production) is pointed out as one of the negative properties, it also contributes to improved adhesive binding between the slag aggregate and the binder, resulting in excellent *stripping resistance*. Grains of a hard, rough, highly angular, and vesicular surface texture result in a high polished stone value, which contributes to good wearing course frictional resistance.[104] However, BF slag aggregate requires a higher binder content compared to conventional natural aggregate.[104,112] This causes a longer retention time for drying during hot mix asphalt manufacturing and a lack of consistency in the physical properties due to variability in the production process. These can create further problems during asphalt mix design and production, such as *flushing* due to a high binder content or *ravelling* due to a low binder content.[104] Finally, BF slag asphalt mixes can be recommended for use on light traffic pavements and in areas where hard braking and acceleration are expected, such as for *slurry seals*.[104]

The utilization of steel slag in the asphalt industry is more common than that of BF slag, particularly in areas with developed steel industries. Although this application is now well known, during the first construction of asphalt surface layers with steel slag, cracks and surface popouts occurred due to slag instability caused by free oxides of Ca and Mg, and overall because of the lack of careful material selection and improper construction.[58,121] The expansive nature of steel slag is usually addressed by sufficient weathering before its use, as previously discussed. However, fracturing of the calcium carbonate product layers caused by coarse steel slag aggregate weathering is responsible for the poor freeze–thaw cycle damage resistance of asphalt mixtures.[122] Hence, for use in asphalt mixtures, weathering is not the only criterion. In particular, expansion is limited if finer grain size fractions are used. Furthermore, owing to the process of asphalt mix production itself, where the aggregate undergoes phases of washing, sieving, and drying, instantaneous hydration and expansion of slag occur.[123,124] Bituminous film coating of steel slag aggregate (wrapping the steel slag grains with a layer of bitumen during asphalt mix production) also limits its potential for expansion.[124] The presence of free CaO in steel slag is usually highlighted as an undesirable property but it has also some positive effects if used in asphalt mixtures. Free CaO increases the resistance to stripping of bituminous binders and improves adhesion between the aggregate and the binder,[121,125] which contribute to the greater durability of the wearing courses. This is also attributed to the fact that steel slag, being a by-product of steel production, cannot contain clay, which would reduce adhesion.[106] The binder-to-aggregate adhesion is also improved by the sharp edges, proper grain shape, and rough surface texture of the slag, which also increase the *coefficient of friction* in bituminous surface courses better than any natural aggregate, thereby increasing the shear strength and resistance to rutting.[106] However, an increased need for binders is caused by the sharp edges and rough surfaces of the steel slag aggregate. Replacing basalt with steel slag could increase the positive texture significantly; abrasion losses and lateral and vertical displacements also decrease because of a stronger interlocking structure, indicating that steel slag can increase the ability of microsurfacing to resist traffic loading.[126] Enhanced surface characteristics are also the reason that steel slag asphalt mixtures are mostly used in places where high abrasion resistance and good adhesion are required, such as for roads in industrial facilities, parking areas exposed to heavy traffic loads, and intersections. The angular grain shape in hot mix asphalts containing 100% steel slag causes susceptibility to high void spaces and bulking problems, which in turn results in susceptibility to over-asphalting during production and subsequent flushing due to in-service traffic compaction.[127] Furthermore, these mixtures require higher optimum proportions of bitumen compared to asphalts with natural aggregate.[128] Therefore, slag is recommended as a replacement for finer or coarser aggregate fractions for applications in wearing asphalt layers. The results presented by Tao *et al.*[129] showed that the addition of steel slag filler as a mineral filler to asphalt binder enhanced the deformation resistance of the binder but slightly decreased its low-temperature cracking resistance.

Asphalt mixtures with steel slag aggregate have a high specific gravity because of the high specific gravity and density of the slag,[121,123,125,127] which is an advantage during winter road maintenance. The high stability of mixtures containing steel slag aggregate and excellent heat retention contribute to patch durability during winter patching. In addition, the electrical conductivity of steel slag asphalt mixtures is better (*i.e.*, greater) than limestone mixtures, which can be useful for the particular requirements of winter maintenance (*e.g.*, in the maintenance of parking garages, sidewalks, driveways, highway bridges and airport runways).[130] The feasibility of adding steel slag to asphalt mixtures as a thermal aggregate to increase the thermal conductivity for electrical–thermal snow melting was researched by Jiao *et al.*[131] The thermal conductivity reached a peak for steel slag volume replacement of 60% and steel slag asphalt mixtures played a positive role in the snow melting process of an electrical–thermal pavement system compared with ordinary asphalt mixtures.[131] Moreover, EAF slag aggregate in asphalt mixtures increases crack initiation resistance, instead of becoming brittle at subzero conditions.[132]

Steel slag asphalt mixtures show 20–80% higher values of modulus of elasticity at 29 °C and lower values of *Poisson's coefficient* compared to standard aggregate mixtures, which indicate greater asphalt mixture stability and resistance to plastic deformation.[123,124] However, the high specific gravity of steel slag aggregate also results in a 15–25% higher weight, which increases the cost of transportation of fresh asphalt mixtures.[123]

Replacing limestone with slag aggregate at 75 vol% gives significant improvements in the mechanical properties of asphalt concrete; however, the optimal replacement percentage was 25 vol%.[133] Slag addition to asphalt concrete wearing courses gives better skid resistance compared to ordinary asphalt concrete.[128]

Observation of a highway with a surface course made of BOF slag asphalt mixture after 2 years of service revealed excellent performance despite the emergence of cracks and ruts.[134] However, some disadvantages are observed for BOF slag asphalt mixtures during construction. Increasing the BOF slag content leads to the need for a longer stabilization time to reach thermal equilibrium because the heat inside the pavement takes longer to be released.[135] This is a result of BOF slag's porosity, which makes it capable of accumulating heat. This accumulated heat is then convected upward to replenish energy in the upper layer of the pavement, decreasing the pavement cooling rate. The high porosity of BOF slag can be a useful property in porous asphalt because it can reduce traffic noise and improve surface water drainage, which contributes to higher traffic safety.[125] BOF slag utilization in stone mastic asphalt gave increased pavement wear resistance; thus, it can endure higher traffic loads and temperature influences. Moreover, it has better resistance to *permanent deformation* at high temperatures and better crack resistance at low temperatures.[136]

Phosphorus (P) slag is produced as a by-product of the electric furnace method of yellow P production, where silica and carbon are added to remove impurities during the smelting process.[137] In the process of yellow P production, around 8 to10 tonnes of slag are produced per 1 tonne of yellow P.[138] For P slag-modified asphalt, rutting resistance and fatigue performance at high temperatures are optimal when the content of P slag powder is around 10 wt%.[138]

The addition of Cu slag to asphalt mixtures with reclaimed asphalt pavement aggregate reduces the *flow* and resilient modulus to values similar to those of conventional mixtures owing to the shape of the material. The optimal combination of these aggregates is in the range of 45% to 55% over the total aggregate volume of the mixture.[139]

7.6.5 Pervious Concrete Pavement

New paving solutions are focused on reducing the impact of transportation infrastructure on the environment and generally on human life. In that sense, studies are oriented toward finding new materials and techniques

Figure 7.5 Ordinary (left) and pervious concrete (right). (Photo: Ivana Barišić).

for reducing natural material exploitation, traffic noise, and the urban heat island effect (increased air temperature within urban areas due to highly concentrated infrastructure and limited green areas), as well as making surface drainage more effective and improving runoff water quality. One of the solutions that is currently being highly investigated and used is constructing pervious concrete pavements (Figure 7.5), whose material has been known since 1852[140] but was patented later in 1980.[141]

Pervious concrete is a standard concrete pavement designed to have reduced levels, or even the complete absence, of fine aggregate, which contributes to natural aggregate preservation. It is designed to have high porosity (*i.e.*, void content between 11% and 35%[142,143]) and permeability (typically approximately 2–6 mm s^{-1} [143,144]), but consequently relatively low strengths (*e.g.*, compressive and flexural strengths vary between 3.5 MPa and 28 MPa and between 1 MP and 3.8 MPa, respectively[144]). Thus, most of the studies currently are focused on finding an efficient way of increasing strength while preserving the porosity and permeability. The two main approaches used are binder modification and aggregate modification. Concerning binder modification, alkali-activated binders (*e.g.*, alkali-activated slag, metakaolin, or their combination) are a good solution to improve the strength of pervious concrete.[145]

In pervious concrete pavement systems, steel slag is mainly used as a substitute aggregate because it leads to a higher density compared to that of pervious concrete mixtures with dolomite aggregate, although the flexural strength and dynamic modulus of elasticity are lower.[146] Based on analysis of pavement surface characteristics, steel slag as an aggregate ensures higher abrasion resistance but has poor acoustic characteristics compared to those of mixtures with dolomite aggregate.[147] However, the results presented by Yeih *et al.*[148] showed higher compressive strength, greater *permeability coefficient*, and lower weight loss in soundness tests of EAF slag compared to natural river gravel pervious concrete mixtures. The increased mechanical performance of pervious concrete containing steel slag aggregate can be explained by physical and chemical factors.[149] The pores, angularities, and rough surface texture of steel slag aggregate physically and chemically improve the aggregate-to-cement-paste bonding while the cementitious ingredients on its surface are hydrated. Finally, besides the possibility of improving the mechanical characteristics, the use of steel slag aggregate in pervious concrete intended for sidewalk and bicycle lane service can reduce costs by 18%.[150]

The use of alkali-activated slag cement in EAF slag aggregate mixtures results in higher mechanical strength and a lower *British pendulum number*, indicating better anti-skid performance than that of pervious concrete made with Portland cement.[151] For pervious concrete with steel slag aggregate, attention should be paid to the selection of a suitable binder type and binder content. Although a higher binder content increases the unit weight and the 28 day compressive strength, the *connected porosity* and the water permeability coefficient decrease dramatically;[152] however, sulfoaluminate cement and calcium aluminate cement should not be used with EAF slag because the mechanical strength does not reach the expected level.[153]

Similar benefits to those of steel slag aggregate application in pervious concrete can be achieved by incorporating Cu slag aggregate. Replacing up to 60 vol% of dolomite aggregate with Cu slag led to improvements in the compressive, flexural, and splitting tensile strengths, as well as the porosity and permeability rate, while the bonding strengths from the pull-off adhesion test were similar to those of dolomite aggregate mixes.[154]

7.7 Concluding Remarks

Civil engineering essentially relies on the exploitation of natural resources but the ever-growing demand for materials in the construction industry cannot be fully met by natural resources or traditional materials. Hence, there is a need to develop alternative materials and innovative techniques to meet the increasing requirements of the construction industry. The use of slags instead of conventional materials presents a solution to this problem: it preserves natural resources and energy and avoids expensive and/or potentially harmful waste disposal. The other advantages of using slags are as follows: less energy being consumed because these materials already exist, reduced pollution, and smaller amounts of these waste materials going to landfills. However, slags are diverse materials and each type of slag needs to be investigated separately before its final application in construction. As for any other waste and alternative materials, detailed and comprehensive laboratory and field investigations are required depending on the type of slag and area of application before its commercial industrial use. In particular, engineering properties need to be tested and a multidisciplinary approach should be applied, along with taking into consideration its environmental influence and long-term consequences.

List of abbreviations

ABFS	Air-cooled blast furnace slag
BF	Blast furnace
BOF	Basic oxygen furnace
CBC	Cement bound granular material for pavement base course construction
CBR	California bearing ratio
EAF	Electric arc furnace
GBFS	Granulated blast furnace slag, a by-product of iron production
GGBS	Ground granulated blast furnace slag, a by-product of iron production
ITZ	Interfacial transition zone
LM	Ladle metallurgy
MDD	Maximum dry density
OHF	Open-hearth furnace
OMC	Optimum moisture content
UBC	Unbound base course within pavement structure

Glossary

Absorption ability of concrete/mortar to absorb water. It is measured in kg m^{-2}.

Accelerators materials used to shorten the set time of binder (cement) or accelerate the binder setting.

Aggregate granular material used in the production of different construction materials (mortar, concrete, asphalt) or used as a compacted layer for various pavement applications (embankment, subgrade, base course).

Alkali activator any substance that provides alkali cations, raises the pH of the reaction mixture, and facilitates dissolution.

Alkali–silica reaction chemical reaction that occurs between alkali cations and hydroxyl ions in the pore solution of hydrated cement paste and certain reactive silica phases present in the aggregates used in concrete. As a result of this reaction, many cracks appear on the surface of the hardened cement composite (mortar or concrete).

Angle of internal friction physical property of earth materials or the slope of a linear representation of the shear strength of earth materials. It represents the interlocking and frictional resistance of aggregates.

Bearing capacity the maximum load that can be applied on a material before failure or uncontrolled deformations occur. It is measured in kN m^{-2}.

Binder material that holds or draws other materials together to form a cohesive whole mechanically, chemically, by adhesion or cohesion.

British pendulum number the measure of how pavement surface course will perform under wet conditions. It is the value of skid resistance obtained from the British pendulum test, a portable dynamic pendulum device that measures the energy absorbed by a rubber pad as it slides over a test surface.

California bearing ratio (CBR) numerical value used to evaluate the subgrade strength primarily of roads, pavements, and foundations. It is an indirect measure of soil strength based on resistance to penetration. It compares the bearing capacity of a material with that of a well-graded crushed stone and it is expressed in %.

Carbonation the result of interaction between carbonate ions and cement hydrates. Its degree presents the value of how much of the original cement is carbonated.

Cement clinker nodules (diameters, 5–25 mm) of sintered material produced by heating a homogeneous mixture of raw materials in a kiln to a sintering temperature of approximately 1450 °C for modern cements.

Coefficient of friction the ratio of the force required to move two sliding surfaces over each other and the force holding them together. It is a ratio of the frictional force resisting the motion of two surfaces in contact (vehicle tire and road surface) to the normal force pressing the two surfaces together. If lost, vehicles will slide on the pavement surface.

Cohesion the binding force that holds together molecules or like particles within a soil. Cohesive soils are fine-grained, low-strength, and easily deformable soils that have a tendency for particles to adhere. The soil is classified as cohesive if the amount of fines (silt and clay-sized material) exceeds 50% by weight.

Compressive strength compressive stress that, under a gradually applied load, a given solid material can sustain without fracture. It is measured in MPa or N mm^{-2}.

Concrete composite made of fine aggregate (sand), coarse aggregate, binder, and water.

Concrete strength class compressive strength of 28 day-old concrete expressed by two numbers, the first denoting the compressive strength of concrete determined on a cylinder 15 cm in diameter and 30 cm high and the second denoting the compressive strength of concrete determined on a 15 cm edge cube.

Connected porosity the ratio of the connected pore volume to the total volume. Connected pores have high permeability and can transport water easily.

Consistency measure of the fluidity and/or wetness of the mortar/concrete and gives a measure of the deformability of the fresh mortar/concrete when subjected to a certain type of stress.

Curing regime during which adequate moisture, temperature, and time are provided to building materials to allow it to achieve the desired properties for its intended use.

Deep stabilization add bearing capacity to areas of soft soils and reduce settlement so structures can be built on it. Generally, stabilization is a method for reinforcing weak material (soil, aggregate...) by mixing it with a binder used for road and foundation works.

Diffusion coefficients rate-determining parameter of chloride ingress in concrete. It is measured in m^2 s^{-1}.

Dry density density of the soil when it is taken in the dry state. It is the weight of a unit volume of a dry sample of soil. It is measured in kg m^{-3}.

Durability ability to last a long time without significant deterioration and requiring minimum maintenance.

Early strength strength of concrete/mortar determined at its early age (usually up to 7 days of concrete age). It is measured in MPa or N mm^{-2}.

Fineness particle size of cement affects hydration rate and thus the rate of strength gain. Depending on the method of testing, it is expressed in cm^2 g^{-1} or as the percentage of material retained on a sieve of a certain size.

Fire resistance property of a material to withstand fire and continue to perform its given function for a specified period of time.

Flexural strength maximum bending stress that can be applied to that material before it yields. It is measured in MPa or N mm^{-2}.

Flow characteristic measuring the flexibility of asphalt mixtures. It is measured in mm.

Free swelling the increase in the volume of a soil, without any external constraints. It is measured in %.

Freeze–thaw cycle resistance property of a material to withstand cyclic changes in freezing and thawing and continue to perform its given function.

Flushing or bleeding type of asphalt surface disintegration when a film of asphalt binder occurs on the pavement surface creating a shiny, glass-like reflecting surface that can become sticky.

Grindability measure of the ability of a clinker to resist grinding forces. It is expressed qualitatively.

Heat of hydration heat generated when water and binder react, measured in kJ kg^{-1}.

Interfacial transition zone (ITZ) region of the cement paste around the aggregate particles.

Latent hydraulic properties hidden hydraulic properties, being expressed when the material (binder) is appropriately activated.

Maximum dry density (MDD) the dry density obtained by the compaction of soil at its optimum moisture content.

Mechanical properties physical properties that a material exhibits upon the application of forces. Examples of mechanical properties are the modulus of elasticity, tensile strength, compressive strength, flexural strength.

Microbleeding occurrence in concrete/mortar which leads to an accumulation of water under the aggregate.

Microstructure material structures seen at the micrometer level.

Modulus of elasticity measure of the stress–strain relationship, an important parameter in the evaluation of the deformation response of concrete under working loads. It is measured in MPa or N mm^{-2}.

Mortar composite made of fine aggregate (sand), binder, and water.

Optimum moisture content (OMC) water content at which a maximum dry unit weight can be achieved after a given compaction effort.

Permanent deformation irreversible deformation. On asphalt pavement, it can be represented, for example by rutting.

Permeability coefficient describes how easily a liquid will move through a soil or other material (pervious concrete, porous asphalt).

Pillar strength maximum resistance of a pillar to axial compression. Pillar is a column created by stabilizing the *in situ* material with binder such as cement and lime within deep soil stabilization (see term deep stabilization). It is a foundations method.

Plasticity generally, it is the property of a material to be deformed repeatedly without rupture by the action of a force, and remain deformed after the removal of force. In the context of roadworks, it is the property of soil to possess plasticity, that is, get deformed without rupture/breaking under external force. It is calculated in terms of soil moisture content.

Plasticity index range of moisture contents over which the soil deforms plastically. It is a measure of the plasticity of a soil and according to its value the amount and type of clay present in a soil can be determined.

Poisson's coefficient the material characteristic expressed as the ratio of the change in a sample width per unit width to the change in the length per unit length.

Porosity measure of the volume of voids in concrete. It is measured in %.

Pozzolanic activity material ability to react with calcium hydroxide.

Ravelling progressive disintegration of an asphalt layer from the surface downward as a result of the dislodgement of aggregate particles (gravel, sand, and crushed stone). It is the deterioration of pavement by the loss of asphalt and rocks. It reduces skid resistance, roughens the road surface, and exposes the layers underneath to further deterioration.

Reflective cracks cracks in an asphalt surface over an existing crack or joint. It occurs due to breaks or cracks in underlying layers because of movement at the original crack.

Resilient modulus measure of subgrade material stiffness. It is an estimate of materials modulus of elasticity (E). While the modulus of elasticity is stress divided by strain for a slowly applied load, resilient modulus is stress divided by strain for rapidly applied loads – like those experienced by pavements. It is measured in MPa or N mm^{-2}.

Setting time time required for stiffening of cement paste to a defined consistency, measured in min.

Shear strength shear is a type of stress in which an applied force causes a structure to 'slide' in two or more ways. Thus, shear strength is the external load (shear forces) a soil can sustain without structural failure. It is measured in MPa or N mm^{-2}.

Shrinkage time dependent deformation which reduces the volume of concrete, without the impact of external forces. It is measured in %.

Slurry seals asphalt maintenance treatment that prevent water from damaging asphalt, creates a new wear surface, and provides a fresh look for asphalt roadway. It is a mixture of aggregate, asphalt emulsion, water, and mineral fillers.

Soil (geotechnical meaning) sediment or other unconsolidated accumulation or deposit of earth material produced naturally from the disintegration (physical and chemical) of rocks and it may or may not contain organic matter. It is a natural, loose material that can be easily excavated by construction equipment. Subgrade is an undisturbed supporting soil beneath pavement.

Soundness of cement ability of a hardened paste to retain its volume after setting. A cement is said to be unsound (*i.e.*, having lack of soundness) if it is subjected to delayed destructive expansion. It is measured in mm.

Stiffness extent to which an object (in here referred to pavement wearing (surface) course) resists deformation in response to an applied force (usually represented by traffic load).

Stripping resistance ability of asphalt to resist the separation of the asphalt binder film from the aggregate surface due to the action of water.

Swelling potential the swelling tendencies of expansive soils. It is defined as the ratio of increase in thickness to the initial thickness of the soil sample – see free swelling.

Tensile strength tensile stress that, under a gradually applied load, a given solid material can sustain without fracture. It is measured in MPa or N mm^{-2}.

Ultimate strength strength of concrete/mortar determined at its age of at least 28 days, final strength. It is measured in MPa or N mm^{-2}.

Ultrasound pulse velocity measurement that can be utilized for the determination of concrete uniformity, cracks or voids' presence, changes in properties with time. It is measured in m s^{-1}.

Vertical strain ratio between the deformation or displacement and the original length. It is the deformation or displacement of material that results from an applied vertical stress.

Volume changes changes in volume of hardened cement/concrete such as shrinkage and swelling. It is measured in %.

Workability property of freshly mixed concrete which determines the ease and homogeneity with which it can be mixed, placed, consolidated and finished without segregation and bleeding. Depending on the method of testing, it is expressed in mm or sec or as a coefficient.

Acknowledgements

We are grateful to Jan Elsen (KU Leuven, Belgium) and Yiannis Pontikes (KU Leuven, Belgium) for their thorough reviews, which helped to improve substantially the original version of the chapter. The time devoted by editors in evaluating this work is highly appreciated.

References

1. *European List of Waste*, https://eur-lex.europa.eu/legal-content/EN/TXT/PDF/?uri=CELEX:02000D0532-20150601&from=EN, accessed February 2021.
2. *National Slag Association Report MF 182-6*, http://www.nationalslag.org/sites/nationalslag/files/documents/nsa_182-6_properties_and_uses_slag.pdf, accessed February 2021.
3. D. D. Higgins, presented at 3rd European Slag Conference – Proceedings "Manufacturing and Processing of Iron and Steel Slags", Keyworth, UK, October, 2002.
4. H. Motz, presented at 3rd European Slag Conference – Proceedings "Manufacturing and Processing of Iron and Steel Slags", Keyworth, UK, October, 2002.
5. D. Padovani and B. Corcoran, presented at Cemtech Conference, Istanbul, Turkey, September, 2004.
6. *Properties and Uses of Iron and Steel Slags*, http://www.nationalslag.org/sites/nationalslag/files/documents/nsa_182-6_properties_and_uses_slag.pdf, accessed February 2021.
7. A. O. Purdon, *J. Soc. Chem. Ind.*, 1940, **59**, 191.
8. *Statistics*, 2018, https://www.euroslag.com/products/statistics/statistics-2018, accessed February 2021.
9. *Statistics*, https://www.slg.jp/e/index.htm, accessed February 2021.
10. *Material, Energy and Greenhouse Gas Savings*, https://www.slagcement.org/sustainability/decreasedembodiedenergy.aspx, accessed February 2021.
11. EN 197-1, *Cement. Composition, Specifications and Conformity Criteria for Common Cements*, CEN, Brussels, 2011.
12. EN 14216, *Cement. Composition, Specifications and Conformity Criteria for Very Low Heat Special Cements*, CEN, Brussels, 2004.
13. EN 15743, *Supersulfated Cement. Composition, Specifications and Conformity Criteria*, CEN, Brussels, 2010.
14. R. M. Ferreira, G. Liu, L. Nilsson and O. E. Gjorv, presented at 4th International Conference on Concrete under Severe Conditions: Environment and Loading, Seoul, Korea, June, 2004.
15. H. S. Kim, J. W. Park, Y. J. An, J. S. Bae and C. Han, *Mater. Trans.*, 2011, **52**, 210.
16. D. S. Nandagawali and N. R. Dhamge, *Int. J. Eng. Sci. Innov. Technol.*, 2014, **3**, 473.
17. P. Bhatt, T. J. MacGinley and B. S. Choo, *Reinforced Concrete Design to Eurocodes: Design Theory and Examples*, CRC Press, 4th edn, 2014.
18. P. D. Bust and P. T. McCabe, presented at Proceedings of the International Conference on Contemporary Ergonomics (CE2005), CRC Press, Hatfield, UK, 2005.
19. E. Tsakiridis, G. D. Papadimitriou, S. Tsivilis and C. Koroneos, *J. Hazard. Mater.*, 2008, **152**, 805.
20. Y. L. Chen, J. E. Chang and M. S. Ko, *Sustainability*, 2017, **9**, 1585.

21. G. Bernardo, M. Marroccoli, M. Nobili, A. Telesca and G. L. Valenti, *Resour., Conserv. Recycl.*, 2007, **52**, 95.
22. R. I. Iacobescu, G. N. Angelopoulos, P. T. Jones, B. Blanpain and Y. Pontikes, *J. Cleaner Prod.*, 2016, **112**, 872.
23. V. Konovalov, A. Fedorov and A. Goncharov, presented at 14th International Congress for Applied Mineralogy (ICAM2019), Springer Proceedings in Earth and Environmental Sciences, Belgorod, Russia, September, 2019.
24. *Slag Cement*, http://www.nationalslag.org/sites/nationalslag/files/documents/nsa_181-4_slag_cement.pdf, accessed February 2021.
25. Z. Osmanović, N. Haračić, J. Zelić and A. Lihić, presented at 2nd International Symposium on Environment and Morality, Adiyaman University, Adiyaman, Turkey, October, 2014.
26. O. Karahan and A. Yakupoglu, *Adv. Cem. Res.*, 2011, **23**, 289.
27. W. A. Chaparro, D. Martínez Bastidas and J. H. Bautista Ruíz, *Rev. Fac. Ing., Univ. Antioquia*, 2012, **62**, 189.
28. W. Chen and H. J. H. Brouwers, *J. Mater. Sci.*, 2007, **42**, 428.
29. T. Bakharev, J. G. Sanjayan and Y. B. Cheng, *Cem. Concr. Res.*, 1999, **29**, 113.
30. O. Burciaga-Diaz, M. R. Diaz-Guillen, A. Fuentes and A. I. Escalante-Garcia, *Constr. Build. Mater.*, 2013, **44**, 607.
31. N. T. Abdel-Ghani, H. A. Elsayed and S. AbdelMoied, *HBRC J.*, 2018, **14**, 159.
32. M. Palacios and F. Puertas, *Cem. Concr. Res.*, 2007, **37**, 691.
33. S. A. Bernal, R. Mejía de Gutierrez and E. D. Rodríguez, *Ing. Compet.*, 2013, **15**, 211.
34. C. Cartwright, F. Rajabipour and A. Radlińska, *J. Mater. Civ. Eng.*, 2015, **27**, B4014007.
35. P. Awoyera and A. Adesina, *Case Stud. Constr. Mater.*, 2019, **11**, e00268.
36. PAS 8820, *Construction Materials. Alkali-activated Cementitious Material and Concrete. Specification*, BSI, 2016.
37. BS 8500 Concrete, *Complementary British Standard to BS EN 206. Method of Specifying and Guidance for the Specifier*, BSI, 2019.
38. EN 206 Concrete, *Specification, Performance, Production and Conformity*, CEN, Brussels, 2002.
39. L. Chao, S. Henghu and L. Longtu, *Cem. Concr. Res.*, 2010, **40**, 1341.
40. R. D. Hooton, *Can. J. Civ. Eng.*, 2000, **27**, 754.
41. R. Palod, S. V. Deo and G. D. Ramtekkar, *J. Mater. Cycles Waste Manage.*, 2019, **21**, 1361.
42. M. L. Berndt, *Constr. Build. Mater.*, 2009, **23**, 2606.
43. F. Sajedi, H. A. Razak, H. B. Mahmud and P. Shafigh, *Constr. Build. Mater.*, 2012, **31**, 188.
44. S. J. Barnet, M. N. Soutsos, S. G. Millard and J. H. Bungey, *Cem. Concr. Res.*, 2006, **36**, 434.
45. K. Y. Yeaua and E. K. Kim, *Cem. Concr. Res.*, 2005, **35**, 1391.
46. H. W. Song and V. Saraswathy, *J. Hazard. Mater.*, 2006, **B138**, 226.

47. H. Huang, G. Ye and D. Damidot, *Cem. Concr. Res.*, 2014, **60**, 68.
48. A. Bhat and S. K. Tengli, *Int. J. Res. Appl. Sci. Eng. Technol.*, 2018, **6**, 4578.
49. E. Lubloy, K. Kopeczko, G. L. Balasz, I. M. Szilajyi and J. Madarasz, *J. Therm. Anal. Calorim.*, 2016, **125**, 271.
50. Q. Li, Z. Li and G. Yuan, *Constr. Build. Mater.*, 2012, **35**, 687.
51. C. C. Onn, K. H. Mo, M. K. H. Radwan, W. H. Liew, C. G. Ng and S. Yusoff, *Sustainability*, 2019, **11**, 7194.
52. E. Gruyaert, P. Van den Heede and N. De Belie, *Cem. Concr. Compos.*, 2013, **35**, 39.
53. J. Liu and R. Guo, *Adv. Civ. Eng.*, 2018, 1426037.
54. F. Liu, M. Z. Chen, F. Z. Li, Q. L. Li, S. P. Wu and Y. Sang, *Mater. Res. Innov.*, 2015, **19**, S1-150.
55. B. S. Cho, Y. U. Kim, D. B. Kim and S. J. Choi, *Adv. Civ. Eng.*, 2018, 1–7.
56. *Concrete for Sustainable Future*, https://ukcsma.co.uk/mpa-cms/wp-content/uploads/2016/08/ict_additions_sustainable_concrete.doc.pdf, accessed February 2021.
57. *Steel Slag Utilization in Asphalt Mixtures*, http://www.nationalslag.org/sites/nationalslag/files/documents/nsa_186-1_steel_slag_utilization_in_asphalt_mixes.pdf, accessed February 2021.
58. N. O. Da Silveira, M. V. A. M. E Silva, E. J. Agrizzi, M. F. De Lana and R. L. D. De Mendonca, presented at 4th European Slag Conference – "Slags-Providing Solutions for Global Construction and Other Markets", Oulu, Finland, June, 2005, vol. 3, p. 145.
59. B. Pang, Z. Zhou and H. Xu, *Constr. Build. Mater.*, 2015, **84**, 454.
60. M. C. Nataraja, P. G. D. Kumar, A. S. Manu and M. C. Sanjay, *Int. J. Civ. Struct. Eng. Res.*, 2013, **2**, 59.
61. M. Nadeem and A. D. Pofale, *Int. J. Eng. Res. Appl.*, 2012, **2**, 1258.
62. M. C. Pacheco, J. M. Avila, M. Etxebarria, J. M. Meneses and I. Berridi, presented at 6th European Slag Conference – Proceedings "Ferrous Slag – Resource Development for an Environmentally Sustainable World", Madrid, Spain, October, 2010.
63. A. . S. Ouda and G. A. Abdel-Gawwad, *HBRC J.*, 2015, **13**, 255.
64. M. Valcuende, F. Benito, C. Parra and I. Minano, *Constr. Build. Mater.*, 2015, **76**, 1.
65. T. Humam and R. Siddique, *Leonardo J. Sci.*, 2013, **23**, 53.
66. K. Takanashi, K. Watanabe, K. Niitani and K. Hosotani, *JFE Tech. Rep.*, 2018, **23**, 102.
67. P. Jariyathitipong, K. Hosotani, T. Fujii and T. Ayano, presented at 3rd International Conference on Sustainable Construction Materials and Technologies, Kyoto, Japan, August, 2013.
68. H. El Nouhy, *et al.*, *Key Eng. Mater.*, 2014, **600**, 227.
69. H. Qasrawi, F. Shalabi and I. Asi, *Constr. Build. Mater.*, 2009, **23**, 1118.
70. I. Arribas, A. Santamaría, E. Ruiz, V. Ortega-López and J. M. Manso, *Constr. Build. Mater.*, 2015, **90**, 68.
71. A. Sharath, M. N. Vathsala and G. Narayana, *Int. J. Res. Eng. Technol.*, 2015, **4**, 99.

72. S. J. Choi, Y. U. Kim, T. G. Oh and B. S. Cho, *Materials*, 2020, **13**, 940.
73. M. Maslehuddin, A. M. Sharif, M. Shameem, M. Ibrahim and M. S. Barry, *Constr. Build. Mater.*, 2003, **17**, 105.
74. N. Faraone, G. Tonello, E. Furlani and S. Maschio, *Chemosphere*, 2009, **77**, 1152.
75. G. I. Sezer and M. Gulderen, *Indian J. Eng. Mater. Sci.*, 2015, **22**, 339.
76. I. Netinger, D. Bjegović and G. Vrhovac, *Mater. Struct.*, 2011, **44**, 1565.
77. A. A. Ramezanianpour, M. Nikravan and M. Jafari Nadoushan, presented at 9th International Congress on Advances in Civil Engineering, Trabzon, Turkey, September, 2010.
78. I. Netinger, M. Jelčić Rukavina and A. Mladenovič, *Procedia Eng.*, 2013, **62**, 745.
79. G. A. Khoury, Y. Anderberg, K. Both, J. Fellinger, N. P. Hoj and C. Majorana, *Fib Bulletin 38: Fire Design of Concrete Structures – Materials, Structures and Modelling*, International Federation for Structural Concrete (fib), Lausanne, Switzerland, 2007.
80. T. Uygunoğlu and İ. B. Topçu, *Constr. Build. Mater.*, 2009, **23**, 3063.
81. N. Saikia, G. Cornelis, O. Cizer, C. Vandecasteele, D. Van Gemert, K. Van Balen and T. Van Gerven, *J. Mater. Cycles Waste Manage.*, 2012, **14**, 102.
82. K. S. Al-Jabri, M. Hisada, A. H. Al-Saidy and S. K. Al-Oraimi, *Constr. Build. Mater.*, 2009, **23**, 2132.
83. M. Knežević, M. Korać, Ž. Kamberović and M. Ristić, *J. Miner. Metall.*, 2010, **16**, 195.
84. A. Alwaeli, *Radiat. Phys. Chem.*, 2013, **83**, 54.
85. S. H. G. Mosavinezhad and S. E. Nabavi, *KSCE J. Civ. Eng.*, 2012, **16**, 989.
86. H. Jamil, M. Ishak, I. Mohamad Pauzi, S. Suhairy, A. Azhar, S. Shaharudin, I. Mohd Zaidi, A. A. Abul and A. A. R. Abdul, *IOP Conf. Ser.: Mater. Sci. Eng.*, 2018, **298**, 012014.
87. R. S. Edwin, E. Ngii, R. Talanipa, F. Masud and R. Sriyani, *IOP Conf. Ser.: Mater. Sci. Eng.*, 2019, **615**, 012014.
88. A. K. Saha and P. K. Sarker, *Int. J. Concr. Struct. Mater.*, 2018, **12**, 1–12.
89. H. L. Robinson, presented at 3rd European Slag Conference – Proceedings " Manufacturing and processing of Iron and Steel Slags", Keyworth, UK, October, 2003.
90. N. Saikia, G. Cornelis, G. Mertens, J. Elsen, K. Van Balen, T. Van Gerven and C. Vandecasteele, *J. Hazard. Mater.*, 2008, **154**, 766.
91. EN 12457, *Characterisation of Waste. Leaching. Compliance Test for Leaching of Granular Waste Materials and Sludges. One Stage Batch Test at a Liquid to Solid Ratio of 2 L kg^{-1} for Materials with High Solid Content and with particle Size below 4 Mm (Without or with Size Reduction)*, CEN, Brussels, 2002.
92. NEN 7345, *Leaching Characteristics of Solid Eartly and Stony Building and Waste Materials-leaching Tests-determination of the Leaching of Inorganic Components from Buildings and Monolitic Waste Materials with the Diffusion Test*, CEN, Brussels, 1995.

93. D. D. Higgins, *Soil Stabilisation with Ground Granulated Blastfurnace Slag*, https://www.ecocem.co.uk/wp-content/uploads/2016/08/ECL042_Soil_Stabilisation.pdf, accessed February 2021.
94. M. Mäkikyrö, presented at 4th European Slag Conference – Proceedings "Slags – Providing Solutions for Global Construction and other Markets", Oulu, Finland, June, 2005.
95. M. Corrêa-Silva, T. Miranda, M. Rouainia, N. Araújo, S. Glendinning and N. Cristelo, *J. Cleaner Prod.*, 2020, **267**, 122017.
96. M. Salimi and A. Ghorbani, *Appl. Clay Sci.*, 2020, **184**, 105390.
97. C. A. Mozejko and F. M. Francisca, *Constr. Build. Mater.*, 2020, **239**, 117901
98. F. I. Shalabi, I. M. Asi and H. Y. Qasrawi, *J. King Saud Univ. Eng. Sci.*, 2017, **29**, 394.
99. J. M. Manso, V. Ortega-López, J. A. Polanco and J. Setién, *Constr. Build. Mater.*, 2013, **40**, 126.
100. V. Ortega-Lópeza, J. M. Manso, I. I. Cuesta and J. J. González, *Constr. Build. Mater.*, 2014, **68**, 455.
101. O. Amini and M. Ghasemi, *Constr. Build. Mater.*, 2019, **223**, 409.
102. M. Mahedi, B. Cetin and A. Y. Dayioglu, *Waste Manage.*, 2019, **95**, 334.
103. G. W. Josephson, F. Sillers and D. G. Runner, *Iron Blast-furnace Slag: Production, Processing, Properties and Uses*, United States Department of the Interior Bureau of Mines, Washington, USA, 1949.
104. *Blast Furnace Slag, User Guidelines for Waste and Byproduct Materials in Pavement Construction*, https://www.fhwa.dot.gov/publications/research/infrastructure/structures/97148/bfs.cfm, accessed February 2021.
105. A. Tikkakoski, K. Kujala and M. Mäkikyrö, presented at 4th European Slag Conference – Proceedings "Slags – Providing Solutions for Global Construction and Other Markets", Oulu, Finland, June, 2005.
106. National Slag Association, *Steel Slag A premier Construction Aggregate*, 2014, http://nationalslag.org/.
107. H. Motz and J. Geiseler, *Waste Manage.*, 2001, **21**, 285.
108. M. Shiha, S. El-Badawy and A. Gabr, *Constr. Build. Mater.*, 2020, **248**, 118710.
109. A. E. A. E.-M. Behiry, *Ain Shams Eng. J.*, 2013, **4**, 43.
110. İ. Gökalp, V. E. Uz, M. Saltan and E. Tutumluer, *Transp. Geotech.*, 2018, **14**, 61.
111. G. Wang, Y. Wang and Z. Gao, *J. Hazard. Mater.*, 2010, **184**, 555.
112. S. P. Singh, D. P. Tripathy and P. G. Ranjith, *Waste Manage.*, 2008, **28**, 1331.
113. P. Y. Mahieux, J. E. Aubert and G. Escadeillas, *Constr. Build. Mater.*, 2009, **23**, 742.
114. F. Autelitano and F. Giuliani, *Constr. Build. Mater.*, 2016, **113**, 280.
115. J. Liu, B. Yu and Q. Wang, *J. Cleaner Prod.*, 2020, **269**, 121733.
116. I. Barišić, S. Dimter and T. Rukavina, *Composites, Part B*, 2014, **58**, 386.
117. I. Barišić, T. Dokšanović and H. Draganić, *Constr. Build. Mater.*, 2015, **83**, 299.
118. I. Barišić, B. Marković and M. Zagvozda, *Int. J. Pavement Eng.*, 2019, **20**, 448.

119. W. Li, L. Lang, Z. Lin, Z. Wang and F. Zhang, *Constr. Build. Mater.*, 2017, **134**, 540.
120. L. P. De Bock and H. Van de Bergh, presented at International RILEM Conference on the Use of Recycled Materials in Building and Structures, Spain, November, 2014.
121. *Steel Slag, User Guidelines for Byproducts and Secondary Use Materials in pavement Construction*, https://www.fhwa.dot.gov/publications/research/infrastructure/structures/97148/ssa1.cfm, accessed February 2021.
122. Z. Chen, Z. Gong, Y. Jiao, Y. Wang, K. Shi and J. Wu, *Constr. Build. Mater.*, 2020, **251**, 118987.
123. J. Emery, *Slag Utilization in pavement Construction, Extending Aggregate Resources*, American Society for Testing and Materials, 1982.
124. J. Emery, Steel slag utilization in asphalt mixes, *National Slag Association Report MF 186-1*, 1984.
125. D.-H. Shen, C.-M. Wu and J.-C. Du, *Constr. Build. Mater.*, 2009, **23**, 453.
126. P. Cui, S. Wu, Y. Xiao, C. Yang and F. Wang, *Constr. Build. Mater.*, 2020, **239**, 117870.
127. N. A. Ali, J. S. S. Chan, T. Papagiannakis, E. G. Theriault and A. T. Bergan, *Effects of Aggregates and Mineral Fillers on Asphalt Mixture Performance*, American Society for Testing and Materials, 1992.
128. F. Kehagia, *Waste Manage. Res.*, 2009, **27**, 288.
129. G. Tao, Y. Xiao, L. Yang, P. Cui, D. Kong and Y. Xue, *Constr. Build. Mater.*, 2019, **201**, 439.
130. P. Ahmedzade and B. Sengoz, *J. Hazard. Mater.*, 2009, **165**, 300.
131. W. Jiao, X. Li, A. Sha, Z. Liu, W. Jiang and L. Hu, *J. Cleaner Prod.*, 2020, **261**, 121197.
132. S. M. Motevalizadeh, R. Sedghi and H. Rooholamini, *Constr. Build. Mater.*, 2020, **240**, 117965.
133. I. Asi, H. Qasrawi, H. Qasrawi and F. I. Shalabi, Use of steel slag aggregate in asphalt concrete mixes, *Can. J. Civ. Eng.*, 2007, **34**, 902.
134. Y. Xue, S. Wu, H. Hou and J. Zha, *J. Hazard. Mater.*, 2006, **B138**, 261.
135. L.-S. Huang, G.-L. Zou, H.-L. Luo and C.-C. Chao, *Int. J. Pavement Res. Technol.*, 2013, **6**, 386.
136. S. Wu, Y. Xue, Q. Ye and Y. Chen, *Build. Environ.*, 2007, **42**, 2580.
137. M. Criado, X. Ke, J. L. Provis and S. A. Bernal, Alternative inorganic binders based on alkali-activated metallurgical slags, *Sustainable and Nonconventional Construction Materials Using Inorganic Bonded Fiber Composites*, 2017.
138. H. Yu, X. Zhu, G. Qian, X. Gong and X. Nie, *Constr. Build. Mater.*, 2020, **256**.
139. A. C. Raposeiras, D. Movilla-Quesada, R. Bilbao-Novoa, C. Cifuentes, G. Ferrer-Norambuena and D. Castro-Fresno, *Constr. Build. Mater.*, 2018, **190**, 427.
140. N. Ghafoori and S. Dutta, *J. Mater. Civ. Eng.*, 1995, 7, 286.

141. H. Hodson, Method of producing and distributing a pervious concrete product, *US Pat.*, 1980, 4225357.
142. B. J. Putman and A. I. Neptune, *Constr. Build. Mater.*, 2011, **25**, 3480.
143. V. R. Schaefer, K. Wang, M. T. Suleiman and J. T. Kevern, *Mix Design Development for Pervious Concrete in Cold Weather Climates*, Centre for Transportation Research and Education, Iowa State University, Ames, USA, 2006.
144. P. D. Tennis, M. L. Leming and D. J. Akers, *Pervious Concrete Pavements, EB302.02*, Portland Cement Association and National Ready Mixed Concrete Association, Silver Spring, Maryland, USA, 2004.
145. Z. Sun, X. Lin and A. Vollpracht, *Constr. Build. Mater.*, 2018, **189**, 797.
146. K. Ćosić, L. Korat, V. Ducman and I. Netinger, *Constr. Build. Mater.*, 2015, **78**, 69.
147. I. Netinger Grubeša, I. Barišić, T. Keser and M. Vračević, *Road Mater. Pavement Des.*, 2018, **20**, 727.
148. W. Yeih, T. C. Fu, J. J. Chang and R. Huang, *Constr. Build. Mater.*, 2015, **93**, 737.
149. X. Chen, G. Wang, Q. Dong, X. Zhao and Y. Wang, *J. Cleaner Prod.*, 2020, **254**, 120149.
150. G. Wang, X. Chen, Q. Dong, J. Yuan and Q. Hong, *J. Cleaner Prod.*, 2020, **262**, 121208.
151. J. J. Chang, W. Yeih, T. J. Chung and R. Huang, *Constr. Build. Mater.*, 2016, **109**, 34.
152. G. Zhang, S. Wang, B. Wang, Y. Zhao, M. Kang and P. Wang, *Constr. Build. Mater.*, 2020, **257**, 119543.
153. W. Yeih and J. J. Chang, *Constr. Build. Mater.*, 2019, **197**, 813.
154. A. R. Lori, A. Hassani and R. Sedghi, *Constr. Build. Mater.*, 2019, **197**, 130.

CHAPTER 8

Environmental Applications of Slag

HELENA I. GOMES*[a], WILLIAM M. MAYES[b] AND REBECCA FERRARI[a]

[a]Food, Water, Waste Research Group, Faculty of Engineering, University of Nottingham, University Park, Nottingham, NG7 2RD, UK; [b]Department of Geography, Geology and Environment, University of Hull, Cottingham Road, Hull, HU6 7RX, UK
*E-mail: helena.gomes@nottingham.ac.uk

8.1 Introduction

Current patterns of natural resource consumption are incompatible with the planet's boundaries.[1] Increasing awareness and focus have been dedicated to implementing a circular economy, and particularly the need to further valorize wastes and by-products such as slags with environmental benefits.[2,3] Several obstacles to resource recovery persist, particularly in the regulatory framework, regarding ownership and responsibilities, and the lack of incentives to consider alternatives or investigate the potential to seek financial gain from legacy slag residues.[4]

Different environmental applications can add value to ferrous and non-ferrous slags, such as removing nutrients or trace elements, controlling unwanted industrial emissions, or use as acid-neutralizing agents, soil conditioners, fertilizers, or soil liming material.[5] Despite the abundant literature focusing on batch studies, limited pilot or full-scale applications have been implemented or commercialized.

Chemistry in the Environment Series No. 2
Metallurgical Slags: Environmental Geochemistry and Resource Potential
Edited by Nadine M. Piatak and Vojtěch Ettler

Published by the Royal Society of Chemistry, www.rsc.org

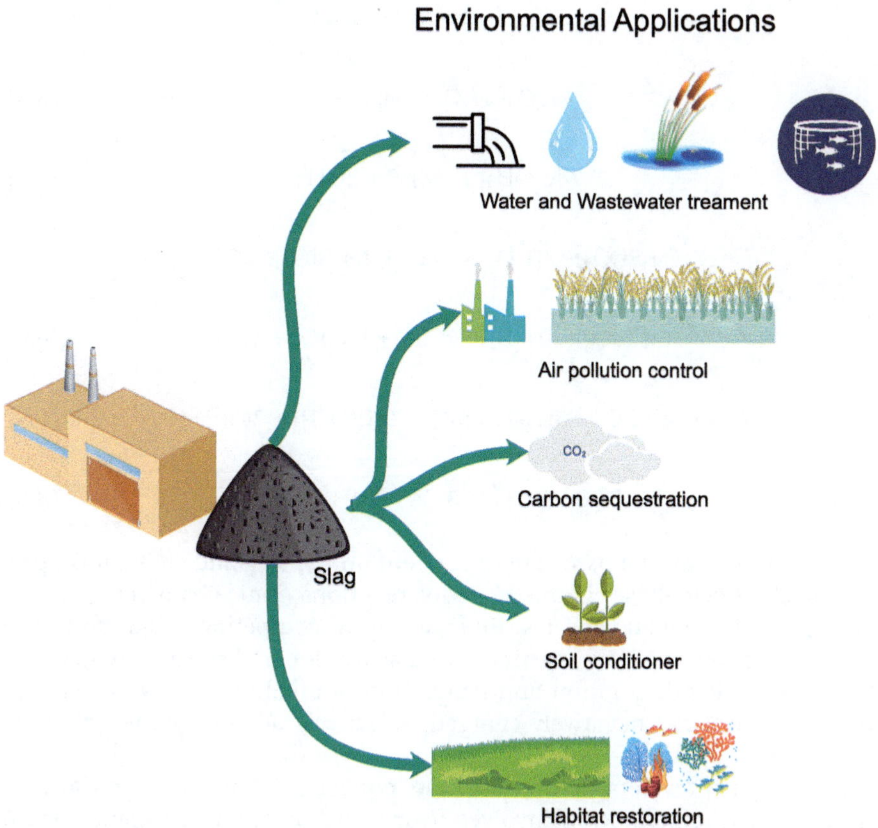

Figure 8.1 Environmental applications of ferrous and non-ferrous slags.

This chapter aims to review and highlight environmental applications of ferrous and non-ferrous slags, focusing on the latest developments in the literature, considering the last five years. Several applications are described, such as water and wastewater treatment, air pollution control, carbon sequestration, soil conditioning, and habitat restoration (Figure 8.1), assessing the advantages and disadvantages, as well as current knowledge gaps and obstacles to full-scale implementation.

8.2 Water and Wastewater Treatment

Both ferrous and non-ferrous slags have been used in different water and wastewater treatment applications, namely as adsorbents, catalysts, and photocatalysts. The mechanisms of contaminant removal from waters are driven by the dissolution of the ferrous slag phases, in most cases generating alkalinity due to the release of hydroxide ions (OH^-), according to the following reactions:[6,7]

$$CaO + H_2O \leftrightarrows Ca(OH)_2 \tag{8.1}$$

$$Ca(OH)_2 \leftrightarrows Ca^{2+} + 2OH^- \quad (8.2)$$

$$MgO + H_2O \leftrightarrows Mg(OH)_2 \quad (8.3)$$

$$Mg(OH)_2 \leftrightarrows Mg^{2+} + 2OH^- \quad (8.4)$$

$$Ca_3Si_2O_7 + 7H_2O \rightarrow 3Ca^{2+} + 2H_4SiO_4 + 6OH^- \quad (8.5)$$

$$Ca_2SiO_4 + 4H_2O \rightarrow 2Ca^{2+} + H_4SiO_4 + 4OH^- \quad (8.6)$$

$$Ca_2Al(AlSi)O_7 + 5H_2O \leftrightarrows SiO_2 + 2Al(OH)_3 + 2Ca^{2+} + 4OH^- \quad (8.7)$$

$$Ca_3Si_2O_7 + 4H^+ \leftrightarrows 3\,Ca^{2+} + 2H_2O + 2SiO_2 \quad (8.8)$$

The removal mechanisms are dependent on the slag and element properties, such as acid–base chemistry, redox reactions, ionic character, pH, concentration of aqueous species, ionic strength, competing adsorbates, and adsorbent properties (*e.g.*, particle size distribution and surface area). Given the ferrous slag dissolution and generation of alkalinity [eqns (8.1)–(8.8)], the slag surface is negatively charged, selectively adsorbing the positively charged metal ions.

Sorption is prevalent at low adsorbate concentrations and high slag surface areas. Chemisorption involves hydrogen bonds and covalent bonds, particularly inner-sphere complexation (ions bind directly to the surface with no intervening water molecules). Physisorption encompasses surface adsorption by van der Waals bonds, absorption by penetration into the solid network, and outer-sphere complexation. Physisorption bonds are weak, reversible, electrostatic, and intermolecular, while chemisorption bonds are stronger, making it hard to remove the chemisorbed species. Sorption of anion species can also occur, and it is more relevant at pH closer to the anion pK_a value (negative logarithm of the acid dissociation constant or K_a value). However, sorption and slag-based adsorbents' performance and applications are significantly affected by the slag's chemical and shape heterogeneity.[8]

Iron (*i.e.*, blast furnace) and steel slags are primarily used as adsorbents (Figure 8.2) or as reagents for the co-precipitation of contaminants, while copper (Cu) slag finds applications in advanced chemical oxidation processes [chemical treatments to degrade organic and inorganic contaminants in water and wastewater by oxidation with hydroxyl radicals ($^{\bullet}OH$)] in both bulk and modified form.[9] An increase in the amount of crystalline phases in slags, such as gehlenite ($Ca_2Al_2SiO_7$), larnite (Ca_2SiO_4), bredigite ($Ca_5MgSi_3O_{12}$), merwinite [$Ca_3Mg(SiO_4)_2$], and åkermanite (Ca_2MgSiO_7), significantly increases the adsorption capacity.[8]

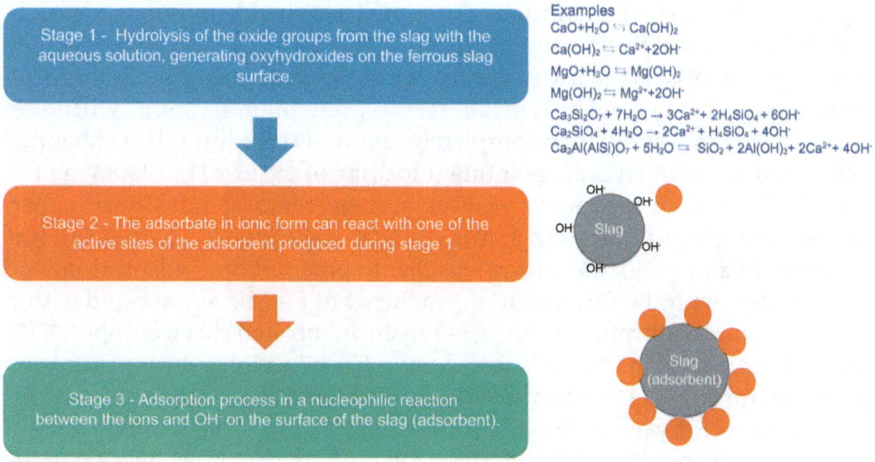

Figure 8.2 Stages of adsorption.[8]

The use of slags in water and wastewater treatment has the advantage of reducing the volume of waste material deposited and simultaneously diminishing resource consumption of primary materials used in the treatment. This bulk use of slags in wastewater treatment systems as a substrate with excellent hydraulic efficiency (in addition to any chemical benefits) has been trialed since the 1960s.[10] If the resulting treatment has a lower cost or higher efficiency than using standard chemicals, it can make water resources available for which current treatment is not viable.[9]

In the following subsections, particular slag applications for drinking and wastewater treatment will be detailed, focusing on removal mechanisms and applications from the last five years.

8.2.1 Drinking Water

The use of slags has been dominant in wastewater treatment. However, there is evidence that these materials can also be used in emerging applications to produce drinking water, despite the reluctance to use industrial by-products or waste materials for this. This could be particularly useful in areas under water stress and where recycling of effluents is a strategic priority, and also to minimize the environmental footprint of drinking water treatment (minimizing chemicals consumption).

An integrated semi-pilot system was tested in South Africa to produce drinking water from acid mine drainage (AMD) using basic oxygen furnace (BOF) slag, lime/soda ash, and reverse osmosis.[11] Alkalinity generated by BOF slag allowed the pH of AMD to increase up to ≥8, removing 99% of metals and 75% of sulfate by precipitation.[11] The treatment process produced water complying with the requirements of the South African National Standard (SANS) 241 Drinking Water Specifications, while simultaneously recovering gypsum ($CaSO_4{\cdot}2H_2O$) and brucite [$Mg(OH)_2$] as valuable minerals to potentially offset some of the treatment costs.[11]

Another example is the use of granulated blast furnace slag as a filter material (supported in a gravel bed) in laboratory-scale column tests, at a filtration rate of 0.32 $m^3\ h^{-1}\ m^{-2}$, which completely removed turbidity, total suspended solids, and color for an average influent loading of 28.26 NTU, 128.85 mg L^{-1}, and 177.05 PCU, respectively.[12] Lower average removal rates were apparent for hardness, sulfate (SO_4^{2-}), nitrate (NO_3^-), chloride (Cl^-), and iron (Fe) (between 28 and 94%).[12] Removal was due to precipitation, adsorption, and ion exchange with the formation of products both at the surface and within the slag media, confirmed by Fourier-transform infrared spectroscopy (FTIR) and X-ray diffraction (XRD).[12] Sorption rates of anions during filtration decreased for $NO_3^- > Cl^- > SO_4^{2-}$, while at a higher filtration rate (0.64 $m^3\ h^{-1}\ m^{-2}$), the sorption was decreased for $Cl^- > NO_3^- > SO_4^{2-}$. The removal rates of Cl^- and SO_4^{2-} were lower at higher filtration rates due to the exhaustion of the slag's exchange capacity. Results showed that intermittent washing helped improve the water quality for various filtration and loading rates, and also that after backwashing there was considerable improvement in Cl^- removal.[12]

None of these applications have considered the leaching of potentially toxic elements in the treated water, such as vanadium (V), which is classified as an emerging pollutant and listed on the United States Environmental Protection Agency (US EPA) Drinking Water Contaminant Candidate List (CCL4) due to its carcinogenic effects and is often overlooked in environmental monitoring.[13] Other potentially toxic elements, such as arsenic (As) and chromium (Cr), should also be monitored. Slag applications for drinking water must assess this vital issue.

8.2.2 Wastewater

8.2.2.1 Nutrient Removal

Excess nutrients, such as nitrogen (N) and phosphorus (P), in wastewater are one of the leading causes of eutrophication, which can lead to excessive growth of plants and algae in aquatic ecosystems and may ultimately result in oxygen depletion of the water body. The use of slag for nutrient removal can be advantageous in reducing chemical costs and minimizing sludge production compared with chemical precipitation,[14] and has been extensively studied.[10] Despite the abundant research on P removal by slags, limited research has focused on further valorization of the saturated slags and if phosphate could be released in the soil when using these slags as soil amendments (see Section 8.5), or if any other processes of nutrient recovery could be implemented.

The mechanism for nitrogen removal is ion exchange adsorption between metal ions (*e.g.*, Al^{3+}, Fe^{2+}, and Mn^{4+}) and NH_4^+, and the reaction is as follows:[15]

$$Z\text{–}M^+ + NH_4^+ \rightarrow Z\text{–}NH_4^+ + M^+ \quad (8.9)$$

where M^+ represents a metal ion and Z represents the unreactive part of the slag.

On the other hand, for PO_4^{3-} removal, there is a series of reactions between the dissolved metal ions including Ca^{2+}, Al^{3+}, Fe^{3+}, and Fe^{2+} from the slags:[15]

$$3Ca^{2+} + 2PO_4^{3-} \rightarrow Ca_3(PO_4)_2\downarrow \quad (8.10)$$

$$Al^{3+} + PO_4^{3-} \rightarrow AlPO_4\downarrow \quad (8.11)$$

$$3Fe^{2+} + 2PO_4^{3-} \rightarrow Fe_3(PO_4)_2\downarrow \quad (8.12)$$

$$Fe^{3+} + PO_4^{3-} \rightarrow FePO_4\downarrow \quad (8.13)$$

8.2.2.1.1 Laboratory Tests for Sorption of Reactive Materials. Most research on nutrient removal is based on small-scale laboratory tests (batch and column experiments, Figure 8.3), where different slags and experimental conditions are assessed to optimize removal through adsorption and precipitation. Although these tests provide useful information about the mechanisms of nutrient removal and potential setbacks of the treatment process, further research at pilot scale is needed to assess how influent variability, environmental conditions (*e.g.*, hydrology, climatic conditions), and time influence the treatment effectiveness.

In these experiments, slag is milled and/or size separated, and a small amount (a few grams) is used in batch tests to determine equilibrium isotherms and adsorption kinetics models (*e.g.*, Langmuir, Freundlich, or pseudo-first or second-order), and maximum adsorption capacity. Typical variables are initial nutrient concentration, slag particle size, adsorbent dosage, and equilibrium concentration of nutrients to assess the removal from a synthetic water solution. Column tests are flow-through experiments that allow using higher amounts of slag in a column, simulating conditions closer to a working filter to assess breakthrough curves and optimize removal for higher volumes of water.

Phosphorus removal efficiency higher than 95% was achieved in batch and column experiments using BOF steel slag aggregates and synthetic wastewater [tap water with different phosphate (PO_4^{3-}) concentrations].[16] The primary P removal mechanism identified was not adsorption, but calcium phosphate [$Ca_3(PO_4)_2$] precipitation, which depends on the hydration and dissociation of calcium oxide (CaO) and other Ca phases in the slag and further dissolution of calcium hydroxide [$Ca(OH)_2$] in water, releasing Ca^{2+} and OH^-. However, once Ca phases are exhausted or effectively non-reactive due to armoring, P removal will depend on Ca^{2+} in the influent to facilitate calcium phosphate precipitation [eqn (8.10)].[16] Piatak *et al.*[7] identified the secondary Ca phosphate phases associated with phosphate removal from slag, namely brushite ($CaHPO_4 \cdot 2H_2O$) and octacalcium phosphate

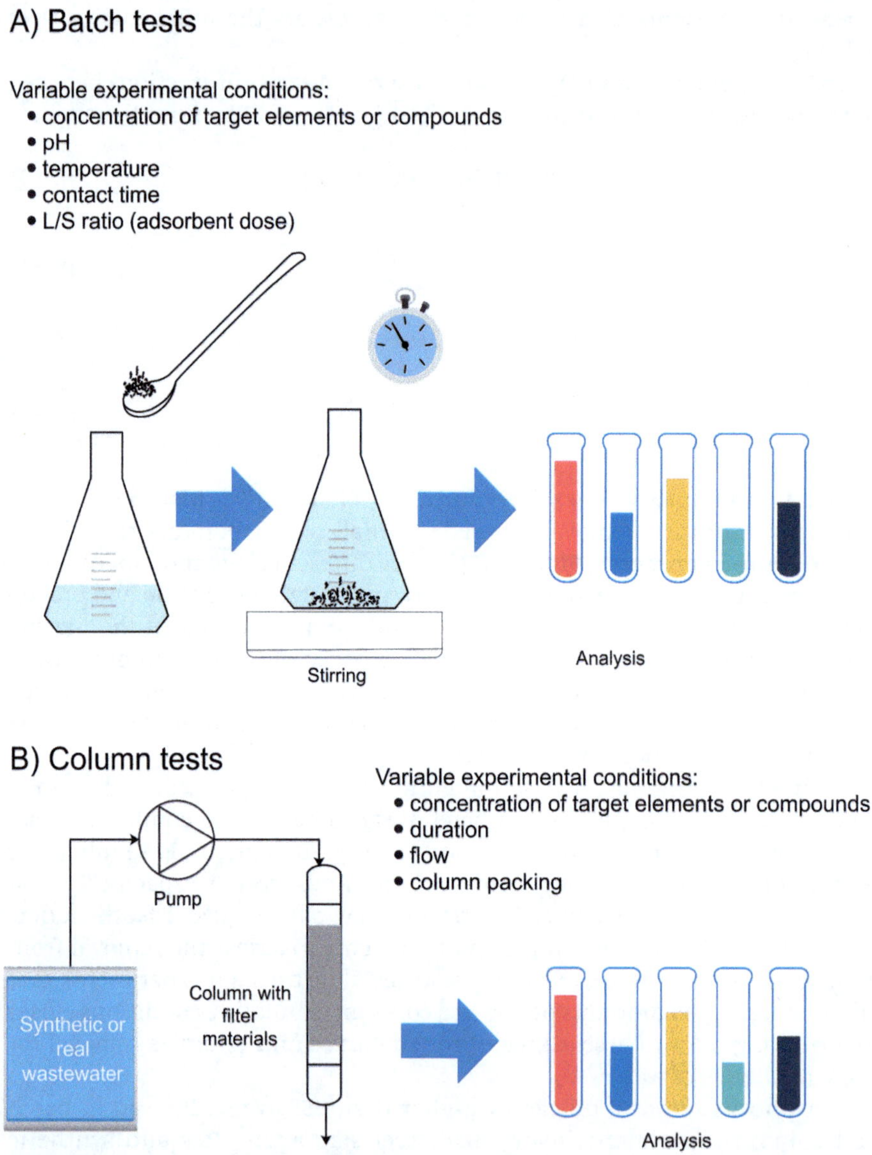

Figure 8.3 Laboratory tests used on slag. (A) Batch tests and (B) column tests for filter materials.

[$Ca_8H_2(PO_4)_6 \cdot 5H_2O$], but also monetite ($CaHPO_4$), and apatite phases such as hydroxyapatite [$Ca_5(PO_4)_3(OH)$] and carbonate-hydroxyapatite [$Ca_5(PO_4,CO_3)_3(OH)$] (Figure 8.4).

Blast furnace (BF) slag was also tested to remove phosphate from synthetic wastewater at different concentrations, showing that a high $CaO:SiO_2$ ratio in the slag reduces its phosphate removal capacity.[17] The use of BF slag also

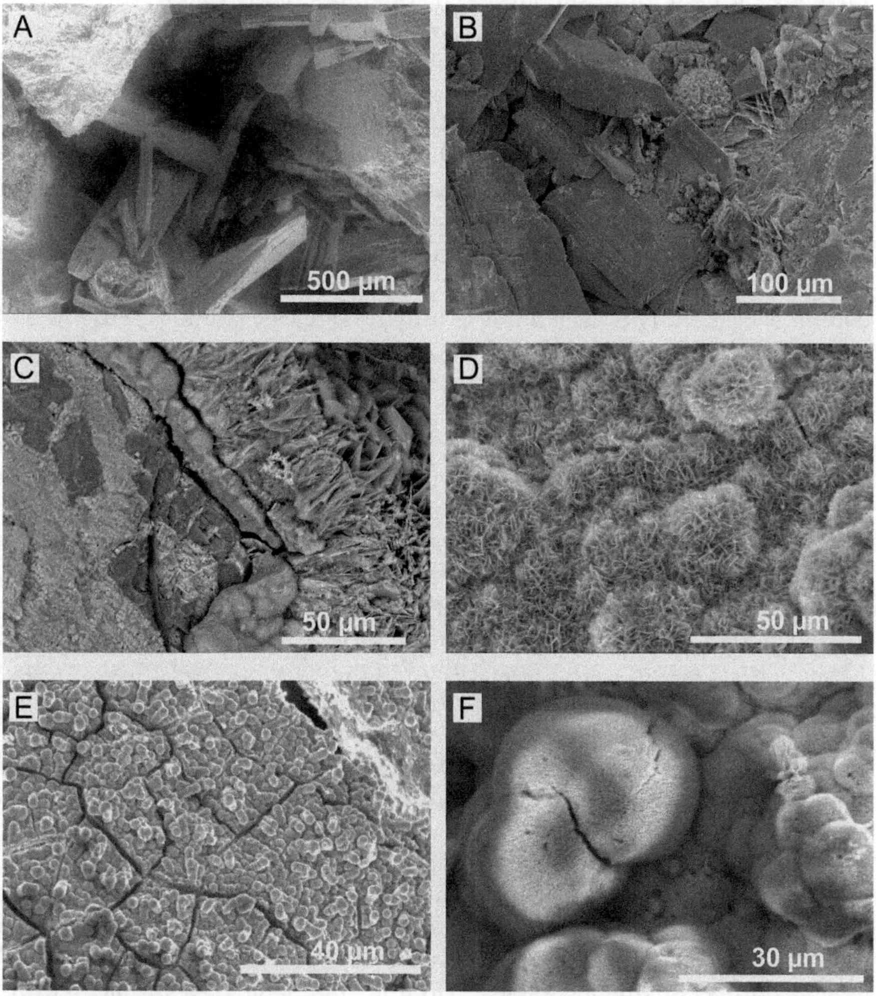

Figure 8.4 Backscattered scanning electron photomicrographs of reacted ferrous slags showing the formation of secondary phases after phosphate removal column experiments. Slags are from the Chicago–Gary area, Illinois and Indiana, USA. (A) Tabular brushite on reacted legacy slag. (B) Tabular Ca phosphate and finer-grained radiating rosette-shaped plates of Fe, Ca, and phosphate on reacted legacy slag. (C) Layers of Ca phosphate coatings, some with trace amounts of Mn and Mg, on reacted legacy slag. (D) Radiating rosette-shaped plates of hydroxyapatite on reacted BF slag (E) Worm-like morphology of Ca, Mg, and phosphate on reacted BF slag. (F) Fibrous nodule morphology of Ca, Zn, and phosphate on reacted BF slag. Reproduced from ref. 7, https://doi.org/10.3390/min9080468, under the terms of the CC BY 4.0 license, https://creativecommons.org/licenses/by/4.0/.

releases trace metals (Al, Cd, Co, and Hg) into the treated water, requiring further treatment for metal removal before reuse or discharge.[17]

Steel slag coated with sodium hydroxide (NaOH) was used to enhance precipitation–coagulation for P removal in batch and column tests. The coating process consists of stirring the slag with 1 M NaOH for 2 h, followed by drying at 80 °C for 48 h, 105 °C for 2 h, and 550 °C for 1 h. The coated slag was cooled down at room temperature and finally dried at 105 °C for 24 h. The slag's capacity to maintain high pH decreases with time.[18] Despite keeping the pH higher than uncoated slag, and subsequently showing a higher P removal efficiency, coated slag provided an excess of aluminum (Al) from the slag and sodium (Na) from the NaOH coating. Therefore, a return process where the column's effluent is returned to the initial wastewater tank to reuse Al and Na as a coagulant was introduced, allowing a longer lifetime of the slag with higher P removal, as well as a lower concentration of cations in the effluent.[18] Phosphorus removal efficiency increased as the slag grain size decreased. However, the coarse-grained slag coated by NaOH showed higher removal efficiency than the fine-grained slag without coating, which can be attributed to the higher pH level and faster coagulation caused by the NaOH coating.[18]

Another approach explored for P removal focused on combining deep-sea bacterium (*Alteromonas* 522-1) with BOF slag (Bio-BOFS) in batch tests.[19] With different experimental conditions (initial solution pH value of 5.0–9.0, P concentration of 5–30 mg L^{-1}, salinity of 0.5–3.5%, and temperature of 15–35 °C), the combined approach allowed higher P removal efficiency than BOF slag (>90%) as well as a neutral final solution (pH 7.8–8.0).[19] This combined approach avoided cementing of the slag and expanded the particle size. This approach could increase the longevity and retention capacity of slag for use in artificial marine construction for long-term P removal from shallow coastal environments.[19]

Combinations of dual filters consisting of BF slag, argon oxygen decarburization (AOD) slag, and electric arc furnace (EAF) slag were evaluated in column experiments with domestic wastewater for 24 days.[20] The column with only EAF slag had the highest P removal (93–98%). Phosphorus was removed as amorphous calcium phosphate.[20] Ground granulated blast furnace (GGBF) slag was also identified as a low-cost material for treating graywater (wastewater from households excluding that from toilets), focusing on phosphate removal.[21]

Removal of P from stormwater by EAF slag filters was also demonstrated through batch tests.[22] The primary mechanism of P removal in batch tests identified by X-ray absorption near edge structure (XANES) spectroscopy was the formation of hydroxyapatite precipitate (37–87% of total P), along with monetite [$Ca(PO_3OH)$], amorphous iron phosphate ($FePO_4$), amorphous aluminum phosphate ($AlPO_4$), crystalline and semicrystalline $AlPO_4$, and phosphate sorbed to aluminum hydroxide [$Al(OH)_3$].[22]

Research has also progressed on the synthesis of materials for P removal, such as porous calcium silicate hydrate derived from carbide slag (by-product from the production of acetylene by hydrolysis of calcium carbide (CaC_2)

and mainly composed of calcium hydroxide [$Ca(OH)_2$]).[23] This porous material has shown 95% removal of P in 4 h, a removal capacity up to 8.66 mg P g^{-1} after being reused four times, and the dominant reaction products were $Ca_3(PO_4)_2$ and $Ca_5(PO_4)_3OH$.[23]

8.2.2.1.2 Constructed Wetlands. Constructed wetlands are wastewater treatments that mimic the processes occurring in natural wetlands for water purification. These systems can treat municipal or industrial wastewater, AMD, graywater, or stormwater runoff. The main means of contaminant removal are biological processes (bacterial activity) in the rhizosphere. Nutrients and metals are mainly removed by plant uptake and phytoaccumulation. Wetlands are complex systems where several processes can remove contaminants, such as precipitation, reduction and oxidation, ion exchange, adsorption, volatilization and phytovolatilization, photochemical oxidation, sedimentation, and biofiltration. Slag has been extensively used as an active filter material in constructed wetlands, both at laboratory, pilot, and full scale (Table 8.1), using different designs (Figure 8.5). The use of iron and steel slags as a reactive substrate in constructed wetlands offers a cost-effective alternative to conventional wastewater treatment, especially for nutrient-rich wastewater.[24]

A pilot-scale constructed wetland at the Center for Alternative Wastewater Treatment, Ontario (Canada), was designed to remove nutrients, pathogens, and other contaminants from domestic wastewater, complemented by a CO_2 dosing system for neutralization of the effluent before discharge.[25] The system had a top-dosed, downward flow, sealed, and unvegetated BOF slag cell to remove P under fully saturated conditions. The mean hydraulic retention time was 6.3 days, with an average influent wastewater flow rate of 68 L day^{-1}.[25] Phosphorus was removed through adsorption on calcite ($CaCO_3$) as well as precipitates of β-tricalcium phosphate [$Ca_3(PO_4)_2$] and hydroxyapatite identified by XANES.[25] Under high pH conditions (pH > 9), HPO_4^{2-} and PO_4^{3-} were the dominant species and are attracted to positively charged surfaces, such as calcite, present in the BOF slag as a weathering product.[25] Elemental maps and energy-dispersive X-ray (EDX) spectra obtained from polished sections of samples indicated an accumulation of P on the slag grain boundaries. Disinfection of wastewater was also achieved due to the alkaline pH generated by BOF slag.[25]

Vertical flow wetlands for treating polluted urban river water showed superior steel slag performance (70.9–76.0% of total P) compared with gravel (48.6–65.0%) over 220 days.[26] In this study, chemisorption was considered the rate-controlling step and it was affected by different organic matter fractions, such that citric acid inhibited the P sorption, whereas oxalic acid promoted the combination of P with metal ions, especially at high concentrations.[26]

Steel slag and zeolite were selected as the filter medium for treatment of rural landfill leachate in a constructed wetland.[27] The removal performance

Table 8.1 Summary of the use of slag in constructed wetlands for P removal.

Type of constructed wetland	Active filter media	Type of wastewater	Duration of experiments (days)	Inlet P concentration ($mg\ L^{-1}$)	Hydraulic retention time (days)	Sorption capacity ($mg\ g^{-1}$)	Maximum P removal (%)	Authors and ref.
Pilot-scale indoor constructed wetland (capacity for 50–100 L day^{-1})	Basic oxygen furnace (BOF) slag	Domestic wastewater	240	6.97 (average)	6.5–14	—	95–99	Hussain *et al.*[25]
Vertical flow constructed wetland (V = 197 L)	BOF slag	Urban river polluted water	220	2–7.5	—	3.15	76	Ge et al.[26]
Rural household constructed wetlands (V = 0.8 L)	Steel slag	Domestic wastewater	150	3.4–13.9	5	—	>80	Lu et al.[101]
Subsurface flow constructed wetland (V = 1400 L)	Coarse gravel, zeolite and slag, soil and sand	Rural landfill leachate	50	0.2–0.6	—	—	—	He *et al.*[27]
Laboratory-scale constructed wetland (V = 130 L)	BOF, sand, fine gravel	Stormwater runoff	365	—	—	—	88 ± 3	Lucas and Babatunde[28]
Batch tests and small-scale horizontal flow wetland (V = 8 L)	Rapid cooled basic oxygen furnace slag (RC-BOFS)	Hydroponic wastewater	240	12.4 ± 2.2	3.2	3.6	94.2	Park *et al.*[102]

Batch and column tests, and field-scale	BOF slag	Peat extraction runoff	0.04 83 1825	0.008–0.250 0.670 0.025–0.050	— — —	— — —	98–99 (batch) 85 (column) 0 (field-scale)	Postila *et al.*[29]
Pilot-scale constructed wetland (total volume 8000 L) with free water surface, subsurface vertical flow, and subsurface horizontal flow	Iron and steel slags	Agricultural and domestic wastewater	60	3.1	—	—	81	Yuan *et al.*[24]
Horizontal subsurface flow constructed wetlands (V = 96 L)	Ti-bearing blast furnace slag Converter steel slag	Simulated wastewater	6	4.24 ± 0.87	6	—	98	Xu *et al.*[15]

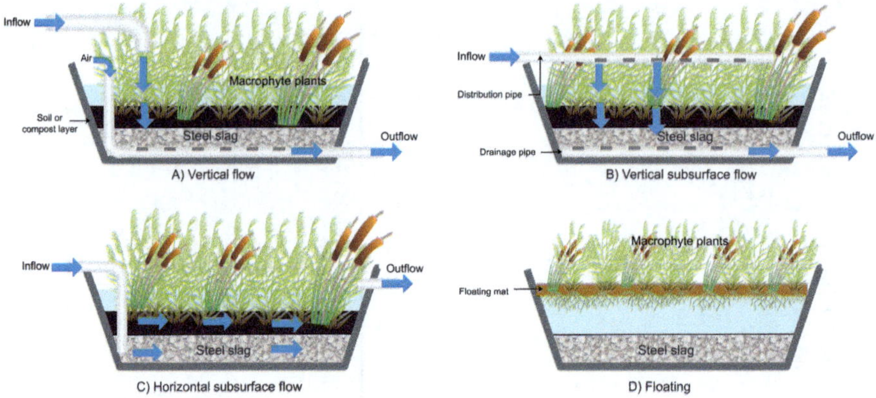

Figure 8.5 Use of slag in different designs for constructed wetlands: A) vertical flow; B) vertical subsurface flow; C) horizontal subsurface flow; D) floating wetland.

of the up-flow system demonstrated that it was a suitable alternative to conventional materials for treating this leachate, with 84.0–99.9% removal of NH_3–N and 80.3–92.1% removal of total N.[27]

Constructed wetlands for stormwater runoff treatment using BOF slag as the filter medium showed effective total N (>95%) and phosphate (>90%) removal.[28] Overall, the most effective configuration was the system with BF slag as the primary media, and that received a regular inflow of stormwater runoff (wet conditions).[28]

Most studies on constructed wetlands are performed at the laboratory scale, with only a few at pilot scale,[24,25,29,30] so there is a need to upscale and include variability in the flow and composition to better assess treatment performance, as well as long-term monitoring. For example, for low phosphorus concentrations in a field setup, steel slag could not retain PO_4^{3-}.[29] This could be due to the low pH, large grain size, and short retention time compared to laboratory-based experiments.[29] Furthermore, the use of steel slag as a substrate in constructed wetlands can produce effluents with high pH values and high electric conductivity, which need to be considered in the design and to be treated before discharge into water bodies.[16]

8.2.3 Aquaculture Amendment

Slags have also been used as amendments in aquaculture for improving water quality.[31] Desulfurization slag or ladle furnace slag was tested in a field-scale study in Pingtung County, Taiwan, to remove metals, ammonia (NH_3), and phosphate from fish pond water.[31] The NH_3 removal mechanism was physical adsorption and the mechanisms for phosphate removal included adsorption and chemical precipitation.[31] Ammonia was adsorbed as NH_4^+–N on the negative charges on the slag surface, obtaining a maximum of 77% removal, following a pseudo-second-order kinetic model. Maximum PO_4^{3-} removal was 99% and resulted in chemical adsorption with Ca present in the slag and also precipitation as hydroxyapatite. Algal growth and biomass accumulation could also be enhanced by the slag dissolution as it releases trace elements and hydroxide ions, which are beneficial to algal growth (*Chlorella* sp.).[31] Other benefits of the use of slag include an increase in water alkalinity and subsequent pH control.

8.2.4 Removal of Trace Elements

Ferrous slags have also been extensively studied and tested to remove trace elements, particularly metals from aqueous solutions. Different mechanisms are involved in the removal of metals, such as adsorption (physisorption and chemisorption), precipitation and co-precipitation (*e.g.*, hydroxides, sulfides, carbonates), and ion exchange.[14] As the slag surface is negatively charged, it selectively adsorbs the positively charged metal ions. Zhou and Haynes[32] suggested that precipitation and co-precipitation are the primary mechanisms for metal removal at high concentrations, whereas adsorption is predominant at lower metal concentrations.

Research focused on adsorption showed that water-quenched BF slag could be used for the removal of Cu, zinc (Zn), and cadmium (Cd) ions from aqueous solutions due to the high $(SiO_4)^{4-}$ content, which can promote ion exchange and adsorption, and calcium oxide (CaO), which has a high affinity for metal ions.[33] The removal efficiency was higher for $Cu^{2+} > Cd^{2+} > Zn^{2+}$, and adsorption was selective due to the difference in the adsorbent structure, ionic radius, and free energy of the hydrated ions.[33]

Steel slag was also used in batch tests to remove nickel (Ni) by adsorption.[34] At the optimal pH of 5, the slag had a maximum adsorption capacity of 36.49 mg g^{-1} for Ni^{2+} and adsorption increased with contact time, reaching saturation after 30 min.[34] Carbon steel slag, stainless steel slag, and Fe-modified stainless steel slag were used to remove As from synthetic and real groundwater.[35] The stainless steel slag was soaked in different concentrations of $FeCl_3$ solution and shaken at 160 rpm for 24 h at room temperature (25 °C) to produce the Fe-modified stainless steel slag. The precipitate was separated by centrifugation at 4000 rpm for 10 min and washed with deionized water to obtain pH 6.5–7.5. Finally, the sample was dried at 110 °C for 24 h and sealed in a desiccator before use. Under highly alkaline conditions (pH = 11), the mechanism of As removal by carbon steel slag and stainless steel was chemisorption, including chemical precipitation and coordination reaction.[35] Surface precipitation by electrostatic interaction and specific adsorption by Fe oxides were key processes for As removal. In particular, As forms inner-sphere bidentate surface complexes with goethite (α-FeOOH).[35] Arsenic(III) removal may also be attributed to the formation of calcium arsenite ($CaHAsO_3$) under high pH as follows:

$$HAsO_3^{2-} + Ca^{2+} + nH_2O \rightarrow CaHAsO_3 \cdot nH_2O \tag{8.14}$$

Regarding As(V), the main species are $HAsO_4^{2-}$ and AsO_4^{3-}, especially at pH > 9, which may react with Ca^{2+} as shown by the following equations:

$$HAsO_4^{2-} + Ca^{2+} + nH_2O \rightarrow CaHAsO_4{\cdot}nH_2O \tag{8.15}$$

$$2AsO_4^{3-} + 3Ca^{2+} + nH_2O \rightarrow Ca_3(AsO_4)_2{\cdot}nH_2O \tag{8.16}$$

Under weakly alkaline conditions (pH = 9), electrostatic interaction and specific adsorption are the mechanisms of As removal by Fe-modified stainless steel slag, showing that pH is a critical parameter for As removal.[35]

Besides the removal of nutrients covered in Section 8.2.2.1.2, constructed wetlands can also use slags as filter media for trace element removal. Steel slag was used as the filter medium in a pilot horizontal flow constructed wetland (volume of 0.69 m^3) for metal removal from landfill leachate, reaching 97% Cu, 95% Zn, and 91% Pb average removal over 15 months.[30] Recent batch studies showed that precipitation (over adsorption) is the main mechanism involved in the removal of Pb and Cd from natural waters by steel slag.[36] Both metals precipitated as the hydroxides $Pb(OH)_2$ and $Cd(OH)_2$, respectively.

AMD can be treated using constructed wetlands with BOF slag as the medium, replacing other neutralizing agents such as lime and sodium hydroxide (NaOH).[37] Iron was mainly removed as $Fe(OH)_3$, goethite, and jarosite [$KFe^{3+}{}_3(OH)_6(SO_4)_2$], whereas Al was removed as basaluminite [$Al_4(SO_4)(OH)_{10}{\cdot}4H_2O$], boehmite [$\gamma$-AlO(OH)], jurbanite [$AlSO_4(OH){\cdot}5H_2O$], gibbsite [$Al(OH)_3$], and diaspore [AlO(OH)].[37] Aluminum and iron precipitated as iron (oxy)-hydroxides and aluminum (oxy)-hydroxides.[37] Manganese precipitated as rhodochrosite ($MnCO_3$) and manganite (MnOOH), Ca and sulfate were removed as gypsum ($CaSO_4{\cdot}2H_2O$), while magnesium (Mg) was removed as brucite [$Mg(OH)_2$] and dolomite [$CaMg(CO_3)_2$].[37] The presence of FeO and Fe_3O_4 in the slag contributed to magnetite ($FeO{\cdot}Fe_2O_3$) formation as Fe^{2+} co-precipitated with Fe^{3+} from AMD. The authors targeted gypsum and magnetite ($FeO{\cdot}Fe_2O_3$) for recovery from the treated AMD, which can potentially and theoretically offset some treatment costs as sulfate concentrations in AMD are compatible with those for commercial gypsum recovery. However, these precipitates were only identified by XRD and their purity and separation processes were not specified.

However, a disadvantage of this technology is that most of the alkali and alkaline earth metals as well as sulfate from the AMD remained in solution, which implies that a polishing technology such as reverse osmosis needs to be coupled to produce water of drinking quality.[37] BOF slag can be a viable replacement for lime in AMD remediation.[38] A pilot study demonstrated the possibility of upscaling, treating 200–1000 L day^{-1}, with removals of 97% Al, 87% Fe, 100% Mn, and 87% SO_4^{2-}.[39]

Another mechanism for metal removal is based on redox processes. Hexavalent chromium [Cr(VI)] removal was achieved by homogeneous photocatalytic BF slag reduction with Fenton reagent in aqueous solutions.[40]

The Fenton reaction [eqn (8.17)] is a common oxidation reaction induced by iron complexes, generating $^{\bullet}OH$ radicals from H_2O_2 by Fe(II) salts:[41]

$$Fe^{2+} + H_2O_2 \rightarrow Fe^{3+} + OH^- + {}^{\bullet}OH \quad (8.17)$$

Han *et al.*[42] showed that it is possible to remove Cr(VI) from wastewater by Fe^{2+} released from BOF slag, with Fe^{2+} release being the rate-limiting step of Cr(VI) removal.[42] The Fe^{2+} is primarily derived from the dissolution of wüstite (FeO) and magnetite. The mechanism of removal was identified using an XRD and a scanning electron microscope (SEM) equipped with an energy-dispersive spectrometer (EDS) system and consists of several consecutive steps. The first is BOF slag dissolution releasing Fe^{2+} and Ca^{2+}, followed by Cr(VI) reduction to Cr(III) by Fe^{2+} and gypsum precipitation. The adsorption and crystallization of gypsum on the BOF slag surface hindered the contact of H^+ with the slag.[42]

Legacy waste metallurgical slag was used as an adsorbent for cerium (Ce^{3+}) ions in aqueous solution.[43] The mechanism of removal is based on the incorporation of Ce^{3+} ions into the structure of the aluminosilicate matrix through ion exchange reactions:[43]

$$3Ca^{2+} \leftrightarrow 2Ce^{3+} \quad (8.18)$$

$$3Mg^{2+} \leftrightarrow 2Ce^{3+} \quad (8.19)$$

Copper slags have been applied to the removal and immobilization of As from Cu smelting wastewaters.[44] Copper slag acted as an *in situ* Fe source and Si donor through dissolution:

$$Fe_2SiO_4 + 2H_2SO_4 \leftrightarrows 2FeSO_4 + H_4SiO_4 \quad (8.20)$$

$$Fe_3O_4 + 4H_2SO_4 \leftrightarrows FeSO_4 + Fe_2(SO_4)_3 + 4H_2O \quad (8.21)$$

When combined with hydrogen peroxide (H_2O_2) addition, this facilitated co-precipitation of silica gel and sulfate complexes, promoting the nucleation and growth of amorphous ferric arsenate ($FeAsO_4$, a scorodite precursor) and scorodite ($FeAsO{\cdot}2H_2O$), a non-hazardous and environmentally stable hydrated iron-arsenate mineral.[44] The overall reaction for As removal was suggested to be the dissolution of fayalite (Fe_2SiO_4) and magnetite as follows:[44]

$$Fe_2SiO_4 + 2H_3AsO_4 + H_2O_2 \rightarrow H_4SiO_4 + 2FeAsO_4 + 2H_2O \quad (8.22)$$

$$2Fe_3O_4 + 6H_3AsO_4 + H_2O_2 \rightarrow 6FeAsO_4 + 10H_2O \quad (8.23)$$

In a parallel study by the same authors, further attachment of silica gel on the scrodite surface resulted in a precipitate with good stability and very low As leachability.[45]

Slags have also been used in the development of innovative materials for trace metal removal. For example, crosslinked magnetic chitosan particles from steel slag and shrimp shells have been applied to remove metals such as Cu and Ni.[46] Chitosan is obtained through alkaline *N*-deacetylation of chitin from shrimp shells and steel slag is used to produce magnetite (Fe_3O_4). These materials showed high adsorption capacity for both metal ions (126.58 mg g^{-1} for Cu^{2+} and 66.23 mg g^{-1} for Ni^{2+}) and excellent regeneration stability (>83% after five cycles of the adsorption–regeneration process).[46] The magnetization of adsorbents and the application of magnetic separation techniques can be an effective way of resolving problems associated with separation and filtration in water treatment trains.[46] Another material resulted from the co-pyrolysis (300–700 °C) of rice straw and steel slag with CO_2 co-feed, giving a carbon–slag composite for As adsorption and removal from aqueous solution (99.3%).[47] Further applications for these materials, that is AMD or industrial wastewater treatment, have not yet been studied.

Research has substantially progressed in developing better adsorbents from slags, with activation processes (*e.g.*, thermal, mechanochemical, alkali activation, and geopolymerisation) to increase the specific surface area and density of adsorption sites on the slag surfaces to overcome issues like their chemical and shape heterogeneity.[8] Functionalized adsorbents (*e.g.*, geopolymers, zeolites, and layered double hydroxides) from silicate and aluminosilicate phases can be the next step in using slags for adsorption. However, further research is needed to integrate the engineered properties of slags with the full recovery and stabilization of dissolved and/or removed metals, as well as regeneration of the slag-based adsorbents.[8]

8.2.5 Removal of Sulfate

Sulfate removal mechanisms with slags are similar to those for other contaminants. The main mechanism for sulfate removal using steel slag is ion exchange with the hydroxyl group present in slag.[12] Precipitation is also an important mechanism. Sulfate removal from AMD by steel slag is attributed to the acid-neutralizing capacity and formation of gypsum.[48] Using BOF slag in batch tests, the efficiency of sulfate removal from AMD increased with temperature (25–45 °C), time (0–24 h), and solid loading (2–12 g BOF slag/50 mL AMD).[49] The maximum sulfate removal was 65% when 12 g of BOF slag per 100 mL of AMD was used at 45 °C with 24 h residence time.[49]

Blast furnace slag combined with metakaolin was used for developing a geopolymer to sorb sulfate from AMD, with a maximum sorption capacity of up to 119 mg g^{-1}, as demonstrated in both batch and column tests.[50] This can be an attractive sorbent for wastewater treatment for the mining industry or for industrial water applications that need low sulfate levels (*e.g.*, to minimize corrosion).[50]

8.2.6 Degradation of Emerging Pollutants and Organic Contaminants

The Fenton reaction [eqn (8.17)] has been used to tackle recalcitrant and persistent organic contaminants, mixed contaminants (inorganic and organic), and emerging pollutants[51] (*e.g.*, surfactants, flame retardants, pharmaceuticals and personal care products, gasoline additives and their degradation products, biocides, polar pesticides and their degradation products, and various proven or suspected endocrine-disrupting compounds, currently not included in routine environmental monitoring programs). The Fenton reaction can be converted into a photocatalytic process (photo-Fenton reaction) when at least one of the ligands is hydroxide, and the photoreduction of Fe^{3+} complexes generates a second $^{\bullet}OH$ radical.[41] Heterogeneous (solid–liquid) Fenton processes are highly efficient over a wide range of operational pH and show increased durability compared with the homogeneous Fenton reaction.[9] Montoya-Bautista *et al.*[52] recently published an extensive review of the catalyst and photocatalyst applications of slags for water treatment.

Metallurgical slags, both ferrous and base metal, showed potential to be used as Fenton photocatalysts in sunlight-assisted advanced oxidation processes at neutral pH.[53] Several Fe oxides commonly found in slag [*e.g.*, magnetite, goethite, maghemite (γ-Fe_2O_3), and hematite (Fe_2O_3)] have specific optical properties and are suitable for this application.[53] In addition, wüstite, zincite (ZnO), and quartz and its polymorphs (SiO_2) as constituents of iron slag can increase O_3 decomposition and enhance reaction rates with H_2O_2 leading to the generation of more $^{\bullet}OH$ radicals in Fenton and ozonation processes.[54]

Copper and steel slags were used as Fenton-type photocatalysts for degradation of diclofenac (a non-steroidal anti-inflammatory drug that can be found in waters as an emerging contaminant),[53] with the Cu slag showing higher photocatalytic activity as well as higher $^{\bullet}OH$ radical production. The oxidation experiments were performed in 250 mL Erlenmeyer flasks, stirred at 250 rpm and 35 °C. The flasks were placed inside a sunlight emission simulation chamber (SUNTEST CPS + ATLAS) with a filtered Xe arc lamp and an air-cooled Xe lamp system with 5–6% photon emissions between 290 and 400 nm. Complete removal of diclofenac was achieved after 90 min, with initial concentrations of 30 and 120 mg L^{-1}, whereas total mineralization took 5 hours.[53] Table 8.2 presents a summary of advanced oxidation processes where slags were used to degrade emerging pollutants in different treatments.

Steel converter slag was successfully used as a Fenton catalyst for the degradation of the herbicide alachlor after pre-treatment with salicylic acid and methanol to dissolve calcium silicate phases in steel slag and thus to improve the catalytic properties.[55] The pre-treatment (1) provided more solid–liquid interfaces for the heterogeneous Fenton-like reaction, (2) promoted the release of Fe from slag, and (3) prevented the rapid increase of the solution pH.[55]

Phosphorous slag, a waste aluminosilicate from phosphoric acid production, has been used successfully as an aluminosilicate precursor to synthesize an alkali-activated mesoporous adsorbent.[56] Milled P slag was mixed with

Table 8.2 Summary of the use of slag in water treatment processes to target emerging contaminants.

Type	Target contaminant	Treatment process	Type of slag	L/S ratio	Time (min)	Maximal removal (%)	Authors and ref.
Dyes and pigments	Brilliant red X-3B (azo dye)	Persulfate activated with steel slag (photocatalysis)	Steel slag	67–250	120	79.2	Wang *et al.*[103]
	Methylene blue	Absorption	Steel slag	0.1	120	99.9	Le *et al.*[34]
	Methylene blue and acid blue 29	Photodegradation (Fenton-like catalyst)	Magnetically activated electric arc furnace (EAF) slag	—	180	94 and 98	Nasuha *et al.*[104]
	Azo dye – Reactive red 24	Fenton and ozonation processes	Iron slag	—	60	85	Van *et al.*[54]
	Reactive black 5	Fenton process	EAF slag	5	180	94	Nasuha *et al.*[105]
Pesticides	Alachlor	Fenton process	Salicylic acid–methanol-modified steel converter slag	0.2	240	75.7	Chang *et al.*[55]
	Chlorpyrifos and its metabolite 3,5,6-trichloro-2-pyridinol	Subsurface flow constructed wetland	Coal boiler slag (melted bottom ash)	—	11 520	99	Yu *et al.*[106]

Pharmaceuticals	Diclofenac	Fenton – photocatalysis	Copper and steel slag	—	300	87	Arzate-Salgado *et al.*[53]
	Sulfamethazine	Fenton process	Steel slag modified with salicylic acid–methanol (SAM) and cobalt nitrate ($Co(NO_3)_2$)	0.14	360	96.7	Cheng *et al.*[107]
Preservatives	Paraben	Fenton process with activated persulfate	Basic oxygen furnace (BOF) slag	—	90	90	Matthaiou *et al.*[108]
Other	Petroleum hydrocarbons	Fenton process	Steel slag	0.03–0.08	240	64	Heidari *et al.*[109]
	Bisphenol A	Advanced oxidation process – solid activator of peroxydisulfate	Steel slag	0.33	60	70	Xu *et al.*[110]

a range of silica fume loadings and NaOH solution concentrations, before being cast in cylindrical molds. Ambient and hydrothermal curing (80 °C, saturated atmosphere, 48 h) followed by sequential boiling then washing was used to activate and purify the adsorbent. The cylindrical pellets were classified as mesoporous as determined by Brunauer–Emmett–Teller (BET) analysis. The adsorbent was applied to remove two cationic toxic dyes, basic violet (BV) and malachite green oxalate (MGO), from water with adsorption capacities of 46.57 mg g^{-1} and 46.36 mg g^{-1}, respectively.[56] These adsorption capacities are lower than others reported in the literature for synthetic adsorption materials,[57,58] but this process reuses a waste material, which can be more sustainable. The optimal adsorbent formulation based on the response surface method (RSM) was found to be when P slags were combined with 30 wt% of silica fume and 15 mol L^{-1} of NaOH.[56] Following seven cycles of sorption and desorption, an 11% reduction in overall capacity was observed.[56] The high removal efficiency was attributed to the high Si : Al ratio and highly ordered structure of Si–O–Si bonds, with a lower negative charge density. High OH^- selectivity resulted in better adsorption of cationic elements, as well as cationic dyes.

Copper slag can contain between 30 and 40 wt% Fe and up to 2 wt% Cu and is, therefore, a suitable feedstock for flotation recovery. The resulting flotation waste from copper slag (FWCS) has found use as an Fe source and precursor for the synthesis of micro-electrolysis materials (MEM) *via* a carbothermal reduction route for the degradation of organic contaminants in wastewater.[59] FWCS was mixed with different amounts of coal (15, 20, 25, 30, and 35 wt%), water, and sodium carboxymethyl binder, and shaped into 10 mm pellets, before being oven-dried (105 °C, 2 h).[59] A range of roasting temperatures (1000 to 1300 °C) were employed. Batch experiments investigated the effect of formulation, roasting temperature, and processing conditions on methyl orange (MO) decolorization (as measured by UV-vis). An increase in the roasting temperature was shown to promote the formation of zero-valent iron (ZVI) microparticles within the MEM, which were also shown to contain elemental Cu, which could enhance MO degradation. Close to 100% decolorization was observed within 20 min.[59] Mechanistic analysis indicated a redox reaction between the MEM and MO, destroying the azo bond, with the generation of sulfanilic acid.[59]

The indirect use of slags to synthesize photocatalyst materials has also progressed recently. Some examples are the synthesis of titanium dioxide (TiO_2) using titanium-bearing EAF molten slag[60] or supported on calcium silicate from BF slag,[61] as well as alkali-activated magnesium slag-based cementitious nanocomposites.[62]

Other advanced oxidation/electrochemical oxidation processes used a magnetic steel slag particle electrode in a novel electrochemical reactor for advanced treatment of oilfield wastewater.[63] Steel slag, clay, and a pore-forming agent were used as the primary raw materials for the electrode preparation. This treatment removed more than 85% of total organic carbon in 2 h, which was considered effective.[63]

8.3 Air Pollution Control

Ferrous and non-ferrous slags can also be applied in air pollution control processes, in particular as amendments or capping materials to control greenhouse gas emissions (GHG) and other fugitive emissions from rice paddy soils and solid waste landfills, but also as reactive materials to treat industrial exhaust gases. The phases in the slag, in particular Fe oxides, can react with the gases, generating stable solid products and avoiding atmospheric emissions.

Methane (CH_4) and nitrous oxide (N_2O) are two major greenhouse gases emitted from rice paddy soils. Estimates of global methane emissions from paddy soils range from 31 to 112 Tg a^{-1}, accounting for up to 19% of total emissions, while 11% of global agricultural nitrous oxide emissions are from rice fields.[64] Steel slag amendment of rice paddy soil has an emission mitigation potential of up to 60% in subtropical rice fields, with negative N_2O emissions.[65] Iron (Fe^{3+}) oxides resulting from the oxidation of Fe^{2+} from steel slag during dry periods in the rice fields play a major role in the mitigation of CH_4 emissions from paddy soil as they are thermodynamically more favorable electron acceptors than H_2/acetate for non-aerobic degradation of organic matter, thereby suppressing the production of CH_4 by methanogens:[65]

$$CH_2O + 4H^+ \rightarrow CH_4 + H_2O \tag{8.24}$$

$$CH_2O + 4Fe^3 + 2H_2O \rightarrow HCO_3 + 5H^+ + 4Fe^{2+} + 4e^- \tag{8.25}$$

Ferric Fe ions ultimately reduce CH_4 emissions *via* competition with the methanogens for electron donors and the sequential inhibitory effects. Silica oxides mitigate methane emissions by enlargement of the aerenchyma (soft tissue in aquatic plants that allows the exchange of gases between the shoot and the root), which increases oxygen transportation from the atmosphere to the root region of the plant, thus enhancing rhizospheric CH_4 oxidation (Figure 8.6).[65] Addition of steel slag increases the soil redox potential, which results in lower cumulative CH_4 emissions, especially in saturated or flooded soils.[65] Kumar *et al.*[65] showed that leaching of potentially toxic elements from the slag to the paddy soil was low enough not to pose any environmental risk, and an 18.3–19.6% decrease in inorganic As concentrations in rice grains was measured compared to the unamended control.[65]

The high Fe oxide concentrations in steel slag could also have mitigated the total N_2O emissions during the season by suppressing microbial activity. In steel slag amended paddy soils, higher C : N ratios and nitrification inhibition slowed down the N_2 cycling, thus affecting N_2O emissions.[65] Besides, the high amounts of ferric Fe might have also stimulated the reduction of nitrite to N_2O.[65]

Different slags [EAF slag, ladle furnace slag, and Linz–Donawitz (LD) converter slag] were used as amendments to reduce methane emissions from rice paddy fields.[66] Both bench-scale and greenhouse experiments showed

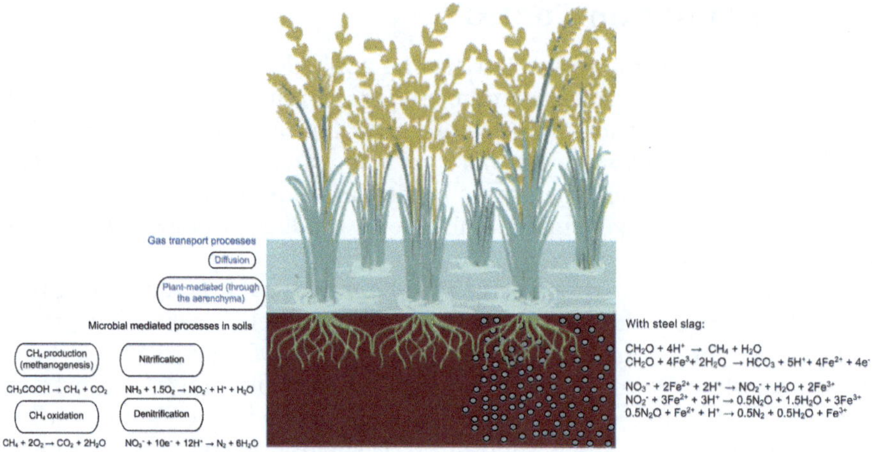

Figure 8.6 Mechanisms for mitigating greenhouse gas (GHG) emissions from rice paddy soils.[65,84]

markedly decreased CH_4 production rates in rice fields.[66] LD slag amendment at a rate of 2.0 Mg ha^{-1} in flooded rice fields significantly ($p < 0.05$) enhanced grain yield by 10.3–15.2% and mitigated CH_4 emissions by 17.8–24.0%.[66]

BOF slag with biochar was also tested to mitigate fugitive landfill emissions, including CH_4, CO_2, and H_2S, from municipal solid waste landfills.[67] Batch experiments showed carbonation in the range of 53–68 mg CO_2 g^{-1} BOF slag after 24 h.[67] Slags have also demonstrated promising potential in sequestering H_2S from landfill emissions. In particular, GGBF slag, in laboratory column tests, showed that the H_2S concentration could be reduced to below the olfactory threshold (0.02 mg m^{-3}) at an application rate of 30%.[68] The mechanism of sequestration is based on alkaline compounds (*e.g.*, CaO) absorbing H_2S to become sulfide ions [eqn (8.26)], some of which would subsequently be oxidized by metal oxides (*e.g.*, Fe_2O_3) to form elemental sulfur, metal sulfate, or metal thiosulfate through redox reactions, like the example shown in eqn (8.27):[68]

$$H_2S + CaO \rightarrow S^{2-} + Ca^{2+} + H_2O \quad (8.26)$$

$$Fe_2O_3 + 3H_2S \rightarrow 2FeS + 3H_2O + \tfrac{1}{2}S_2 \quad (8.27)$$

The direct treatment of industrial exhaust gases using slag-based catalysts has also been investigated. Copper slag has been used as an Mn ion support to synthesize Mn–Cu slag nanocatalysts for SO_2 oxidation owing to its resistance to acid solutions, mechanical stability, and high temperature resistance, which are critical for the proposed application environments.[69] As well as acting as a support for Mn ions, the Fe, Cu, and Mn oxides and hydroxides present in the Cu slag surface also contributed to the autoxidation of S^{6+}

oxides, ultimately generating H_2SO_4, critically at low temperatures (10–30 °C) and without the need for additional oxidizing agents such as H_2O_2.[69]

Copper slag has also been used to oxidize elemental mercury (Hg) in coal combustion flue gases in a bench-scale experimental setup with a tubular furnace. Here, iron oxide (Fe_2O_3) and copper oxide (CuO) present in the slag were responsible for the surface-mediated oxidation of Hg^0 in the presence of HCl in the flue gas, with good performance between 200 and 300 °C (optimal performance of 93.8% Hg^0 oxidation at 250 °C).[70]

8.4 Carbon Sequestration

Slags are the product of high-temperature processes and their mineralogy (see Chapter 3) is usually dominated by a range of silicate and oxide minerals.[5,6] The dominant phases in iron and steel slags (*e.g.*, larnite and gehlenite) can react with atmospheric carbon dioxide (CO_2) that has dissolved in water to form thermodynamically stable carbonate minerals [eqns (8.28) and (8.29)].

$$Ca_2SiO_{4(s)} + 2H_2CO_{3(aq)} \rightarrow 2CaCO_{3(s)} + 2H_2O_{(l)} + SiO_{2(s)} \tag{8.28}$$

$$Ca_2Al_2SiO_{7(s)} + 2H_2CO_{3(aq)} \rightarrow 2CaCO_{3(s)} + Al_2O_{3(s)} + 2H_2O_{(l)} + SiO_{2(s)} \tag{8.29}$$

The potential of iron and steel slags to act as sinks for atmospheric carbon through mineral carbonation has been demonstrated for several years[71] and is a much more rapid process than with many natural silicate minerals.[72] A large number of studies have investigated the carbon capture potential of slags under a range of environmental conditions. These have primarily been laboratory-based studies that show carbon capture potential can be enhanced at high partial pressures of CO_2 and high temperature (Table 8.3). These direct carbonation applications require active energy input for optimizing carbonation conditions and contrast with more passive or indirect approaches to carbonation under ambient environmental conditions. The latter has only recently been investigated at legacy iron and steel disposal sites where integrated hydrogeochemical and geophysical data were used to compare long-term carbon capture (*i.e.*, from mass balance of secondary carbonates) with carbonation potential (*i.e.*, based on slag composition and volume).[73,74] The data show that only a tiny portion of the potential atmospheric carbon capture is realized in ambient disposal conditions, typically less than 3%.[73,74] Low rates of diffusion of atmospheric CO_2 through capped waste heaps and passivation of slags due to secondary mineral formation are thought to minimize carbonation in such settings.[74,75] Efforts to enhance weathering, such as weathering slag in shallow heaps with controlled throughflow of water to limit secondary armoring, could provide a way forward for realizing the value of mineral carbonation at slag disposal sites.[66] Sites actively producing and handling slags would be best suited to integrating carbon capture as part of

Table 8.3 Recent carbonation experiments and carbonation potential of iron and steel slag. Adapted from ref. 73, https://pubs.acs.org/doi/10.1021/acs.est.8b01883, under the terms of the ACS AuthorChoice/Editors' Choice *via* Creative Commons CC-BY agreement, https://pubs.acs.org/page/policy/authorchoice_ccby_termsofuse.html.

Type of slag[a]	Type of carbonation	Ca extraction agent	Particle size	Liquid/solid ratio (kg/kg)	Temp.	CO_2 pressure	Reaction time	Maximum carbonation degree	Authors and ref.
EAF	Direct	Water	33.9 μm	0.3	100 °C	10 bar	24 h	280 g CO_2 kg^{-1} slag	Baciocchi *et al.*[111]
BOF	Direct	Water	14.3 μm	5	100 °C	10 bar	24 h	403 g CO_2 kg^{-1} slag	Baciocchi *et al.*[111]
BOF	Direct	Water	<125 μm	—	80 °C	20 bar	16 h	150 g CO_2 kg^{-1} slag	Quaghebeur *et al.*[112]
BOF	Direct	Water	125–250 μm	—	80 °C	20 bar	16 h	90 g CO_2 kg^{-1} slag	Quaghebeur *et al.*[112]
BOF	Direct	Water	250–500 μm	—	80 °C	20 bar	16 h	50 g CO_2 kg^{-1} slag	Quaghebeur *et al.*[112]
BOF	Direct	Water	500–1000 μm	—	80 °C	20 bar	16 h	40 g CO_2 kg^{-1} slag	Quaghebeur *et al.*[112]
SS	Direct	Water	<2 mm	10	30 ± 2 °C	6 bar	3 h	82 g CO_2 kg^{-1} slag	Rushendra *et al.*[113]
BOF	Direct	Water	<0.5 mm	5	100 °C	245 bar	24 h	71%	Su *et al.*[114]
EAF	Direct	Water	24 μm	4	Ambient	10.68 bar	10 min	46 g CO_2 kg^{-1} slag	Ghacham *et al.*[115]
EAF	Direct	Water	24 μm	10	Ambient	10.68 bar	10 min	52 g CO_2 kg^{-1} slag	Ghacham *et al.*[115]
BOF	Direct	Water	<44 μm	20	50 °C	1 bar	120 min	57%	Pan *et al.*[116]
BOF	Direct	Water	850–1000 μm	5	50 °C	5 bar	4 h	4.7 g CO_2 kg^{-1} slag	Polettini *et al.*[117]
BOF	Direct	Water	<38 μm	5	50 °C	5 bar	4 h	465 g CO_2 kg^{-1} slag	Polettini *et al.*[117]
BOF	Direct	Water	25–37 μm and 53–75 μm	30	25 °C and 90 °C	1 bar	240 h	45–70%	Yadav and Mehra[118]
AOD	Direct	Water	1.5–80 μm	0.4	Ambient temp	7.5 bar	>6 h	50–52%	Wang *et al.*[119]
SS	Direct	Water	100–1000 μm	10	10–60 °C	0.5 bar	24 h	30-75 g CO_2 kg^{-1} slag	Nielsen *et al.*[120]

[a]EAF – electric arc furnace, BOF – basic oxygen furnace, AOD – argon oxygen decarburization slag, SS – steel slag

their process trains, while recently closed steelworks would provide candidates for mineral carbonation to be integrated into post-closure restoration activities, where there may be co-benefits for minimizing leachate generation and improving bulk slag reuse potential.[76]

8.5 Soil Conditioner

Ferrous slags have also been used extensively as soil conditioners (*i.e.*, liming materials for soil pH correction) and fertilizers [*i.e.*, as rich sources of Ca and Mg].[14] In recent years, ferrous slags have been used to neutralize acidity in acid sulfate soils after mixing with other materials to reduce leachate pH.[77] Mechanical stabilization of clay soil was also achieved with GGBF slag activated by olivine (Mg_2SiO_4) and NaOH by mixing, compaction, and curing processes.[78]

Sewage sludge stabilized with steel slag was studied in a pot experiment using three soils with different pH to assess the effects on soil properties and winter wheat growth.[79] Results showed that the mixture of sewage sludge (75%, wet weight basis) stabilized with steel slag (25%, wet weight basis) could be used in agriculture, increasing biomass yield and nutrient uptake by crops, and applied at rates based on the sludge's Kjeldahl-N [sum of N bound in organic substances, ammonia (NH_3), and ammonium (NH_4^+)] content and the crop's demand for N.[79] It also contributed to improving the pH of acidic soils and the fertility of soils, mainly in terms of N and P.[79] The potential environmental impacts of these soil amendments include increased metal concentrations in soil, as well as soil salinization, nitrate leaching, phosphorus buildup, and boron phytotoxicity in the case of repeated heavy applications of the mixture to soil.[79]

A steel slag-based fertilizer was used in a greenhouse experiment with rice paddy soils to improve soil Si availability and crop productivity.[80] The results showed that application of steel slag reduced soil acidity, increased plant-available Si, promoted rice growth (dry weight of rice straw and grain), and inhibited Cd transport to rice grains in the soil–plant system.[80] No accumulation of Cd or Pb was observed in the paddy soil.[80] Iron and steel slags were also used for Na removal and nutrient supply for the recovery of the paddy fields damaged by the tsunami in Japan in 2011.[81] Positive effects of slag application on desalination, pH conditioning, and nutrient supply (Si) have been found on both regular and affected paddy fields.[81]

The use of iron and steel slags as environmental amendments (*e.g.*, filtration media, soil amendment) is recommended after ≥10 times dilution to avoid potential toxic and ecotoxicological effects due to alkaline leachate and the presence of potentially toxic elements.[82] Recently, restrictions on the use of steel slag as a soil amendment have been implemented in China (Cleaner Production Promotion Law) precisely due to the presence of potentially toxic elements, in particular, Cr, V, and As.[83]

An extensive review of the use of ferrous slags in agriculture highlights how slag amendment can increase crop yield and mitigate GHG emissions.[84] The potential risks of over-liming of soils and metal accumulation in the food chain (Cr, V, As, Ba, and Pb), potentially affecting human health, are also featured. There is a need for standardized and regulated soil application, mainly through process verification, assessment of environmental effects, acquisition of public approval, and also agreement and understanding from relevant authorities and stakeholders.[84]

8.6 Habitat Restoration

8.6.1 Calcareous Grasslands

Development of vegetation cover over slag deposits can aid land reclamation efforts, limit water ingress into waste heaps, and minimize the risk of fugitive dust generation.[85] In addition, the ecological importance of plant communities spontaneously colonizing slag deposits has long been recognized. For iron and steel slags, the dominant Ca silicate phases create a nutrient-poor substrate, with Ca limiting P availability, giving rise to a low-growing sward, which lends itself to floristic diversity. In some cases, this can be accelerated with soil amendments and the introduction of suitable native species.[86] The importance of such sites for biodiversity is reflected in conservation designations that have arisen at former slag disposal sites. A recent review of iron and steel slag disposal sites in the UK showed that 30% of historical disposal sites overlap with a range of conservation designations ranging from local to international significance. In numerous cases, the slag deposits themselves are cited in site notifications for a variety of reasons, including the presence of orchid-rich grassland and provision of habitats for declining amphibians as well as ground-nesting wading birds.[87] Such calcareous grasslands have been declining in extent in many jurisdictions and, as such, slag disposal areas can provide important regional inventories of such habitats.[88]

For base metal slags, skeletal, nutrient-poor substrates can limit plant growth in a similar fashion to iron and steel slags. There are, however, additional issues of metal availability that can both constrain plant growth and permit the colonization of metallophyte species tolerant to the conditions, such as *Minuartia verna* (spring sandwort), *Silene vulgaris* (bladder campion), and *Viola calaminaria* (zinc pansy).[89] In Europe, such calaminarian grassland habitats, as they are known, are rare and, while being largely anthropogenic in origin, they are protected under Annex I of the European Union Habitats Directive.[90] Early studies of grassland communities on metallurgical slags focused on revegetating slag heaps as part of phytostabilization and the addition of amendments to potentially accelerate the process.[91] Recent studies have assessed broader ecosystem value and functioning. Studies on Pb–Zn slag dumps in Upper Silesia (Poland) highlight the tolerance of cryptogam species (*e.g.*, lichens, bryophytes, and liverworts) to extreme metal concentrations in slag substrates.[92] These non-vascular plants serve as crucial pioneer communities for the natural revegetation of metal-rich slags and could even be targeted for subsequent seeding with vascular plants.[92] A wide-ranging study of aphids on Central European calaminarian grasslands found no significant difference in aphid diversity on metal-enriched sites, although community structure differed, which was related to the plant community structure.[93] The benefits of slag disposal sites for a range of insect pollinators, including bees, have also been demonstrated at sites in southern Belgium.[94] These studies of mature legacy sites are encouraging in showing the ecological enhancement that is possible at metallurgical slag disposal sites. Future research efforts to optimize the early stages of plant colonization will be beneficial to both encourage development of communities of conservation value and limit environmental risks at a time when the risk of leaching and dust generation is most pronounced.

8.6.2 Marine Habitats

Artificial marine reefs are essential for coastal rehabilitation and providing habitats for marine biota. Green artificial reef concrete, made with granulated blast furnace slag and steel slag, showed favorable physical properties and was suitable for the attachment growth of algae while simultaneously minimizing the use of primary raw materials.[95] However, no leaching tests were done on the concrete and further research is needed to ensure that both alkalinity and metals are not leaching from the slags.

In southwestern Taiwan, an *in situ* experiment deployed several BOF slags in a sandy bed off the Fangshan coast.[96] The results showed that the BOF slag was a potential substrate for attachment growth of coral reefs and subsequent coral rehabilitation.[96] EAF slag was used for coral nursery and transplantation into artificial lagoons in Sabah Al-Ahmad Sea City, Kuwait, serving as an artificial reef ecosystem.[97] Coral nubbins were glued individually to discs made of powdered EAF slag and marine-grade cement. The nubbins had a 56% survival rate and a mean skeletal extension of 10.6–13.4 mm in 10 months.[97]

Marine forest restoration of coastal algae grounds that have been damaged or lost due to sea desertification using steel slag has been tested from laboratory scale to actual field-scale in Japan.[98] Steel slag can act as both the substrate and a ferrous fertilizer for seaweeds (*Laminaria religiosa*), restoring algae grounds in coastal areas, which helps enhance coastal fisheries, but is also crucial from the standpoint of blue carbon (CO_2 removed from the atmosphere by coastal ecosystems).[98]

The use of slag as a basal medium for artificial tidal flats and seagrass beds showed that biodiversity, structure, and function could be restored in a degraded ecosystem.[99] Some slags can also mitigate the intense and rapid release of P from basal media due to adsorption and co-precipitation of phosphate with Fe, Ca, and Mg (major elements in slags) – two critical roles that impact the P cycle in sediments, one of which involves removal of P, and the other consists of the release of P at the sediment–seawater interface.[99]

Advances have also been made in developing high-density geopolymer concrete with steel furnace slag aggregate for coastal protection structures.[100] Steel slag aggregate offers higher bulk density to concrete, leading to adequate strength gain in ambient curing, and allowing diffusion of the free lime from the slag into the geopolymer matrix to eliminate the delayed hydration and expansion of the aggregate.[100] This high-density geopolymer concrete results in better breakwater stability and reduced material consumption, and can be utilized in breakwater armor unit fabrication.[100]

8.7 Concluding Remarks

Reuse of ferrous and non-ferrous slags in environmental applications can be a sustainable alternative to the use of primary raw materials, especially in water and wastewater treatment, contributing to the circular economy. Most studies on the environmental applications of slags are at bench scale (batch and column), so there is a need to develop research to upscale these applications in pilot studies, as well as include variability and long-term monitoring to better assess the performance. It is also crucial to demonstrate that potential environmental impacts, such as the generation of alkaline leachate and leaching of potentially toxic elements, are minimized. Another untapped area of potential is the recovery and valorization of metals and nutrients from saturated slags used as adsorbents.

Acknowledgements

The authors gratefully acknowledge the editors and the reviewer (Cristian Barca, Aix-Marseille University, France) for their insightful comments that helped improve this chapter. H.I. Gomes acknowledges the University of Nottingham for her funding through an Anne McLaren Fellowship. W.M. Mayes acknowledges the funding from the UK Natural Environment Research Council under grant NE/P019943/2.

References

1. B. Oberle, S. Bringezu, S. Hatfield-Dodds, S. Hellweg, H. Schandl, J. Clement, L. Cabernard, N. Che, D. Chen and H. Droz-Georget, *Global Resources Outlook 2019: Natural Resources for the Future We Want*, A Report of the International Resource Panel, United Nations Environment Programme, Nairobi, Kenya, 2019.
2. A. P. M. Velenturf, S. A. Archer, H. I. Gomes, B. Christgen, A. J. Lag-Brotons and P. Purnell, *Sci. Total Environ.*, 2019, **689**, 963.
3. H. I. Gomes, M. Rogerson, R. Courtney and W. M. Mayes, in *Resource Recovery from Wastes*, RSC Green Chemistry, 2020, pp. 168–191.
4. P. Deutz, H. Baxter, D. Gibbs, W. M. Mayes and H. I. Gomes, *Geoforum*, 2017, **85**, 336.

5. N. M. Piatak, M. B. Parsons and R. R. Seal, *Appl. Geochem.*, 2015, **57**, 236.
6. H. I. Gomes, W. M. Mayes, M. Rogerson, D. I. Stewart and I. T. Burke, *J. Cleaner Prod.*, 2016, **112**, 3571.
7. N. M. Piatak, R. R. Seal, D. A. Hoppe, C. J. Green and P. M. Buszka, *Minerals*, 2019, **9**, 468.
8. J. Manchisi, E. Matinde, N. A. Rowson, M. J. Simmons, G. S. Simate, S. Ndlovu and B. Mwewa, *Sustainability*, 2020, **12**, 2118.
9. B. M. Mercado-Borrayo, J. L. González-Chávez, R. M. Ramírez-Zamora and R. Schouwenaars, *J. Sustain. Metall.*, 2018, **4**, 50.
10. A. R. Lee, *Blastfurnace and Steel Slag, Production Properties and Uses*, 1974.
11. V. Masindi, M. S. Osman and A. M. Abu-Mahfouz, *Desalination*, 2017, **424**, 45.
12. M. S. Anjali, S. Shrihari and B. M. Sunil, *Environ. Technol. Innovation*, 2019, **15**, 100418.
13. H. I. Gomes, W. M. Mayes, P. Whitby and M. Rogerson, *J. Environ. Manage.*, 2019, **243**, 30.
14. S. Hosseini, S. M. Soltani, P. S. Fennell, T. S. Y. Choong and M. K. Aroua, *Environ. Technol. Rev.*, 2016, **5**, 1.
15. R. Xu, Y. Zhang, R. Liu, Y. Cao, G. Wang, L. Ji and Y. Xu, *Environ. Sci. Pollut. Res.*, 2019, **26**, 16229.
16. I. Blanco, P. Molle, L. E. Sáenz De Miera and G. Ansola, *Water Res.*, 2016, **89**, 355.
17. S. Yasipourtehrani, V. Strezov and T. Evans, *Sci. Rep.*, 2019, **9**, 1.
18. T. Park, V. Ampunan, S. Maeng and E. Chung, *Chemosphere*, 2017, **167**, 91.
19. W. Zhou, Z. Huang, C. Sun, H. Zhao and Y. Zhang, *Bioresour. Technol.*, 2016, **214**, 534.
20. M. Zuo, G. Renman, J. P. Gustafsson and W. Klysubun, *Environ. Sci. Pollut. Res.*, 2017, **25**, 7391.
21. D. T. K. James and A. O. Ifelebuegu, *Water Sci. Technol.*, 2018, **2017**, 667.
22. Z. Qin, A. L. Shober, K. G. Scheckel, C. J. Penn and K. C. Turner, *J. Environ. Qual.*, 2018, **47**, 1232.
23. D. Fang, L. Huang, Z. Fang, Q. Zhang, Q. Shen, Y. Li, X. Xu and F. Ji, *Chem. Eng. J.*, 2018, **354**, 1.
24. Z. Yuan, S. Fu, X. Xu and L. A. Wendling, *Environ. Processes*, 2017, **4**, 523.
25. S. I. Hussain, D. W. Blowes, C. J. Ptacek, J. H. Jamieson-Hanes, B. Wootton, G. Balch and J. Higgins, *Environ. Eng. Sci.*, 2015, **32**, 34.
26. Y. Ge, X. C. Wang, M. Dzakpasu, Y. Zheng, Y. Zhao and J. Xiong, *Water Sci. Technol.*, 2016, **73**, 2644.
27. H. He, Z. Duan, Z. Wang and B. Yue, *Environ. Sci. Pollut. Res.*, 2017, **24**, 1754.
28. R. Lucas and A. O. Babatunde, *J. Environ. Eng.*, 2017, **143**, 04017015.
29. H. Postila, S. M. Karjalainen and B. Kløve, *Ecol. Eng.*, 2017, **98**, 40.
30. A. Batool, *Environ. Sci. Pollut. Res.*, 2019, **26**, 3150.
31. B.-M. Yang, W.-L. Lai, Y.-M. Chang, Y.-S. Liang and C.-M. Kao, *J. Cleaner Prod.*, 2017, **143**, 131.

32. Y.-F. Zhou and R. J. Haynes, *Crit. Rev. Environ. Sci. Technol.*, 2010, **40**, 90.
33. Z. Wang, G. Huang, C. An, L. Chen and J. Liu, *Desalin. Water Treat.*, 2016, **57**, 22493.
34. V. T. Le, H. S. Le, X. V. Tran, T. K. N. Tran, D. Q. Vo, B. C. Tran, Q. P. Ngo and T. X. T. Le, *Vietnam. J. Sci. Technol. Eng.*, 2017, **59**, 7.
35. W. Shi, H. Li, G. Liao, G. Pei and Y. Lin, *Int. J. Environ. Sci. Technol.*, 2018, **15**, 2337.
36. D. H. Huy, E. Seelen and V. Liem-Nguyen, *J. Water Process Eng.*, 2020, **36**, 101369.
37. V. Masindi, M. S. Osman, R. N. Mbhele and R. Rikhotso, *J. Cleaner Prod.*, 2018, **172**, 289.
38. T. S. Naidu, L. D. Van Dyk, C. M. Sheridan and D. G. Grubb, *J. Hazard., Toxic Radioact. Waste*, 2020, **24**, 04019028.
39. T. S. Naidu, C. M. Sheridan, L. D. Van Dyk and A. Higginson, presented in part at the IMWA 2019 - Mine Water: Technological and Ecological Challenges, Perm, Russia, 2019.
40. D. Li and J. Han, *Pol. J. Environ. Stud.*, 2019, **28**, 37.
41. Z. Stasicka, in *Advances in Inorganic Chemistry*, ed. R. V. Eldik and G. Stochel, Academic Press, 2011, vol. 63, pp. 291–343.
42. C. Han, Y. Jiao, Q. Wu, W. Yang, H. Yang and X. Xue, *J. Environ. Sci.*, 2016, **46**, 63.
43. A. G. Morozova, T. M. Lonzinger, V. A. Skotnikov, J. N. Sahu, G. G. Mikhailov, J. L. Schenk, A. Bhattacharyya and Y. Kapelyushin, *J. Cleaner Prod.*, 2020, **255**, 120286.
44. Y. Li, X. Zhu, X. Qi, B. Shu, X. Zhang, K. Li, Y. Wei, F. Hao and H. Wang, *J. Cleaner Prod.*, 2020, **270**, 122428.
45. Y. Li, X. Zhu, X. Qi, B. Shu, X. Zhang, K. Li, Y. Wei and H. Wang, *Chem. Eng. J.*, 2020, **394**, 124833.
46. L. V. Thuan, T. B. Chau, T. T. K. Ngan, T. X. Vu, D. D. Nguyen, M.-H. Nguyen, D. T. T. Thao, N. To Hoai and L. H. Sinh, *Environ. Technol.*, 2018, **39**, 1745.
47. T. Lee, J. I. Oh, H. Yi, Y. F. Tsang and E. E. Kwon, *Chem. Eng. J.*, 2019, **361**, 1630.
48. K. R. Reddy, A. Gopakumar and J. K. Chetri, *Rev. Environ. Sci. Biotechnol.*, 2019, **18**, 127.
49. F. Ntuli and N. P. Magwa, *IOP Conf. Ser.: Earth Environ. Sci.*, 2018, **191**, 012116.
50. H. Runtti, T. Luukkonen, M. Niskanen, S. Tuomikoski, T. Kangas, P. Tynjälä, E.-T. Tolonen, M. Sarkkinen, K. Kemppainen, J. Rämö and U. Lassi, *J. Hazard. Mater.*, 2016, **317**, 373.
51. V. Geissen, H. Mol, E. Klumpp, G. Umlauf, M. Nadal, M. van der Ploeg, S. E. A. T. M. van de Zee and C. J. Ritsema, *Int. Soil Water Conserv. Res.*, 2015, **3**, 57.
52. C. V. Montoya-Bautista, E. Avella, R.-M. Ramírez-Zamora and R. Schouwenaars, *Sustainability*, 2019, **11**, 2470.
53. S.-Y. Arzate-Salgado, A.-A. Morales-Pérez, M. Solís-López and R.-M. Ramírez-Zamora, *Catal. Today*, 2016, **266**, 126.

54. H. T. Van, L. H. Nguyen, T. K. Hoang, T. P. Tran, A. T. Vo, T. T. Pham and X. C. Nguyen, *Sep. Purif. Technol.*, 2019, **224**, 43.
55. M. Cheng, G. Zeng, D. Huang, C. Lai, Y. Liu, P. Xu, C. Zhang, J. Wan, L. Hu, W. Xiong and C. Zhou, *Chem. Eng. J.*, 2017, **327**, 686.
56. A. Salehi and E. Najafi Kani, *Adsorption*, 2018, **24**, 647.
57. A. Awadallah-F, *J. Polym. Environ.*, 2017, **25**, 258.
58. M. Abdel Salam, M. R. Abukhadra and A. Adlii, *ACS Omega*, 2020, **5**, 2766.
59. W. Yu, Y. Sun, M. Lei, S. Chen, T. Qiu and Q. Tang, *J. Hazard. Mater.*, 2019, **361**, 221.
60. Y. Li, Y. Yang, M. Guo and M. Zhang, *RSC Adv.*, 2015, **5**, 1347.
61. J. Shi, Y. Kuwahara, T. An and H. Yamashita, *Catal. Today*, 2017, **281**, 21.
62. Z. Yaojun, K. Le and L. Licai, *Rare Met. Mater. Eng.*, 2015, **44**, 303.
63. Z. Wang, X. He, J. Li, J. Qi, C. Zhao and G. Yang, *J. Ind. Eng. Chem.*, 2018, **58**, 18.
64. E. P. Win, K. K. Win, S. D. Bellingrath-Kimura and A. Z. Oo, *Greenhouse Gases*, 2020, **10**, 88.
65. S. S. Kumar, A. Kumar, S. Singh, S. K. Malyan, S. Baram, J. Sharma, R. Singh and A. Pugazhendhi, *Chemosphere*, 2020, **241**, 124824.
66. H. S. Gwon, M. I. Khan, M. A. Alam, S. Das and P. J. Kim, *J. Hazard. Mater.*, 2018, **353**, 23.
67. K. Reddy, G. Kumar, A. Gopakumar, R. Rai and D. Grubb, presented in part at International Conference on Protection and Restoration of the Environment XIV, Thessaloniki, 2018.
68. C. W. W. Ng, M. Xie and A. K. Leung, *J. Environ. Eng.*, 2017, **143**, 04017016.
69. F. Rabiee and K. Mahanpoor, *Int. J. Ind. Chem.*, 2018, **9**, 27.
70. H. Li, W. Zhang, J. Wang, Z. Yang, L. Li and K. Shih, *Waste Manage.*, 2018, **74**, 253.
71. W. J. J. Huijgen, G.-J. Witkamp and R. N. J. Comans, *Environ. Sci. Technol.*, 2005, **39**, 9676.
72. P. Renforth, *Nat. Commun.*, 2019, **10**, 1401.
73. W. M. Mayes, A. L. Riley, H. I. Gomes, P. Brabham, J. Hamlyn, H. Pullin and P. Renforth, *Environ. Sci. Technol.*, 2018, **52**, 7892.
74. H. Pullin, A. W. Bray, I. T. Burke, D. D. Muir, D. J. Sapsford, W. M. Mayes and P. Renforth, *Environ. Sci. Technol.*, 2019, **53**, 9502.
75. D. I. Stewart, A. W. Bray, G. Udoma, A. J. Hobson, W. M. Mayes, M. Rogerson and I. T. Burke, *Environ. Sci. Pollut. Res.*, 2018, **25**, 9861.
76. W. M. Mayes, S. L. Hull and H. I. Gomes, in *Sustainability and Circular Economy*, ed. A. Stefanakis, Elsevier, 2021.
77. L. A. Wendling, G. B. Douglas and S. Coleman, *Miner. Eng.*, 2012, **39**, 21.
78. M. H. Fasihnikoutalab, S. Pourakbar, R. J. Ball, C. Unluer and N. Cristelo, *Acta Geotech.*, 2019, **15**, 1981.
79. E. Samara, T. Matsi and A. Balidakis, *Waste Manage.*, 2017, **68**, 378.
80. D. Ning, Y. Liang, Z. Liu, J. Xiao and A. Duan, *PLoS One*, 2016, **11**, e0168163.
81. X. Gao, M. Okubo, N. Maruoka, H. Shibata, T. Ito and S. Y. Kitamura, *Miner. Process. Extr. Metall.*, 2015, **124**, 116.

82. L. A. Wendling, M. T. Binet, Z. Yuan, F. Gissi, D. J. Koppel and M. S. Adams, *Environ. Toxicol. Chem.*, 2013, **32**, 2602.
83. J. Guo, Y. Bao and M. Wang, *Waste Manage.*, 2018, **78**, 31.
84. S. Das, S. J. Galgo, M. A. Alam, J. G. Lee, H. Y. Hwang, C. H. Lee and P. J. Kim, *Crit. Rev. Environ. Sci. Technol.*, 2020, 1.
85. S. Siebielec, G. Siebielec, T. Stuczyński, P. Sugier, E. Grzęda and J. Grządziel, *Sci. Total Environ.*, 2018, **636**, 104.
86. H. Ash, R. Gemmell and A. Bradshaw, *J. Appl. Ecol.*, 1994, 74.
87. A. L. Riley, J. M. MacDonald, I. T. Burke, P. Renforth, A. P. Jarvis, K. A. Hudson-Edwards, J. McKie and W. M. Mayes, *J. Geochem. Explor.*, 2020, **219**, 106630.
88. G. Rosquist, *Åtgärdsprogram för sandstäpp, 2015–2019: (Xeric sand calcareous grasslands). Naturvårdsverket*, Swedish Environmental Protection Agency, Stockholm. Sweden, 2017, vol. 45, p. 2122.
89. EEA, *Calaminarian Grasslands of the Violetalia Calaminariae*, https://eunis.eea.europa.eu/habitats/10113, accessed 20 June 2020.
90. EC, *The Habitats Directive*, https://ec.europa.eu/environment/nature/legislation/habitatsdirective/index_en.htm, accessed 25 June 2020.
91. D. W. Shimwell and A. E. Laurie, *Environ. Pollut.*, 1972, **3**, 29.
92. K. Rola, P. Osyczka, M. Nobis and P. Drozd, *Ecol. Eng.*, 2015, **75**, 332.
93. B. Osiadacz and R. Hałaj, *Redia*, 2016, **XCIX**, 3.
94. T. Henneresse and D. Tyteca, *J. Insect Sci.*, 2016, **16**, 104.
95. X. Huang, Z. Wang, Y. Liu, W. Hu and W. Ni, *Constr. Build. Mater.*, 2016, **112**, 24.
96. P. Chen, C. Kang and J. Yu, *OCEANS 2019 – Marseille*, 2019, p. 1, DOI: 10.1109/OCEANSE.2019.8867082.
97. M. Nithyanandan, L. Le Vay, D. K. Raja, R. Kesavan and D. Pereira, *Cogent Environ. Sci.*, 2018, **4**, 1480334.
98. T. Kato, C. Kosugi, E. Kiso and K. Torii, *Nippon Steel & Sumitomo Metal Technical Report*, 2015, vol. 109, p. 79.
99. A. Tsukasaki, M. Suzumura and W. Nishijima, *ISIJ Int.*, 2015, **55**, 183.
100. A. H. Mahmood, S. J. Foster and A. Castel, *Constr. Build. Mater.*, 2020, **248**, 118681.
101. S. Lu, X. Zhang, J. Wang and L. Pei, *J. Cleaner Prod.*, 2016, **127**, 325.
102. J.-H. Park, J. J. Wang, S.-H. Kim, J.-S. Cho, S.-W. Kang, R. D. Delaune and D.-C. Seo, *Chem. Eng. J.*, 2017, **327**, 713.
103. Z. Wang, Y. Guo, Y. Chen, S. Liang, J. Zuo and X. Yu, *J. Environ. Eng.*, 2016, **142**, 04015048.
104. N. Nasuha, S. Ismail and B. H. Hameed, *J. Environ. Manage.*, 2017, **196**, 323.
105. N. Nasuha, S. Ismail and B. Hameed, *J. Taiwan Inst. Chem. Eng.*, 2016, **67**, 235.
106. X. Yu, H. Zhu, B. Yan, Y. Xu, G. Bañuelos, B. Shutes, H. Wen and R. Cheng, *J. Hazard. Mater.*, 2019, **373**, 67.
107. M. Cheng, G. Zeng, D. Huang, C. Lai, Y. Liu, C. Zhang, J. Wan, L. Hu, C. Zhou and W. Xiong, *Water Res.*, 2018, **138**, 7–18.

108. V. Matthaiou, P. Oulego, Z. Frontistis, S. Collado, D. Hela, I. K. Konstantinou, M. Diaz and D. Mantzavinos, *Chem. Eng. J.*, 2019, **360**, 72.
109. B. Heidari, M. Soleimani and N. Mirghaffari, *Water Sci. Technol.*, 2018, **78**, 1159.
110. X. Xu, D. Liu, W. Chen, S. Zong and Y. Liu, *Environ. Chem. Lett.*, 2018, **16**, 1435.
111. R. Baciocchi, G. Costa, M. Di Gianfilippo, A. Polettini, R. Pomi and A. Stramazzo, *J. Hazard. Mater.*, 2015, **283**, 302.
112. M. Quaghebeur, P. Nielsen, L. Horckmans and D. Van Mechelen, *Front. Energy Res.*, 2015, **3**, 52.
113. T. D. Rushendra Revathy, K. Palanivelu and A. Ramachandran, *Environ. Sci. Pollut. Res.*, 2015, 7349–7359.
114. T. H. Su, H. J. Yang, Y. H. Shau, E. Takazawa and Y. C. Lee, *J. Environ. Sci.*, 2016, **41**, 9.
115. A. B. Ghacham, L.-C. Pasquier, E. Cecchi, J.-F. Blais and G. Mercier, *Environ. Sci. Pollut. Res.*, 2016, 17635–17646.
116. S. Y. Pan, H. L. Liu, E. E. Chang, H. Kim, Y. H. Chen and P. C. Chiang, *Chemosphere*, 2016, **154**, 63.
117. A. Polettini, R. Pomi and A. Stramazzo, *Chem. Eng. J.*, 2016, **298**, 26.
118. S. Yadav and A. Mehra, *Waste Manage.*, 2017, **64**, 348.
119. Y.-J. Wang, Y.-N. Zeng, J.-G. Li, Y.-Z. Zhang, Y.-J. Zhang and Q.-Z. Zhao, *J. Cleaner Prod.*, 2020, **256**, 120377.
120. P. Nielsen, M. Boone, L. Horckmans, R. Snellings and M. Quaghebeur, *J. CO2 Util.*, 2020, **36**, 124.

CHAPTER 9

Secondary Metal Recovery from Slags

ANNA POTYSZ*[a] AND ERIC D. VAN HULLEBUSCH[b]

[a]Institute of Geological Sciences, University of Wrocław, Cybulskiego 30, 50-205 Wrocław, Poland; [b]Institut de Physique du Globe de Paris, Université de Paris, CNRS, F-75005 Paris, France
*E-mail: anna.potysz@uwr.edu.pl, apotysz@gmail.com

9.1 Introduction

Pyrometallurgy is a widely applied technology for producing metals that are imperative for a number of industrial applications, including electronics, construction, automotive, medical, and many other sectors. Therefore, the pyrometallurgical sector continues to be a key contributor in metal production, with activity likely to increase in magnitude and importance in the coming decades.[1] Metal extraction and purification lead to the generation of a final product – pure metal. However, despite an improvement in the efficiency of pyrometallurgical extraction processes, some metal losses occur using current methods.[2] Furthermore, the extraction process generates variable volumes of by-products, such as ashes, residues, and slags, with the lattermost accounting for a considerable mass, reaching levels of over hundreds of millions of tonnes per year.[3,4] Generally, two types of pyrometallurgical slags are generated by the smelting industry: non-ferrous slags and ferrous slags; the latter originate from iron and steel production while the former are commonly derived from the production of base metals other than iron itself.[5] While they differ in many respects, they also have some aspects

Chemistry in the Environment Series No. 2
Metallurgical Slags: Environmental Geochemistry and Resource Potential
Edited by Nadine M. Piatak and Vojtěch Ettler

Published by the Royal Society of Chemistry, www.rsc.org

in common: all of these slag types, when appropriately managed, can be transformed from something overlooked or ignored into valuable resources. Ferrous slags have already found their application route in the building and construction industries[6] as additives to cement (Chapter 7) and as soil conditioners, fertilizers, and liming materials (Chapter 8); ferrous slags can also be used for phosphorous entrapment[7] and CO_2 capturing (Chapter 8),[8] whereas base metal slags may be valuable as a metal repository.[9,10]

The amount of slag produced in comparison to the amount of metal produced varies based on the commodity with approximately 0.6 tonnes of slag for every tonne of Pb and 2.2 tonnes of slag for every tonne of Cu produced.[10,11] Since pyrometallurgical production capacity is not anticipated to diminish in the foreseeable future, an associated rise in waste generation can also be predicted, leading to economic and environmental issues regarding their final destination. Limited space for disposal and mounting evidence on the adverse environmental impacts of slags[12–15] (Chapters 4, 5, and 6) have resulted in urgency in the search for alternative uses for slags maximizing their recycling potential.[16] At present, gravity and magnetic separation, as well as flotation and roasting, are applied to improve the recovery of metals initially incorporated in slags. The supergravity method enables separation of liquid metal phases from the remaining slag. Magnetic separation is applied to concentrate magnetically responsive minerals. Flotation is mainly used to recover base metals hosted in sulfides and metals. Therefore, the choice of flotation is reasonable provided that the metal-rich phases are large enough to be liberated.[17] Roasting is achieved through the conversion of metals incorporated within slags into soluble phases such as sulfates and/or oxides at high temperatures ranging from 200 to 600 °C.[18–23] Finally, one of the prospective treatment routes is a metal extraction technique termed "(bio)leaching".

A strong focus has recently been put on the content of metals incorporated in slags, requiring new estimates regarding the amounts of metal resources on a global scale. In fact, metal resources are greater than previously estimated considering the amount of metal now known to be incorporated in secondary sources.[24] Currently, the European Union (EU) has defined a list of critical elements (updated in 2020) based on two criteria: economic importance and the risk to the supply of the selected raw materials.[25] Critical raw materials as specified by the EU include: antimony (Sb), beryllium (Be), cobalt (Co), gallium (Ga), germanium (Ge), indium (In), magnesium (Mg), niobium (Nb), scandium (Sc), platinum group elements (PGEs), phosphorus (P), rare earth elements (REEs), tantalum (Ta), tungsten (W), molybdenum (Mo), nickel (Ni), and vanadium (V), among others.[25] The majority of these critical elements can be encountered in metallurgical slags and other metallurgical wastes in relatively high concentrations, rendering these residues, with previously unappreciated value, potentially profitable.[26,27] The potential of slags in terms of metal recovery still remains under evaluation. This provides researchers the opportunity to pursue novel experimental studies combining interdisciplinary approaches linking aspects related to slag mineralogy, slag properties, and dissolution rates. The ever-increasing number and availability of scientific reports on metal recovery from slags is proof of interest in the subject and that researchers are striving to expand their knowledge on it.[19,28,29] The widespread experimentation with metal extraction from slags and its abundant production suggest that slags may provide a consistent supply of secondary metal resources; the objective now is to improve the process development and overall optimization.

9.2 Recoverable Metals by Slag Type and Mineralogy

Metallurgical slags are defined as being predominantly composed of silicate and oxide phases originating from the smelting of metal-rich ores or secondary sources. Two main types of metallurgical slag are recognized, one from the primary production of ferrous ores, usually iron and steel manufacturing, and one from non-ferrous ores, typically from the recovery of base metals and some precious metals.[10] Chemical and mineralogical features vary from one slag type to another, whereas the presence of metals is a typical attribute of the compositions of both slags.[14,30–33] Ferrous slags mainly contain Ca and Si accompanied by Mg, Al, and Fe.[5] In contrast, the main components of non-ferrous slags are typically Si and Fe with somewhat lower levels of Al and Ca (Chapter 3). Additionally, variable metallurgical process specifications, different melt compositions, and different cooling rates result in different mineralogical compositions in the final product. The phase composition of ferrous slags is commonly characterized by the presence of Ca-rich olivine group ($(Ca,Mg,Mn,Fe)_2SiO_4$) phases, melilite group ($(Ca,Na)_2(Al,Mg,Fe^{2+})(Al,Si)_2O_7$) phases that contain Al or Mg, silicate glass, and oxides with lesser volumetric proportions of calcite and trace amounts of a variety of sulfides, intermetallic compounds, and pure metals. The non-ferrous slags are predominantly composed of silicate group phases [the olivine, pyroxene ($XY(Si,Al)_2O_6$, where X = Ca, Na, Fe^{2+}, or Mg, and Y = Cr, Al, Fe^{3+}, Mg, Mn, or Ti), and melilite groups], silicate glass, spinels, and SiO_2 (*i.e.*, quartz and other polymorphs).[34–36] The metals are predominantly incorporated within volumetrically minor to trace sulfides and intermetallic compounds intimately associated with the volumetrically major silicates and oxides (Figure 3.6, Chapter 3).

Melt generated during the smelting process solidifies and its cooling regime determines the level of crystallinity and the subsequent behavior of these wastes in terms of environmental and recycling industry matters.[37] Slow cooling generates well-formed crystalline phases resulting in slag of generally crystalline character.[35,38] This is in contrast to water-mediated quenching, during which the melt rapidly solidifies and, depending on the melt compositions, may generate an amorphous or glass-rich slag.[39–41]

Element partitioning during smelting determines what elements end up in the slag and are, therefore, potential targets for recovery. Figure 9.1 presents the tendency of individual elements to end up in the slag, metal, or gaseous phase based on thermodynamics. Fundamental knowledge of the thermodynamic properties of minor and trace elements and their behaviors during smelting are important for predicting which may be favorable for recovery.

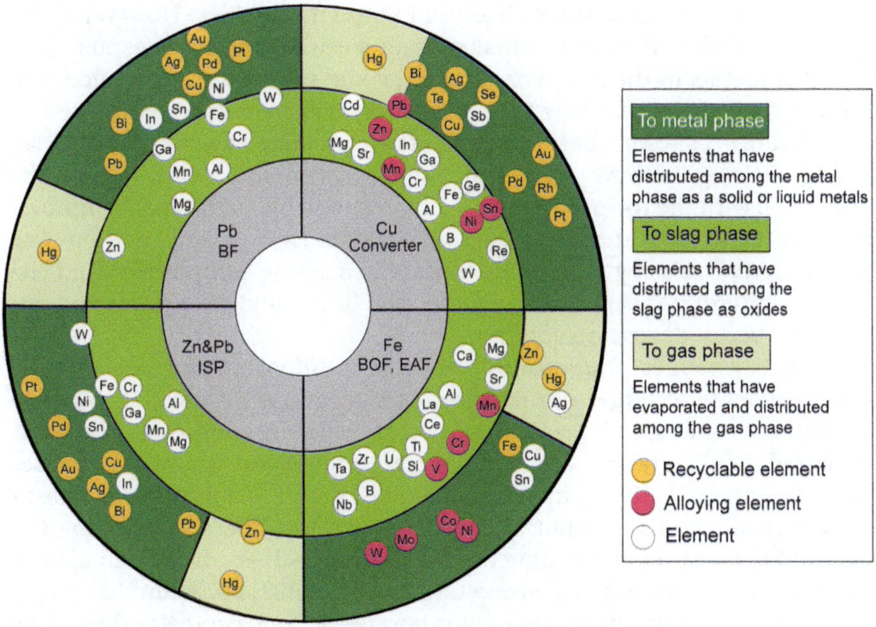

Figure 9.1 Element radar chart presenting distributions of elements among the metal, slag, and gas phases during pyrometallurgical processing. Abbreviations: BF: blast furnace slag, BOF: basic oxygen furnace, EAF: electric arc furnace, ISP: imperial smelting process. Reproduced from ref. 5 with permission from Elsevier, Copyright 2018.

Base metal slags are enriched in metals such as Cu, Zn, Pb, Ni, and Co (Table 3.2, Chapter 3). The quantitatively important enrichment of individual metals in slags is dependent on the type of ore processed, which means that incomplete recovery of the target metal during smelting may result in its transfer to the slag or gas phase. Compiled slag chemistry (Table 3.2, Chapter 3) indicates that Cu slags contain an average of 2 wt% Cu. Enrichment of Zn in Cu slags, with an average of 3 wt%, is lower than that observed in Pb and Zn slags, with averages of nearly 4–5 wt% and 7 wt%, respectively. Lead slags are the main carriers of Pb (with an average of nearly 9 wt%), whereas other base metal slags, on average, contain Pb contents of approximately 1 wt% or less. In contrast, the average Ni content in the various base metal slag types is 0.6 wt% Ni or less, with the highest average and maximum value (0.6 wt% and 1.2 wt%, respectively) for slags resulting from Ni sulfide ore processing (Table 3.2). Other elements in base metals slags occur in strikingly lower amounts; however, Cu slags can still be valuable for Co recovery (up to 2.4 wt%), Pb slags can serve as a secondary source of Ba (up to 19 wt%) and Cd (up to 0.4 wt%), and Ni slags from laterite ore processing incorporate Cr (up to 2.1 wt%). Ferrous slags have different characteristics to base metal slags as a result of processing of Fe oxide ores, which are generally poor in base metals, and consequently ferrous slags are impoverished in base metals as compared to the quantities observed in base metal slags. However, scrap Fe metal added as feed to ironmaking and steelmaking furnaces may be a source of metals in the slag. For example, some steel slags are abundant in Mn and Cr (Table 3.1), likely from a high contribution of scrap to the furnace. Figure 9.1 shows that these elements are predicted to end up in the slag during smelting. Both Mn and Cr are added to steel in order to improve its properties, with the former element being responsible for strength improvement (Mn) and the latter improving steel resistance against corrosion (Cr). In addition, processing conditions that influence slag composition and element partitioning include not only the type of ore and feed processed but also the types of fluxes and additives.[42]

The bulk chemical composition of slags is important for determining which elements to target during secondary recovery. However, in order to define the most adapted recovery method and specific conditions, the partitioning of the targeted elements into the crystalline and amorphous phases has to be known. Sulfides, intermetallic phases, and metal droplets present in base metal slags are important carriers of various metals, including Cu, Zn, and Pb. However, depending on the slag type, other phases can also be important hosts of various recovery-targeted elements. For example, silicate and oxide phases are the predominant hosts of Zn in base metal slags. Zinc is often incorporated into: (a) olivine group silicates (end member willemite, Zn_2SiO_4), (b) spinel group phases (end members franklinite, $ZnFe_2O_4$, or gahnite, $ZnAl_2O_4$), and (c) melilite group phases (end member hardystonite, $Ca_2ZnSi_2O_7$).[32,43,44] The amounts of Zn reported reach as much as 56 wt% in silicates and 32 wt% in oxides (other than ZnO) (Table 3.4, Chapter 3). Copper is often accommodated in spinels and glass, hosting up to 3.3 wt% and 5.5 wt%, respectively (Table 3.4). Lead is mainly found in sulfide and intermetallic compounds; however, it may also be hosted in phases such as pyroxene, feldspars, melilite, and glass, among which feldspars may contain important quantities reaching nearly 32 wt% Pb (Table 3.4). Furthermore, spinel and pyroxene serve as important Co and Cr carriers, reaching up to 4 wt% Co and 36 wt% Cr (Table 3.4).

Ferrous slags incorporate V, Cr, and Mn in quantities that are often similar to, or even higher than, the amounts observed in base metal slags. Silicates and oxides present in ferrous slags are important carriers of Cr, Mn, and V. These elements may also be found in less abundant metals, alloys, or sulfides (Table 3.3).

There is relatively little information about critical and REEs in slags. Figure 9.1 shows critical element distribution in the slag phase during secondary Cu smelting. Simulated secondary Cu smelting with electronic equipment as the input resulted in the generation of Cu slags containing Ge (4.4 mg kg^{-1}), Ga (5280 mg kg^{-1}), and In (4170 mg kg^{-1});[45] however, slags resulting from ore processing are expected to contain lower quantities of these elements.[46] Thus, end-of-life products, such as waste electronic equipment, when added to secondary smelters may contain critical elements and REE that end up in the slag. Blast furnace (BF) slag was found to contain up to 440 mg kg^{-1}

of Nd and 40–170 mg kg^{-1} of La, Ce, and Er.[47] Likewise, Cu slags may also contain Nd (17.9–49.7 mg kg^{-1}), La (60.8–77.6 mg kg^{-1}), and other critical elements.[48] Tin slags were found to have 6.1 wt% and 3.6 wt% of Ta and Nb, respectively.[49] However, the extent to which individual slags can serve as secondary resources of these critical elements remains under evaluation.[47,50–52] Based on recently simulated secondary Cu smelting and refining processes with phase equilibration at high temperature and quenching, Avarmaa *et al.*[45] showed that the distribution of In into Cu slag increased provided that the oxygen partial pressure increased, whereas the opposite behavior was observed for Ga. Thus, amending smelting conditions can improve element partitioning and future recovery.[45] Furthermore, Kayembe *et al.*[50] suggested that during smelting the addition of basic oxides increases the basicity of silicate slag and subsequent depolymerization of the silica network, leading to the production of Ge-bearing matte, hence separating Ge from slag. In particular, basicity ($CaO + MgO/SiO_2 + Al_2O_3$) at 0.7 was favorable for Ge concentration in the form of GeS, FeGe, and Ge, whereas as the basicity increased, the only Ge-bearing phase was one resembling germanite ($Cu_{13}Fe_2Ge_2S_{16}$).[50]

Recovery of Ta and Nb from Sn slag was achieved by chlorination and carbochlorination. Chlorination achieved 65% and 95% recovery at 1000 °C for low-grade concentrate and high-grade concentrate, respectively. Carbochlorination enabled a higher recovery and more pure recovery product from the high-grade concentrate.[49]

Consideration of chemical and mineralogical features is a fundamental aspect when searching for an appropriate recovery method. For example, the presence of magnetic responsive phases is a prerequisite for magnetic separation methods (Section 9.3.1). Sulfide content and distribution are important for flotation (Section 9.3.2). For example, the leachate characteristics of ferrous and non-ferrous slags vary and depend on the slags' chemistry and mineralogy. Ferrous slags generate alkaline leachates resulting from the dissolution of Ca-rich oxides, carbonates, and silicates. In contrast, non-ferrous slags are known for their acid-producing potential, which is mainly determined by their sulfur and sulfide content.[53,54] In terms of leaching and bioleaching process design, this knowledge is absolutely crucial (Sections 9.4.1 and 9.4.2).

9.3 Methods Used to Recover Metals from Slag

9.3.1 Physical Separation

Physical separation techniques applied for metal recovery from slag include gravity, supergravity, and magnetic separation methods as well as flotation. However, prior to applying a recovery method, reduction of the size of the slag particles is required in order to efficiently liberate targeted metal-bearing mineral phases.[55] Primary crushing is achieved using a jaw crusher, whereas secondary crushing is accomplished using a cone crusher, hammer crusher, or impact crusher. Alternatively, in order to simplify the process, a one-stage dry autogenous grinding machine can be applied instead. Such prepared material is suitable for further processing *via* magnetic separation in a cross-belt magnetic separator, drum magnetic separator, or magnetic pulley separator.[55,56] In addition, a single or double deck vibrating screen may be used for size sorting and classification of slag fed. Size reduction and gravity separation commonly result in an improved recovery yield *via* concentration of liberated dense metal-bearing particles, often resulting in a content comparable with that of primary ores. Substantial improvement of metal recovery can then be obtained using size reduction and gravity separation, magnetic separation, or flotation.[56,57]

Magnetic separation is able to concentrate the phases of interest if they are magnetic or paramagnetic. This technique relies on applying an external magnetic field to separate and enrich the desired particles. Conventional magnetic separation works well for magnetically responsive minerals (Fe-bearing minerals), whereas high gradient magnetic separation must be used for weakly magnetically responsive minerals as well as for fine particles. For ferrous slags, this method may be used to separate Fe-bearing particles that are returned to the furnace for reprocessing. Once these particles are removed, CaO- and SiO_2-rich products usable in the cement industry can be generated.[58] Iron removal by magnetic separation from ferrous slags also enhances their application performance for phosphate removal from industrial effluents. Xiong *et al.*[59] demonstrated that better phosphate removal can be achieved using steel slag with a lower Fe and Al content. Furthermore, Menad *et al.*[58] found that dry magnetic separation is efficient for coarse Linz–Donawitz (LD) slag fractions and recovers Fe scraps only, whereas the wet process combining low- and high-intensity magnetic separation enables the recovery of magnetically sensitive particles (Fe oxides) and paramagnetic particles (*e.g.*, ferrite, $M^{2+}Fe_2{}^{3+}O_4$, where M = Zn^{2+}, Fe^{2+}, Mg^{2+}, Mn^{2+}, or Cd^{2+}). However, magnetically responsive phases are not always present in base metal slags.

Slow cooling applied after slag re-melting may influence recovery yields[60] such that slag in the molten state may be cooled slowly to solidify so that mineral particles are increased in size, creating mineralogical properties that are suitable for flotation or magnetic separation. In contrast, fast cooling, depending on the composition, may generate amorphous slag with more homogenous metal distribution, rendering physical separation difficult to perform. In addition, Guo *et al.*[60] demonstrated that through the adjustment of the binary basicity ($CaO:SiO_2$ ratio), conversion of fayalite (Fe_2SiO_4; a common slag silicate) to magnetite (Fe_3O_4) is promoted and subsequently results in a higher magnetic separation efficiency for Cu slag. On the other hand, those authors highlighted that increased magnetite generation increases slag pulp viscosity, which in turn leads to the formation of fine-sized Cu-bearing phases, leading to poor Cu recovery during the slag flotation stage. In such situations, the addition of a sulfide-containing additive may be beneficial because: (i) sulfides play the role of a reducing agent and may reduce magnetite according to eqn (9.1), (ii) the viscosity of the slag decreases,

(iii) and more Cu is captured in a sulfide matte since Cu_2O is susceptible to sulfurization into Cu_2S in the presence of FeS, according to eqn (9.2).[60]

$$FeS + 3Fe_3O_4 \rightarrow 10FeO + SO_2 \quad (9.1)$$

$$FeS + Cu_2O \rightarrow FeO + Cu_2S \quad (9.2)$$

Furthermore, Zhu *et al.*[61] have shown that co-reduction followed by magnetic separation can be applied to treat Cu slag and ferruginous manganese ore. The MnO_2 contained in the ore played a role in the reduction of fayalite and Cu sulfide contained in the slag while facilitating the formation of Fe and Cu alloys, which subsequently allowed their recovery by magnetic separation.[61] Furthermore, reduction of basic oxygen furnace (BOF) slag particles is beneficial for magnetic separation, and, additionally, an increase in the gap between the blades and the drum also improves the process.[62] A study by Li *et al.*[63] concerning the effect of sodium carbonate (Na_2CO_3) dosage and binary basicity ($CaO:SiO_2$ ratio) on the coal-based reduction of Cu slag demonstrated that the addition of CaO at 0.5 basicity and 8% (molar ratio) of sodium carbonate was optimal for Fe and Cu recovery during magnetic separation, generating a 90.5 wt% Cu concentrate. Magnetic separation is also applied in the processing of fine steel slag particles that exhibit low Fe grades and high impurity concentrations. Ma and Houser[62] demonstrated that weak magnetic separation coupled with selective particle size screening leads to the generation of a product with a high Fe grade and low impurity content. Furthermore, magnetic separation is used in multistage treatment scenarios. For example, magnetic separation was beneficial for the recovery of valuable metals (*i.e.*, Fe, Ti, and V) from V-bearing converter slag. First Ca roasting and acid leaching were applied for V recovery from slag and the resulting tailings were subjected to a coal-based direct reduction–magnetic separation following HCl leaching to recover the Fe and Ti.[64]

Supergravity, as a physical method for enhancing mass transfer, can efficiently separate components from complex materials (*e.g.*, electronic waste and tailings) as well as from various slag melts resulting in significant concentration of phases.[65–67] Supergravity is the force of matter greater than that under normal gravity acceleration ($9.80\ m\ s^{-2}$). In the supergravity field (generated by centrifugal force in different temperature-controlled conditions), molecular diffusion and mass transfer processes are faster and liquid phases can flow in and out of porous media and contact with other phases at a rate hundreds of times higher than that in the normal gravity field. High shearing force by supergravity pulls liquid into micrometer to nanometer liquid membranes and then forms a large and new phase interface, greatly improving the micrometer-scale mixing and mass transfer processes.

In a supergravity field treatment, the liquid metal usually concentrates along the direction of supergravity, while the remaining slag phases are moved and aggregated along the opposite direction, thereafter segregating from the metal phase. The supergravity method finds application for separating P- and

Fe-bearing phases in steelmaking slag.[68] At a certain temperature, slags remain in a solid–liquid state including a solid P-bearing phase and melt. As the P-bearing phase has high viscosity, it is hardly separated from the residual melt unless supergravity is applied. Supergravity relies on the reinforced separation of phases in the fluid causing the passage of residual melt while at the same time blocking the passage of the P-bearing phase (precipitating at 1390 °C).[68] Likewise, the supergravity method can also be applied to Cu slag for separating metallic Cu- and Fe-rich phases by reduction of Fe_3O_4 to FeO and forming Fe_2SiO_4 melt. As a consequence, metallic Cu and Cu matte may separate from the Fe-rich slag according to supergravity forces.[69] Furthermore, the same method is applied for Ti recovery from BF slag melt. At a certain temperature, Ti is present as a solid phase (rutile, TiO_2, crystals) in a molten slag that can pass through by supergravity, resulting in efficient Ti separation.[70] It was also noted that during the separation of metals from molten slag, the interface between these two phases occurs perpendicularly to the direction of supergravity and therefore an increase of the gravity coefficient could result in better phase separation and improve the purity of the metal.[71]

9.3.2 Flotation

Flotation is a type of physical and chemical separation process that was initially applied in the 20th century for upgrading sulfide ores and was proclaimed as a revolutionary method in mineral processing.[4,17] This process takes advantage of differences in the hydrophobic and hydrophilic surface properties of minerals driving the separation of targeted and unwanted particles. However, flotation applied to the separation of phases in slags leads to technical problems because the relatively fine-sized sulfides and intermetallic phases commonly targeted often limit their efficient separation from the other phases. Therefore, solid material is ground and then crushed to liberate targeted phases, and size-sorted material enters into the aerated flotation cells. Three general mechanisms are involved in the flotation process: selective attachment of metals and sulfides to the air bubbles, entrainment in water passing the froth, and aggregation, that is, physical entrapment between particles in the froth attached to the air bubbles. Mineral particles resistant to wetting are preferentially surrounded by the air compared to particles susceptible to wetting. Consequently, the former particles are transferred on the foam towards the top of the flotation cell, whereas the latter settle down. Specific chemicals (flotation reagents) are applied to modify the process and improve the recovery of the desired particles. Frothing organic substances are applied to decrease surface tension, which enables abundant foam formation. Collector substances are dedicated to selective adsorption onto specific grain surfaces, altering their hydrophobic properties such that these particles are efficiently recovered in the foam. Modificatory substances are usually inorganic substances used to regulate the efficiency and selectivity of collector substances. As the result of flotation, the mineral particles carrying targeted metals are separated from useless (gangue) particles.[55,60,72,73]

Other phases, such as spinel/ferrite phases, that may contain Cu and Fe display slower floating rates as compared to metallic Cu and sulfides, which consequently affects Cu and Fe recovery yields.

In terms of slag treatment, various collector substances have been studied [*i.e.*, sodium isopropyl xanthane (SIPX), sodium di-ethyl dithiophosphate (DTP) and alkyl hydroxamate, xanthates, dithiophosphates, mercaptobenzothiazole, thionocarbamates, fatty acids, and sulfides and sulfates]. The use of combined proportions of collectors may improve flotation performance by several percent.[17] Additionally, the use of DTP as a co-collector improved the adsorption of coarse Cu-bearing particles.[4] Substances such as FC7245 (sodiumalkyl dithiophosphate, mercaptobenzothiazole), FC4146 (modified thionocarbocarbamate, methyl isobutyl carbinol), and Na_2S were found to be beneficial flotation reagents with FC7245 (sodium alkyl dithiophosphate, mercaptobenzothiazole) reported as a suitable secondary collector after slag re-grinding and flotation.[17] Thus, dual-step flotation accompanied by re-grinding was found to play a crucial role in the improvement of Cu recovery.[17] Anionic collectors, such as amyl, isopropyl, and butyl xanthates, were found to be suitable collectors for Cu sulfides.[74]

The cooling rate of molten slags affects their mineralogical and physical characteristics, impacting flotation behavior. Slow cooling of molten slag during the solidification stage generally results in increased size of Cu-bearing phases because of the diffusion and coagulation of narrow particles.[75] Furthermore, a greater degree of oxidation in the furnace generally results in higher Cu content and a higher proportion of Cu associated with Cu oxides or Cu spinel instead of metallic Cu.[76] Slow cooling rates (1.5 °C min^{-1}) enable the fine particles to coagulate, which results in larger Cu-bearing particles in the slag. The larger particles result in better Cu recovery during flotation as compared to air-cooled slag.[75] Likewise, the use of Na_2CO_3 reduces slag viscosity and consequently improves the crystalline state of Cu slag, increasing the size of Cu-bearing phases and resulting in better flotation efficiency.[77] In addition, close association of target metal-bearing phases with silicates may necessitate better grinding to improve separation.[78]

9.3.3 Roasting

Roasting is a separation method employing thermal gas–solid reactions. Roasting is achieved through the conversion of metals incorporated within slags into soluble phases, such as sulfates and/or oxides, at high temperatures ranging from 200 to 600 °C.[19–23,79] In particular, this process is dedicated to sulfide minerals, which are converted to oxides under exposure to elevated temperature while sulfur is released as gaseous sulfur dioxide [eqn (9.3)]. The latter is often used for sulfuric acid production, whereas the metal oxide is further processed *via* recovery techniques such as leaching. Sulfating roasting converts sulfides into oxides and releases SO_2.

$$2MS_{(s)} + 3O_{2(g)} \rightarrow 2MO_{(s)} + 2SO_{2(g)} \quad \text{where M: metal} \qquad (9.3)$$

Roasting of V slag with CaO results in vanadate (VO_4) formation. Fayalite is more prone to oxidation than V-bearing spinel and lowering the heating rate (2 °C min^{-1}) positively affects V oxidation at 850 °C.[80] Furthermore, addition of sodium hydroxide (NaOH) pellets was found to improve roasting in terms of V recovery (99% instead of 80%) and lower the required temperature from 800 °C to 700 °C. In contrast, the roasting pellet cooling rate was not found to affect V recovery.[81] Additionally, lower roasting temperatures are good for spinel conversion (to inverse V-bearing spinel), whereas higher temperatures (630 °C) would cause a spinel conversion product (V_2O_5) reaction with CaO, leading to $Ca_2V_2O_7$ formation.[82]

The roasting temperature is an important parameter affecting the final metal recovery. Depending on the metal targeted, the temperature has to be appropriately adjusted as sulfation leads to the formation of metal-bearing sulfates, whose decomposition will be specific to the metal incorporated. The decomposition temperatures of Zn, Cu, Co, and Ni sulfates are 600, 650, 735, and 848 °C, respectively.[83] Altundoğan and Tümen[18] revealed that increasing the roasting temperature up to 500–550 °C improved Cu recovery, whereas further temperature increase negatively affected Cu extraction. Moreover, elevated roasting temperatures may cause loss of sulfur and give insufficient time for sulfation to take place.[18] Various roasting durations were reported as optimum for slag treatment (Table 9.1), while temperatures usually oscillate between 400 and 750 °C. However, lower temperatures of 150–250 °C have also been reported as optimal in several cases when applying H_2SO_4 as a roasting agent. Slag particles used for roasting processes usually range from 75 to 750 μm in diameter (Table 9.1).

9.4 Extractive Methods

Extractive leaching can be defined as the compilation of factors that jointly influence the solubility of metallurgical slags and results in the transfer of metals from the solid (slag) to the liquid phase, namely the leachate. The extraction can be carried out by treating the slag with traditional inorganic or organic chemical agents (chemical leaching) or a solution containing microorganisms (bioleaching). Both chemical and biotic leaching processes are influenced by several mechanisms, such as acid attack, complexation, and redox reactions. Acidolysis is one of the most important mechanisms facilitating the dissolution of phases by interaction with hydrogen ions causing metal transfer from solid to liquid. Complexolysis takes place through metal binding to organic molecules on sites immediately following proton detachment. Redox reactions are linked to elemental reduction from an oxidized to a reduced form (or from reduced to oxidized) that often results in its detachment from the mineral lattice. The mechanisms involved in the extraction process depend on the type of chemical or microbial agent being employed as the dissolution catalyst. Furthermore, the rate and intensity of slag dissolution depend on the type of chemical and/or biotic agent acting

Table 9.1 Examples of extractive approaches applied for metallurgical slags.

Type of slag	Elements targeted	Treatment agent	Optimal conditions[a]	Reference
Cu converter slag	Cu, Co, Ni	$FeCl_3$	PS: <150 μm, LS: 10, D: 2.5 h, T: 85 °C, AC: 1.25 stoichiometric amount of $FeCl_3$, AS: 200 rpm	Anand *et al.*[122]
Cu flash smelter slag	Cu, Co, Ni	$FeCl_3$	PS <53 μm, LS: 10, D: 6 h, T: 100 °C, AC: 5 stoichiometric amount of $FeCl_3$, AS: 200 rpm	Anand *et al.*[122]
Cu converter slag	Cu, Co, Ni	Roasting (furnace oil) + $FeCl_3$ leaching	Roasting: PS: 150 μm, AC: 10 wt% furnace oil, T: 750 °C, D: 1 h, D: 2 h, AC: 1.4 stoichiometric $FeCl_3$	Anand *et al.*[23]
Cu converter slag	Cu, Co, Ni	H_2SO_4	PS: 150 μm, LS: 10, AC: 0.35 N H_2SO_4, T: 130 °C, O_2 pressure: 0.59 MPa, D: 4 h, AS: 840 rpm	Anand *et al.*[94]
Cu converter slag	Cu, Co, Ni	Roasting ($(NH_4)_2SO_4$) + H_2O leaching	PS: <75 μm, Roasting AC: 2.5 stoichiometric $(NH_4)_2SO_4$, T: 400 °C, D: 1 h	Sukla *et al.*[123]
Cu converter slag	Cu, Co, Ni	Roasting (H_2SO_4) + H_2O leaching	PS: <75 μm, Roasting, AC: stoichiometric H_2SO_4, T: 150 °C, D: 1 h	Sukla *et al.*[123]
Cu converter slag	Cu, Co, Ni, Zn	Roasting $Fe_2(SO_4)_3$, H_2SO_4, H_2O	PS: <74 μm, Roasting AC: $Fe_2(SO_4)_3$/slag = 1, T: 650 °C, D: 2 h, Leaching: LS: 5, D: 0.5 h, T: 75 °C, AC: 500 g H_2SO_4 kg^{-1}	Altundoğan and Tümen[18]
Cu converter slag	Cu, Zn	$Fe_2(SO_4)_3$, H_2SO_4	AC: 15.7 g L^{-1}, D: 6 h, T: 70 °C, LS: 10	Muravyov *et al.*[108]

(continued)

Table 9.1 (*continued*)

Type of slag	Elements targeted	Treatment agent	Optimal conditions[a]	Reference
Cu converter slag	Cu, Fe, Zn	$H_2O + Cl_2$	PS: 150 µm, LS: 10, D: 1.5 h, AS: 550 rpm, FR: Cl_2 220 mL min^{-1}	Beşe *et al.*[124]
Cu converter slag	Cu, Co, Fe, Zn	H_2SO_4, $K_2Cr_2O_7$	PS: <74 µm, LS: 100, D: 2 h, T: 25 °C, AS: 400 rpm, AC: 1 M H_2SO_4	Altundoğan *et al.*[95]
Cu converter slag	Cu, Co, Fe, Zn	H_2SO_4, $Fe_2(SO_4)_3$ + ultrasound	PS: <125 µm, LS: 25, D: 3 h, T: 65 °C, AC: 0.2 M H_2SO_4, 0.15 M $Fe_2(SO_4)_3$, ultrasound application	Beşe *et al.*[125]
Cu converter slag	Cu, Co	Aging H_2SO_4 + H_2O leaching	PS: <74 µm, Aging: 1.2 stoichiometric H_2SO_4, D: 8 h, T: 75 °C, H_2O:slag: 1, Leaching: LS: 5, T: 80 °C, D: 1–1.2 h	Deng and Ling[19]
Cu converter slag	Cu, Co, Ni	Solvent extraction	Neutralization PC88A: 70%, Organic/aqueous phase ratio: 10, AC: 1 M H_2SO_4	Mishra *et al.*[126]
Cu converter slag, Electric furnace slag	Cu, Co, Ni, Zn	$H_2SO_4 + H_2O$ + oxidative pressure	PS: 50 µm, LS: 2.6, D: 1 h, T: 250 °C, AS: 700 rpm, O_2 pressure 520 kPa, AC H_2SO_4: 5 g L^{-1}, A/S ratio: 0.3	Baghalha *et al.*[90]
Cu converter slag, Flash smelter slag	Cu, Co, Zn	Roasting H_2SO_4 + hot H_2O leaching	RA: H_2SO_4 roasting T: 250 °C, A/S: 3, D: 2 h, Hot water leaching LS: 50, D: 2 h, T: 150 °C	Arslan and Arslan[127]
Cu reverberatory slag, Flash slag	Cu, Fe	Chlorine solution (Cl_2)	PS: <20 µm, T: ambient, D: 5 min., LS: 60, AS: 600 rpm	Herreros[128]
Cu brass melting slag	Cu, Zn, Pb	HCl, H_2SO_4, H_2O_2, NH_4OH	PS: <900 µm, AC: 6 M HCl, LS: 8, D: 12 h, T: 90 °C, AC: 4–16 M H_2SO_4, D: 12 h, T: 70 °C, AC: 6 M HCl + 6.5 M NH_4OH, D: 3 h, T: 70 °C, Addition of 250 mL of H_2O_2 per 1 L of 6 M HCl leachate	Basir[129]

Cu slag	Cu, Co, Fe, Zn	$H_2SO_4 + H_2O_2$	PS: <100 µm, LS: 10, T; 60–80 °C, D: 2 h, AC H_2SO_4: 500 g kg^{-1}, AC H_2O_2 62.5 mL kg^{-1}, Redox: 450 mV + 3 h = 650 mV + 1 h	Banza *et al.*[106]
Cu slag	Cu, Fe, Zn	$H_2O + NH_4Cl$	PS: 75 µm, LS: 2, D: 2 h, T: 320 °C, AS: n.i.	Nadrirov *et al.*[130]
Cu slag	Cu, Co, Ni	H_2SO_4 roasting + H_2O	Roasting AC: 80 wt% H_2SO_4, 50 g slag, T: 200 °C, D: 2 h, Leaching T: 90 °C, D: 0.5 h, LS: 2	Zhang *et al.*[131]
Cu slag	Cu	$Fe_2(SO_4)_3$	PS: 22–840 µm, LS: 50, D: 2 h, T: 60 °C, AS: 240 rpm, AC: Fe^{3+} 11.5 g L^{-1}	Carranza *et al.*[107]
Cu slag	Cu, Co, Zn	H_2SO_4, $NaClO_3$, NaOH, $Ca(OH)_2$	PS: 180 µm, LS H_2SO_4: 880 g kg^{-1}, AC ($NaClO_3$): 2.5 g kg^{-1}, D: 3 h, T: 95 °C, AS: 500 rpm	Yang *et al.*[93]
Cu slag	Cu, Fe	Additives + re-melting + flotation + magnetic separation	0.45 binary basicity, 12% additive, T: 1350 °C, D: 2 h, cooling: 1.5 °C min^{-1} to 900 °C	Guo *et al.*[60]
Cu slag	Cu, Fe	H_2SO_4, KI, $C_2H_7NO_2$, NH_4HF_2, $Na_2S_2O_3$, KSCN, 1,10-phenanthrolinemonohydrate, *p*-nitrophenol	PS: 45–75 µm, T: 200 °C, D: 0.66 h, AS: 600 rpm, AC H_2SO_4: 0.4 M, autoclave 20–220 °C + O_2 pressure	Shi *et al.*[132]
Cu slag	Cd, As, Ni	HCl, H_2SO_4, HNO_3, citric acid	PS: <53 µm, LS: 100, D: 2 h, AS: 200 rpm, AC HCl: 1.5 M	Du *et al.*[133]
Cu slag	Cu, Fe, Ni, Zn	H_2SO_4	AC: 2 M H_2SO_4, LS: 100, MP: 0.26 kW m^{-3}, T: 40 °C, D: 1 h, O_2 2000 mL min^{-1}	Khalid *et al.*[119]

(*continued*)

Table 9.1 (*continued*)

Type of slag	Elements targeted	Treatment agent	Optimal conditions[a]	Reference
Cu slag	Cu, Co	H_2SO_4 + acid baking	PS: 100 µm, AS: 400 rpm, T: 2 h, AC: 120 g L^{-1}, T 60 °C, LS: 2, Acid baking + hot water leaching 200 °C, D: 1 h, LS: 3, Acid baking 200 °C, D: 1 h, LS: 1.5 + thermal decomposition 600 °C, D: 2 h	Bulut *et al.*[134]
Cu slag	Cu, Co, Ni, Zn	H_2O + roasting with pyrite	Roasting: T: 550 °C; D: 1 h, roasting of pre-roasted slag–pyrite mixture (pyrite:slag ratio = 0.25) for 1 h at 550 °C, H_2O leaching 15 rain, LS: 10, T: ambient	Tümen *et al.*[20]
Cu slag	Cu, Zn, Pb, Fe	H_2SO_4, HCl, HNO_3, $C_6H_8O_7$, $C_2H_2O_4$	Amorphous slag: 2 M HCl, LS: 100, D: 24 h, Crystalline slag: 2 M HCl, LS: 100, D: 48 h	Potysz and Kierczak[91]
Cu slag	Cu, Fe	Roasting H_2SO_4 + leaching H_2O	PS: <75 µm, Roasting T: 250 °C, AC: 1.33 stoichiometric H_2SO_4, D: 0.5–2 h, H_2O Leaching T: 30–50 °C, AS: 350 rpm, D: 0.5 h, LS: 5	Dimitrijević *et al.*[22]
Cu slag	Cu, Fe	HCl, H_2O_2 + HCl	PS: <75 µm, LS: 30, AS: 350 rpm, D: 2 h, 3 M H_2O_2 + continuous HCl addition	Dimitrijević *et al.*[135]
Cu slag	Cu, Co, Zn	H_2SO_4, HNO_3	PS: 29–54 µm, D: 24 h, LS: 20, T: 80 °C, AS: 300 rpm, AC: 3 M HNO_3	Kinnunen *et al.*[92]
Cu slag	Cu, Fe, Zn	NH_4Cl (NH_4OH, Cl^-)	PS: <74 µm, LS: 0.39–0.53, D: 4.13–4.56 h, $C_{initial}$ NH_3 17.1–19.9%, $C_{initial}$ Cl^- 160 g L^{-1}	Nadirov[21]

Cu slag	Cu, Fe, Zn	$H_2SO_4 + K_2Cr_2O_7$	PS: <74 µm, LS: 66.6, T: 24.9 °C, D: 2 h, AS: 150 rpm, AC: 0.5 M $K_2Cr_2O_7$ + 1 M H_2SO_4	Nadirov[21]
Cu–Ni slag	Cu, Co, Ni, Zn	Fe^{3+}, H_2SO_4	T: 70 °C, LS: 10%	Fomchenko and Muravyov[109]
Cu water jacket furnace slag	Cu, Co, Zn, Fe, Pb	$HNO_3 + NH_3$	PS: <212 µm, LS: 10, D: 20 h, T: 25 °C, AS: 500 rpm, HNO_3 pH = 0	Tshiongo *et al.*[37]
Brass slag	Cu, Zn	H_2SO_4	PS: <100 µm, LS: 5, D: 0.16–0.5 h, T: 35 °C, AS: 150 rpm, AC: 30–45% H_2SO_4	Ahmed *et al.*[97]
Pb–Bi slag	Bi	$CaCl_2$, HCl	PS: 15 µm, LS: 3–5 (Bi) and 8 (Pb), AC: 100 g L^{-1} $CaCl_2$, LS: 4, T: 50 °C, D: 1 h	Che *et al.*[136]
Zn–Pb slag	Cu, Cd, Ni, Pb, Zn	HNO_3	PS: <4 mm, pH 3	Król *et al.*[137]
Zn slag	Co, Zn	2 M H_2SO_4	LS: 25, AS: 300 rpm, Fe removal: T 80 °C, 1.5 h, pH 4.5, H_2O_2 2.5 wt%, Precipitation: 8 g L^{-1} sodium ethyl xanthate, 1 g L^{-1} K_2MnO_4, pH 4, T: 55 °C, and 5 h	Song *et al.*[138]
Pb–Zn–V slag	V	H_2SO_4	PS: <50 µm, LS: 100, AC: 0.5 M H_2SO_4, T: 70 °C, D: 24 h, AS: 60 rpm	Ettler *et al.*[52]
Cu flash smelting slag	Cu, Pb	$C_6H_8O_7$	PS: <1 mm, AC: 1 M $C_6H_8O_7$, D: 0.5 h, T: 70 °C, LS: 5	Gargul *et al.*[139]
V converter slag	V	Microwave (M) + electric hybrid leaching autoclave (EHA)	D: 0.66 h EHA, D: 0.3 h in M	Tian *et al.*[140]
Cu converter slag, Flash furnace slag	Cu, Fe, Zn	H_2O_2	4 M CH_3COOH, 4 M H_2O_2, MP: 900 W, D: 0.5 h, LS: 25, T: 100 °C	Turan *et al.*[141]

(*continued*)

Table 9.1 (*continued*)

Type of slag	Elements targeted	Treatment agent	Optimal conditions[a]	Reference
Ni slag	Cu, Co, Ni, Zn, Fe	High-pressure oxidative acid leaching	PS: 150 μm, T: 250 °C, acid (HPOXAL): 20%, AS: 700 rpm	Li *et al.*[142,143]
Ni slag	Co, Ni	$SO_{2(aq)}$ leaching	PS: 75–106 μm, T: 35 °C, D: 3 h, AS: 600 rpm	Gbor *et al.*[144]
Steel slag	Cr, V	H_2O (ultrapure)	PS: 1000 μm, LS: 10, T: ambient, AS: 10 rpm	De Windt *et al.*[145]
Steel slag	Ba, Cr	UPW + 0.1 mM HNO_3	Column test, PS: 0–20 mm, 0.1 mM HNO_3	Fällman[146]
Steel slag	Cu, Co	Electrolytic dissolution 2.5 M NH_3 + NH_4^+ and 1.0 M Cl (pH 10.1)	Circulation rate: 15 $dm^3\ h^{-1}$, T: 50 °C, current density 350 A m^{-2}, D: 2–3 h	Rudnik *et al.*[147]

[a]Abbreviations: A/S: agent to slag ratio, AC: agent concentration, AS: agitation speed, d: days, D: duration, FR: flow rate, h: hours, LS: liquid-to-solid ratio, MP: mixing power, PS: particle size, T: treatment temperature.

as well as any accompanying parameters, such as: temperature,[84] liquid-to-solid ratio (pulp density^{-1}),[72,85,86] and the size(s) of solid particles.[4,16,87] A higher liquid-to-solid ratio (*i.e.*, lower pulp density) and smaller particle size cause the ratio of the solution to slag surface area to increase and consequently higher leaching rates can be expected (Figure 9.2). When the slag is subjected to treatment, irreversible chemical changes take place: metals are transferred to the leachate and the initial characteristics of the slag change.[88,89] However, the extent to which the initial material changes entirely depends on the efficiency of the process. Different extraction rates and solid material behaviors can be expected when amorphous or crystalline slags are subjected to the (bio)leaching process (Figure 9.2).[86] Generally, amorphous materials are believed to display greater susceptibility to dissolution as compared to crystalline materials.[37] However, some exceptions to this rule have been observed.[40,90] The generation of a residue with significantly or entirely depleted levels of toxic elements is the desired end result.[91,92] For example, treatment of glassy slag under optimal conditions may lead to the generation of a metal-depleted residue, mainly composed of gypsum, that is more suitable for further use or disposal.[86,91]

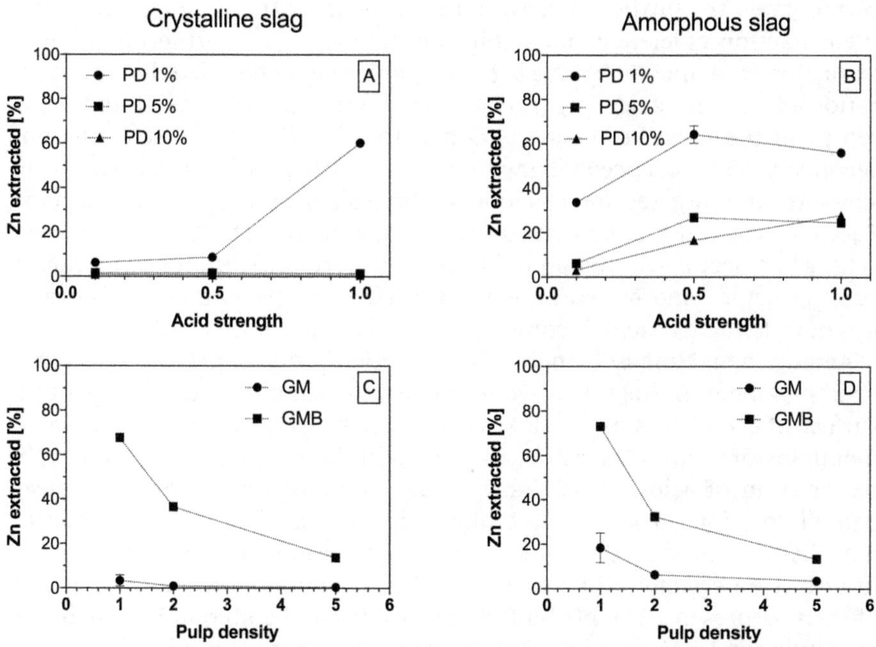

Figure 9.2 Leaching of crystalline and amorphous Cu slags using H_2SO_4 at various strengths (A–B) and various pulp densities (A–B) and bioleaching with Acidithiobacillus thiooxidans and abiotic growth medium (C–D) at different pulp densities. Reproduced from ref. 91 with permission from MDPI. Copyright 2019. PD: pulp density, GM: growth medium, GMB: growth medium with bacteria.

9.4.1 Chemical Leaching

The type of leaching agent is crucial to the efficiency of chemical leaching. Individual slag phases react differently to specific chemical agents, the reaction often being governed by the leaching agent concentration (Figure 9.2) and the abundance and chemistry of phases present in the slag sample. The leaching agents tested for slags include ferric chloride ($FeCl_3$), ferric sulfate ($Fe_2(SO_4)_3$), ammonium sulfate ($(NH_4)_2SO_4$), ammonium chloride (NH_4Cl), chlorine solution, sulfuric acid (H_2SO_4), hydrochloric acid (HCl), ammonium hydroxide (NH_4OH), and nitric acid (HNO_3) (Table 9.1). However, chemical leaching is often accompanied by supplementary treatment, such as: roasting followed by leaching, aging of slag, oxidant addition, high-temperature leaching, or even re-melting (Table 9.1), all of which are meant to improve the extraction efficiency.

Sulfuric acid (H_2SO_4) is one of the most intensively studied agents and has been proven effective for metal extraction from metallurgical slags (Table 9.1). Agent concentration was found to be a crucial factor for determining efficiency and usually an increased extraction yield is observed when the acid concentration increases (Figure 9.2).[93–96] However, an increase of molarity is desirable only to a certain extent (Figure 9.2) because after reaching certain rates any further increase of the acid concentration does not change the extraction efficiency in a significant manner.[91,97,98] Furthermore, metal extraction from metallurgical slags by organic agents revealed that the application of organic acids (*e.g.*, citric, oxalic, and salicylic acids) and EDTA represent the most promising options. However, the crystal structure and secondary reactions occurring with other constituents in the phases in the slag alter the targeted metal's amenability to dissolution in organic acids. Succinic acid can more efficiently recover metals from Cu slags than oxalic acid, while acetic acid was found to have performance approaching that of succinic acid.[99] Another study demonstrated better performance of normality equivalent citric acid as compared to oxalic acid for Cu slag.[91]

Among the factors influencing the extraction process, pH acts as a significant parameter in the slag dissolution system: it modifies the charge of the surface of the phases in the slag; causes acid attack depending on its value; maintains or adjusts the solubility of ions in the solution; and changes the dissociation of acids.[100] Generally, acidic conditions are known to exacerbate dissolution; however, some phases incorporated in slags (*e.g.*, quartz) may dissolve more readily under high pH conditions. Furthermore, the pH in extraction systems will also be modified depending on the composition of the slag present. The pH buffering capacity can be ubiquitously detected in sulfide-rich slag samples, whereas shifts to higher pH can be observed in silicate-rich slags. Moreover, ferrous slags reveal a greater alkaline-generating potential as compared to base metal slags.[101] The alkaline potential of steel slags can be decreased by the application of water washing prior to leaching. For example, a pH drop from the initial value of 11.2 to 8.3 was demonstrated after three water washing cycles with steel slags.[102] On the other hand, the high buffering capacity of steel slags renders them suitable for application as a pH-modifying agent in the leaching system.[103] Furthermore, as the slag particle size decreases, the buffering potential of the slags increases.[86]

A thorough analysis of dissolution processes demonstrated that pH buffering is mainly associated with silicate dissolution and subsequent Si release, indicating that silicate dissolution may play a crucial role in extractive leaching process design.[104,105] On the other hand, the dissolution of silicate phases in an acid medium may cause the presence of elevated Si in the solution, subsequently leading to the formation of silicic acid, denoted as "silica gel" [eqns (9.4)–(9.6)].[19,72,93,106,107] The latter may entrap metals and nearly always makes leachate filtration difficult.[16] Under extremely acidic conditions, a monomeric form of silica occurs, polymerizes, and flocculates. Furthermore, the presence of salts also influences the process, accelerating silica gel growth rates in the following order: $CaCl_2 > MgCl_2 > NaCl$.[96] Some authors have approached the silica gel issue and demonstrated that the addition of oxidants such as sodium chlorate ($NaClO_3$) and calcium hydroxide ($Ca(OH)_2$) as neutralizing agents during extraction is beneficial towards preventing silica gel formation.[93,106] On the other hand, the addition of ferric sulfate ($Fe_2(SO_4)_3$) in the leaching system may inhibit formation of silica gel since it minimizes dissolution of fayalite, a common silicate in base metal slags, and promotes dissolution of Cu-bearing phases.[108] The occurrence of silica gel is known to cause various difficulties in the leaching system, including: (i) the formation of a passivation layer, (ii) the capturing of metal ions present in solution, (iii) the gradual depletion of slag dissolution rates as time progresses, and (iv) the inhibition of the leachate filtration process. For these reasons, this problem should be monitored in the leaching system and the addition of an agent that could limit the formation of silica gel should be considered. Dissolution of silicates such as fayalite may occur in the system according to eqn (9.4), with formation of silica gel according to eqns (9.5)–(9.6).

$$X_2SiO_4 + 2H_2SO_4 \rightarrow 2XSO_4 + H_4SiO_4 \quad \text{where X: element incorporated in silicate} \tag{9.4}$$

$$H_4SiO_4 \rightarrow H_2SiO_3 + H_2O \tag{9.5}$$

$$H_2SiO_3 \rightarrow SiO_2 + H_2O \tag{9.6}$$

9.4.2 Bioleaching

Bacteria and fungi, with their proven role in solid-phase dissolution, have been utilized in bioleaching (Figure 9.3).[72,86,87,102,108,109] Microbial actions are classified as "indirect action" and "direct action", whereby the former happens when phase dissolution occurs as a side effect of microbial activity and the latter refers to activity absolutely attributed to the presence of phases

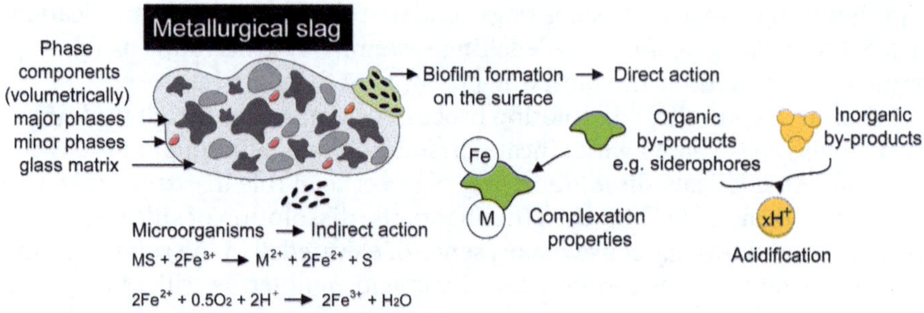

Figure 9.3 (Bio)leaching mechanisms involved in metal extraction from slag phases. Abbreviation: M: metal.

in slags and the subsequent nutritional benefit that can be gained by the microorganisms.[16] Direct action takes place when microorganisms attach to the slag surface and mediate surficial changes through the acquisition of elements from the metal-bearing phases. Indirect dissolution, sometimes called "non-contact leaching", takes place when active microorganisms in the bioleaching system excrete some products that dissolve slag (Figure 9.3). For example, siderophores are well-known bacterial products enhancing the leaching rates of elements.[110] These compounds are low-molecular-weight (approximate sizes range between 500 and 1500 Daltons) organics exhibiting strong binding strength to Fe;[111] however, siderophore-mediated complexation of Cu, Zn, and other elements has also been reported.[112]

The choice of microorganisms is a crucial point for operating a bioleaching reactor. Depending on the element targeted and its partitioning in specific phases, a suitable microorganism has to be selected to perform efficient extraction (Table 9.2). Native (indigenous) microorganisms affiliated with dumping sites are believed to be efficient because these are already adapted to the stress conditions associated with elevated metal contents.[99] The acidophilic group of microorganisms is one of the most intensively studied, owing to its proven efficiency in base metal slag dissolution (Figure 9.3). Biotic and chemical treatment of metallurgical slag can often result in similar efficiency (Figure 9.2), however, chemical leaching is unrivaled in its quick duration as compared to bioleaching. Additionally, Potysz and Kierczak[91] demonstrated that *Acidithiobacillus thiooxidans*-mediated bioleaching of Cu slags resulted in nearly complete dissolution of fayalite in the case of crystalline slag (Figure 9.4B–C) and the transformation of amorphous slag to amorphous silica and secondary gypsum (Figure 9.4F). In contrast, dissolution features observed on slags resulting from abiotic treatment are less pronounced as compared to features resulting from bioleaching (Figure 9.4A–C). Furthermore, maintenance of pH conditions optimal for microbial activity is crucial in bioleaching systems, thus slag neutralization potential has to be taken into consideration during the process design. It has been demonstrated that the use of fine grains of amorphous slag in bioleaching leads to neutralization of the

Table 9.2 Examples of biological approaches applied for metallurgical slags.

Type of slag	Metal targeted	Microorganism used	Optimal conditions[a]	Reference
Cu converter slag	Cu, Co, Ni	*Aspergillus niger*	PS: <150 μm, PD: 1%, D: 48 days, T: 30 °C, pH: 3.5, AC: 12 N HCl addition: 2 mL	Sukla *et al.*[148]
Cu converter slag	Cu, Co, Ni	*Aspergillus niger*	PS: <150 μm, PD: 2%, T: 37 °C, D: 72 h, AS: 120 rpm, pH 6.8.	Sukla *et al.*[99]
Cu converter slag	Cu, Co, Ni	*Thiobacillus ferrooxidans*, *Thiobacillus thiooxidans* (Renamed as *Acidithiobacillus*)	PS: <75 μm, PD: 5%, T: 32 °C, D: 1200–1920 h, pH 2, AS: lack	Mehta *et al.*[149]
Cu converter slag	Cu, Zn	*Acidithiobacillus ferrooxidans*, *Sulfobacillus thermotolerans*, *Leptospirillum ferrooxidans*	T: 40 °C, aeration: 4 min^{-1}, pH: 1.5, PD: 10%, AS: 760 rpm, D: 2.5 h	Muravyov and Fomchenko[84]
Cu slag	Cu, Co, Ni	*Marinobacter* spp., *Acidithiobacillus* spp., *Sulfurospirillum* spp., *Leptospirillum* spp., *Cuniculiplasma* spp., *Sulfobacillus* spp., and *Arcobacter* spp.	Sulfur: 1%, T: 30 °C, AS: 150 rpm, PD: 5%	Kinnunen *et al.*[92]
Cu slag	Cu, Zn	*Acidithiobacillus* spp., *Leptospirillum* spp.	T: 25 °C, PD: 1%, pH: 0.5–1	Vestola *et al.*[150]
Cu slag	Cu, Zn	*Acidithiobacillus ferroxidans*, *A. thiooxidans*, *A. caldus*, *Leptospirillum ferroxidans*, *Sulfobacillus thermotolerans*	Continuously stirred tank reactor, D: 29 days, PD: 5%, PS: <47 μm, pH 1.8, Sulfur: 1%, Fe(II) 4.5 g L^{-1}, AS: 1380 rpm, T: 22 °C	Kaksonen *et al.*[28]
Cu slag	Cu	*At. ferrooxidans*, *L. ferrooxidans* and *At. thiooxidans*	PS 100 μm, pH 15., 15% inoculum, Fe(II) 3 g L^{-1}, PD: 7%, D: 12 days	Panda *et al.*[72]
Cu slag	Cu, Zn, Fe	*Acidithiobacillus thiooxidans*, *Pseudomonas fluorescens*	Crystalline slag PD: 1%, PS: <300 μm, *A. thiooxidans*, T: 30 °C, AS: 100 rpm, Amorphous slag PD: 1%, PS: 1–2 mm, *A. thiooxidans*, T: 30 °C, AS: 100 rpm	Potysz *et al.*[86]

(*continued*)

Table 9.2 (*continued*)

Type of slag	Metal targeted	Microorganism used	Optimal conditions[a]	Reference
Cu slag	Cu, Zn	*Acidithiobacillus ferrivorans*, *Alicyclobacillus cycloheptanicus*, *Al. tolerans and Al. herbarium*	PD: 10%, S: 10 g L^{-1}, D: 37 days, PS: 45 µm, AS: 150 rpm, T: 27 °C	Kaksonen *et al.*[151]
Cu slag	Cu, Co, Zn, Ni	*Acidithiobacillus thiooxidans*, *A. ferrooxidans*, *Leptospirillum ferriphilum*	$pH_{initial}$ = 2.1–2.2, PD 10%, S: 10 g L^{-1}, 5% inoculum, D: 25 days, PS: 45 µm, T: 27 °C, AS: 150 rpm	Kaksonen *et al.*[88]
Cu slag	Cu	*Acidithiobacillus thiooxidans*, *A. ferrooxidans*, *Acidiphilium* spp.	Aerobic conditions, PD: 2%, PS: <63 µm, D: 51 days	Schippers[152]
Cu slag	Co, Mo, V, REE	*Acidithiobacillus thiooxidans*	PS: 250–500 µm, PD: 1%, D: 28 days, Sulfur: 1%, $pH_{initial}$ = 2.5, inoculum 2%	Mikoda *et al.*[51]
Cu slag	Cu, Zn, Pb, Fe	*Acidithiobacillus thiooxidans*	PD: 1%, D: 21 days, $pH_{initial}$ = 3.5, Sulfur: 1%, PS: 250–500 µm	Potysz and Kierczak[91]
Cu slag	Cu, Co, Zn	*Acidithiobacillus ferrooxidans*, *Leptospirillum ferrooxidans* and *Acidithiobacillus thiooxidans*, *Sulfobacillus thermosulfidooxidans* and *Acidithiobacillus caldus*, *Sulfolobus metallicus*, *Metallosphaera sedula* and the genus *Acidianus*	*S. thermosulfidooxidans* + *A. caldus*, T: 55 °C, PD: 10%, PS: 100 µm, D: 48 h	Georgiev *et al.*[153]
Pb–Zn slag	As, Al, Cu, Fe, Mn, Pb, Zn	*Bacillus* spp., *Sporosarcina* spp., *Pseudomonas* spp.	PD: 5%, pH 1.5, T: 65 °C, AS: 120 rpm, D: 6 days, PS: n.i.	Cheng *et al.*[154]
Pb–Zn slag	As, Cu, Fe, Mn, Pb, Zn	*Bacillus* spp., *Sporosarcina* spp., *Pseudomonas* spp.	pH 1.5, PD: 10%, PS: <0.83 mm, T: 65 °C, AS: 120 rpm, D: 6 days	Guo *et al.*[85]

Zn slag	Cu, Fe, Zn	*Bacillus* spp., *Sporosarcina* spp., *Pseudomonas* spp.	PD: 2%, pH 1, T: 65 °C, AS: 120 rpm, PS: 250 μm	Jiang *et al.*[155]
Pb–Zn slag	Ag, As, Cu, Cd, Pb, In	*Acidithiobacillus thiooxidans*, *Leptospirillum ferriphilum*	PD: 10%, PS: 150 μm, T: 30 °C, AS: 120 rpm, D: 1–9 days, pH 1.6–1.9	Wang *et al.*[156]
Cu converter slag	Cu, Fe	*Acidithiobacillus ferrooxidans*	$pH_{initial}$ = 1.8, PD: 71, Fe(II): 7.3 g L^{-1}, PS: 75 μm	Ahmadi *et al.*[157]
Al slag	Ga	*Acidithiobacillus thiooxidans*	PD: 2%, D: 24 h, recovery: lime milk method	Wang *et al.*[120]
Oxygen furnace steel slag	Al, Cr, V	*Acidithiobacillus thiooxidans*, *Acidithiobacillus ferrooxidans*	PS: <75 μm, 2-step hybrid process, D: 3 days	Gomes *et al.*[158]
Steel slag	Mg, Zn, Cu	*Acidithiobacillus thiooxidans*, *Acidithiobacillus ferrooxidans*, *Aspergillus niger*	*A. thiooxidans*, D: 12 days, T: 30 °C, AS: 150 rpm	Hocheng *et al.*[102]
Steel slag	Zn, Fe	*Acidithiobacillus ferrooxidans*	S: 150 rpm, T: 30 °C, $pH_{initial}$ = 1.5, PD 1%, D: 21 days	Bayat *et al.*[159]
Steel slag	Cr, P	*Acidithiobacillus ferrooxidans*	Reductive bioleaching, CO_2 atmosphere, Sulfur: 2.5%, PD: 1%, T: 30 °C, AS: 115 rpm, $pH_{initial}$ = 2.5, D: 12 days	Thalner *et al.*[160]

[a]Abbreviations: AC: agent concentration, AS: agitation speed, D: duration, FR: flow rate, h: hours, PD: pulp density, PS: particle size, T: treatment temperature.

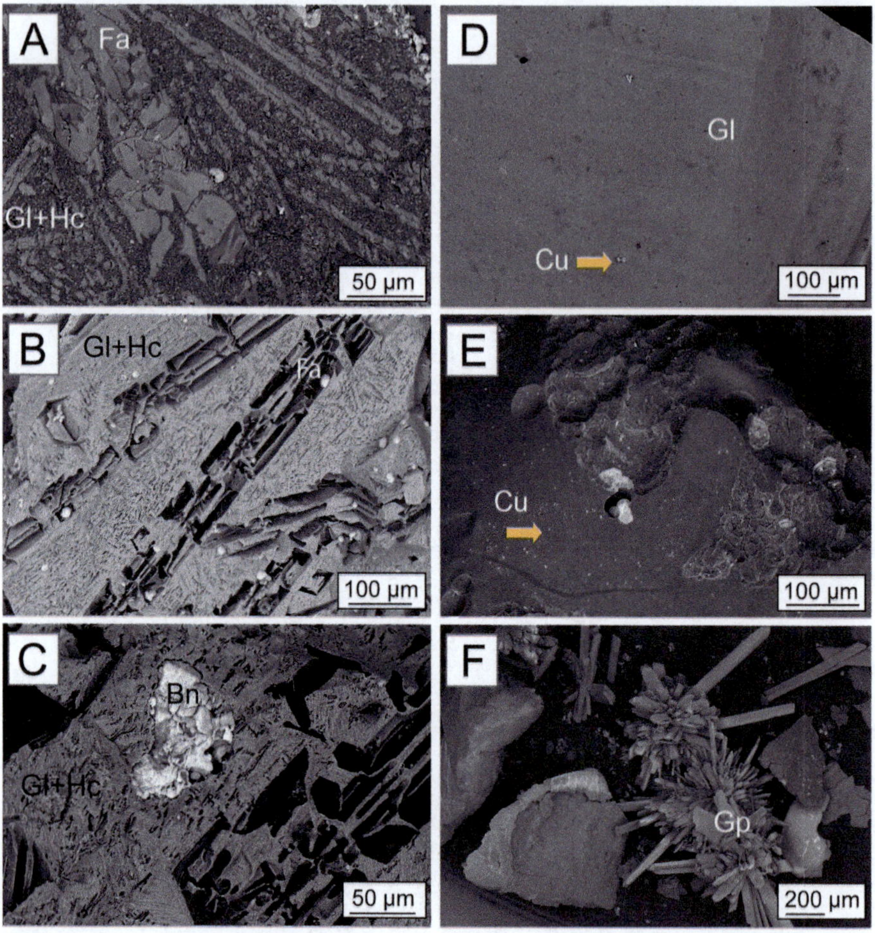

Figure 9.4 Dissolution of crystalline and amorphous slags as the result of bioleaching visualized by scanning electron microscopy. A: Fresh crystalline slag, B: fayalite dissolution resulting from biotic extraction C: fayalite and sulfide dissolution resulting from biotic extraction, D: fresh amorphous slag, E: dissolution of amorphous slag, and F: gypsum formation resulting from biotic extraction. Abbreviations: Fa: fayalite, Gl: glass, Hc: hercynite, Bn: bornite, Cu: metallic Cu, Gp: gypsum. Reproduced from ref. 91, https://doi.org/10.3390/min9090542, under the terms of the CC BY 4.0 license, https://creativecommons.org/licenses/by/4.0/.

solution and creates conditions that are unfavorable for microbial activity, whereas the use of larger grains of amorphous slag neutralized solution to a lower extent, which resulted in maintenance of optimal pH conditions for microbial requirements.[86]

Bioleaching is, at present, a prospective recycling technology for metal removal from metallurgical slags. However, laboratory evaluation remains the central route for future process customization. Leaching scenarios considered at the laboratory scale include batch leaching, semi-open flow-through leaching, and continuously stirred tank reactors.[28,113,114] However, the most desired option for upscaling the leaching scenario is a heap bioleaching operation because it takes place on site, therefore ensuring the large scale of the process and the avoidance of transport to a treatment plant. Heap bioleaching is a specifically designed process where slag is piled on a heap and a fluid solution is sprinkled over it to penetrate the heap from the top downwards.[115] As a result of the contact of the leaching solution with the disposed waste, metals are solubilized and metal-loaded effluent is collected nearby.[115] Based on experimental data, leaching rates can be determined, which can be used as input data for the models predicting material behavior under heap leaching conditions.[115] The physicochemical conditions of a particular layer of the heap may vary, which tends to make this process less homogeneous as compared to batch and tank reactors. The main conditions that will certainly affect the feasibility of heap leaching are: (i) permeability, (ii) optimal solution flow rate, and (iii) maintenance of optimal process parameters, such as pH, aeration rate, and temperature.[104] Permeability dictates the optimal access of the leaching solution towards each waste disposed of in the heap, and excessively compacted material could cause uneven distribution of the solution and possibly its preferential flow, resulting in decreased process efficiency. Temperature is known for its role in activating dissolution so heat loss is undesirable in heap leaching. To avoid this problem, a covering can be applied to the heap. Heap leaching was first applied for Cu leaching at Kennecott mine in 1950.[16] At the present time, only sulfide ores are used for full-scale leaching, whereas low-grade sulfides or concentrates are being explored on a purely demonstrative scale in the hope of discovering optimal operating conditions. Heap leaching, in the context of slag application, requires significant steps forward prior to its commercialization.

9.4.3 Element Recovery from Leachates

Metal recovery refers to the process of acquisition of metal(s) present in the leachate and the conversion into soluble solid forms. Selective recovery is particularly desired in order to get a final product of high purity. The pregnant leachate contains the targeted metal as well as various impurities and major cations released from slags in extremely high quantities, often much higher than that of the targeted metal. On the other hand, the sensitivity of each element and its specific behavior under certain conditions make it possible to selectively separate the desired elements. Physicochemical methods dedicated to metal recovery mainly include: precipitation, electrochemical treatment, electrocoagulation, cementation, ion exchange, and solvent extraction,[116] whereas biosorption is an alternative biological recovery treatment using living (bacteria, fungi, and algae) or dead biomass (seeds, peels, and stalks of different crops) as the sorbent.[117,118] Each method presents its own advantages and limitations. However, the main prerequisite for the use of a chosen method is its selectivity because, apart from the main metal targeted, other valuable elements are also present in the pregnant leachate. In order to make the recovery process sustainable, these other elements should therefore be treated concurrently, which requires an integrated treatment procedure to sequentially and selectively precipitate metals from the solution using a combined approach. For example, Fe in the form of Fe^{2+} exhibits higher solubility in comparison to Fe^{3+}. The release of Fe into the leachate is an impurity and its removal from the leachate through oxidation–precipitation is a pH-dependent process. In order to remove Fe impurities, solvent extraction, ion exchange, and neutralization may be implemented and contribute to the acquisition of concentrated and purified solutions suitable for further processing by appropriate treatment units (*e.g.*, electrowinning).[119] The inhibition of Fe dissolution and avoiding its presence in the subsequent recovery steps improves the pregnant leaching solution downstream treatment process, therefore improving the selectivity and purity of recovered metals.

9.5 Factors Affecting Process Expenses

Metal recovery from slag will aid in limiting the amount of waste deposited at disposal sites, ideally generating less toxic waste after reprocessing, and will keep metals in the value chain of the industrial cycle, making this technology highly important to achieving a circular economy. At present, physical separation methods (*i.e.*, gravity, magnetic separation, and flotation) are applied, whereas (bio)leaching recovery methods are currently under evaluation and remain at the level of an academic debate. A full-scale leaching recovery unit has to be assessed in terms of the expenses, benefits, and, potentially more importantly, environmental consequences. The electricity, heat, and chemicals applied have to be wisely chosen so that both economic and environmental benefits are achieved. Calculations based on laboratory-scale processes may provide insight into the economic contribution of the process before it passes into widespread use. This is particularly true for bioleaching, which has not been employed at full scale to date. For leaching, expenses to be considered include costs related to milling; chemicals required for medium preparation; microbial culture isolation/cultivation; energy spent for incubator shaker operation; maintenance costs; and expenses related to multi-step selective recovery. Conversely, the composition of the leachate should be used for estimation of its value based on metal concentrations and prices at current market values. As already reported, from an economic viewpoint, laboratory leaching is a cost-intensive operation.[16,51,52] The same can be assumed for physical separation methods, which consume a high amount of energy. However, when factors such as: (i) the reduction of space required for slag storage, (ii) the possibility of the application of solid leaching residue in the construction sector, and (iii) a decrease in the environmental footprint related to slag disposal are all considered, the environmental benefit may prevail.

Another important aspect concerning the economics of the recovery processes is the type of element to be recovered. Obviously, base metals can be recovered in higher quantities due to their generally higher initial bulk content in certain types of slag. On the other hand, critical and REEs are also incorporated into slags, in spite of very little attention being given to them so far.[45,46,120] A comparative estimate shows that recovery value can be increased significantly when valuable elements and REEs are considered.[51] That being said, selectivity in the recovery step remains a crucial feature of any recovery method in order to be not only environmentally but also economically viable. Although the economic advantages of any recovery process are a matter of course, undoubtedly its environmental benefit is of primary importance.[121]

9.6 Concluding Remarks

The progressive depletion of primary metal resources (ores) means that the current market is placing an emphasis on the exploitation of lower-grade resources, such as slags. Therefore, the metal industry is seeking to maximize the recycling potential and waste processing, which have become integral parts of their valorization chains. Base metal slags contain valuable and critical elements at various concentrations; often the concentration of these elements is high enough to classify these wastes as resources. In contrast, ferrous slags are generally more suitable for use in the building and construction industries.

Physical recovery methods, including gravity, magnetic separation, flotation, and roasting, represent efficiently applied technologies for improving metal recovery from slags. In contrast, metal (bio)leaching from slags takes place on the laboratory scale and development of suitable protocols is at present one of the top priorities in hydrometallurgy. The recovery methods exhibit distinct advances over traditional disposal of these wastes, which include: (i) the acquisition of elements that would otherwise end their life-cycle in the slag waste piles, (ii) the generation of metal-depleted waste that can potentially be more suitable for disposal or use as solid additive for numerous purposes, (iii) an increase in the amount and accuracy of resource estimates when secondary resources are taken into account, and (iv) the reduction of land area required for the disposal of slags. Metal recovery methods are generally cost-intensive operations, however generation of slags and residues depleted in metals is crucial to decrease slags' footprint in the environment. Therefore, the importance of recovery methods should be viewed through the lens of environmental rather than economic profit.

Acknowledgements

The authors would like to gratefully acknowledge all literature data reports and review studies on metallurgical slags and extraction/recovery processes written by other authors; all publications and references therein have certainly contributed to the development of this chapter. The time devoted by Maxim Muravyov and the editors (Nadine Piatak and Vojtěch Ettler) to evaluating this work as well as their valuable comments and remarks are highly appreciated. The National Science Centre (NCN) in Poland (UMO-2018/31/D/ST10/00738) is acknowledged and greatly appreciated as well. The authors thank Daniel Mitchell and Nadine Piatak for their English proofreading of this manuscript.

References

1. N. J. Themelis, *JOM*, 1994, **46**, 51.
2. R. B. Gordon, *Resour., Conserv. Recycl.*, 2002, **36**, 87.
3. B. Gorai, R. K. Jana and Premchand, *Resour., Conserv. Recycl.*, 2003, **39**, 299.
4. S. Roy, A. Datta and S. Rehani, *Int. J. Miner. Process.*, 2015, **143**, 43.
5. N. M. Piatak, in *Environmental Geochemistry: Site Characterization, Data Analysis and Case Histories*, Elsevier, 2018.
6. C. Shi, C. Meyer and A. Behnood, *Resour., Conserv. Recycl.*, 2008, **52**, 1115.
7. C. Barca, C. Gérente, D. Meyer, F. Chazarenc and Y. Andrès, *Water Res.*, 2012, **46**, 2376.
8. D. Bonenfant, L. Kharoune, S. Sauvé, R. Hausler, P. Niquette, M. Mimeault and M. Kharoune, *Ind. Eng. Chem. Res.*, 2008, **47**, 7610.
9. J.-C. Lee and B. D. Pandey, *Waste Manage.*, 2012, **32**, 3.
10. N. M. Piatak, M. B. Parsons and R. R. Seal, *Appl. Geochem.*, 2015, **57**, 236.
11. S. Sobanska, B. Ledésert, D. Deneele and A. Laboudigue, *C. R. Acad. Sci., Ser. IIa*, 2000, **331**, 271.
12. V. Ettler, *Appl. Geochem.*, 2016, **64**, 56.
13. R. Tyszka, A. Pietranik, J. Kierczak, V. Ettler, M. Mihaljevič and J. Weber, *Appl. Geochem.*, 2012, **27**, 1089.
14. R. Tyszka, A. Pietranik, J. Kierczak, G. Zieliński and J. Darling, *J. Geochem. Explor.*, 2018, **186**, 215.
15. N. M. Piatak and R. R. Seal, *Appl. Geochem.*, 2012, **27**, 623.
16. A. Potysz, E. D. van Hullebusch and J. Kierczak, *J. Environ. Manage.*, 2018, **219**, 138.
17. V. Sibanda, E. Sipunga, G. Danha and T. A. Mamvura, *Heliyon*, 2020, **6**, 03135.
18. H. S. Altundoğan and F. Tümen, *Hydrometallurgy*, 1997, **44**, 261.
19. T. Deng and Y. Ling, *Waste Manage. Res.*, 2007, **25**, 440.
20. F. Tümen and N. T. Bailey, *Hydrometallurgy*, 1990, **25**, 317.
21. R. K. Nadirov, *Trans. Indian Inst. Met.*, 2019, **72**, 603.
22. M. D. Dimitrijevic, D. M. Urosevic, Z. D. Jankovic and S. M. Milic, *Physicochem. Probl. Miner. Process.*, 2016, **52**, 409.
23. S. Anand, R. P. Das and P. K. Jena, *Hydrometallurgy*, 1981, **7**, 243.
24. N. T. Arndt, L. Fontboté, J. W. Hedenquist, S. E. Kesler, J. F. H. Thompson and D. G. Wood, *Geochem. Perspect.*, 2017, **6**, 1.

25. European Commission, *Study on the EU's List of Critical Raw Materials (2020)*, 2020.
26. K. Avarmaa, L. Klemettinen, H. O'Brien and P. Taskinen, *Miner. Eng.*, 2019, **133**, 95.
27. H. Motz and J. Geiseler, *Waste Manage.*, 2001, **21**, 285.
28. A. H. Kaksonen, L. Lavonen, M. Kuusenaho, A. Kolli, H. Närhi, E. Vestola, J. A. Puhakka and O. H. Tuovinen, *Miner. Eng.*, 2011, **24**, 1113.
29. A. Pathak, M. G. Dastidar and T. R. Sreekrishnan, *J. Environ. Manage.*, 2009, **90**, 2343.
30. V. Ettler and Z. Johan, *Appl. Geochem.*, 2014, **40**, 97.
31. N. M. Piatak, R. R. Seal and J. M. Hammarstrom, *Appl. Geochem.*, 2004, **19**, 1039.
32. A. Potysz, J. Kierczak, Y. Fuchs, M. Grybos, G. Guibaud, P. N. L. Lens and E. D. van Hullebusch, *J. Geochem. Explor.*, 2016, **160**, 1.
33. M. Vítková, V. Ettler, Z. Johan, B. Kříbek, O. Šebek and M. Mihaljevič, *Mineral. Mag.*, 2010, **74**, 581.
34. A. Manasse, M. Mellini and C. Viti, *Eur. J. Mineral.*, 2001, **13**, 949.
35. J. Kierczak and A. Pietranik, *Can. Mineral.*, 2011, **49**, 1281.
36. V. Ettler, O. Legendre, F. Bodénan and J.-C. Touray, *Can. Mineral.*, 2001, **39**, 873.
37. N. Tshiongo, R. K. K. Mbaya, K. Maweja and L. C. Tshabalala, *World Acad. Sci. Eng. Technol.*, 2010, **46**, 273.
38. J. Kierczak, C. Néel, J. Puziewicz and H. Bril, *Can. Mineral.*, 2009, **47**, 557.
39. N. Y. Mostafa, S. A. S. El-Hemaly, E. I. Al-Wakeel, S. A. El-Korashy and P. W. Brown, *Cem. Concr. Res.*, 2001, **31**, 899.
40. Y. M. Kuo, J. W. Wang, C. T. Wang and C. H. Tsai, *J. Hazard. Mater.*, 2008, **152**, 994.
41. A. M. Álvarez-Valero, R. Pérez-López and J. M. Nieto, *Am. Mineral.*, 2009, **94**, 1417.
42. R. F. Tylecote, H. A. Ghaznavi and P. J. Boydell, *J. Archaeol. Sci.*, 1977, **4**, 305.
43. J. Puziewicz, K. Zainoun and H. Bril, *Can. Mineral.*, 2007, **45**, 1189.
44. N. M. Piatak and R. R. Seal, *Appl. Geochem.*, 2010, **25**, 302–320.
45. K. Avarmaa, L. Klemettinen, H. O'Brien, P. Taskinen and A. Jokilaakso, *Minerals*, 2019, **9**, 367.
46. A. M. Alfantazi and R. R. Moskalyk, *Miner. Eng.*, 2003, **16**, 893.
47. Abhilash, P. Meshram, S. Sarkar and T. Venugopalan, *Miner. Metall. Process.*, 2017, **34**, 178.
48. B. Mikoda, H. Kucha, A. Potysz and E. Kmiecik, *Appl. Geochem.*, 2019, **101**, 62.
49. I. Gaballah, E. Allain and M. Djona, *Metall. Mater. Trans. B*, 1997, **28**, 359.
50. J. J. N. Kayembe, M. K. Wa Kalenga and D. K. Nyembwe, presented in part at Materials Science and Technology, *2018*, Ohio, USA, November, 2018.
51. B. Mikoda, A. Potysz and E. Kmiecik, *J. Environ. Manage.*, 2019, **236**, 436.

52. V. Ettler, A. Jarošíková, M. Mihaljevič, B. Kříbek, I. Nyambe, F. Kamona and B. Mapani, *J. Geochem. Explor.*, 2020, **218**, 106631.
53. A. Bruynesteyn and R. P. Hackl, *Miner. Environ.*, 1982, **4**, 5.
54. L. E. Eary and M. A. Williamson, presented in part at 7th International Conference on Acid Rock Drainage, ICARD, United States, March, 2006.
55. H. Shen and E. Forssberg, *Waste Manage.*, 2003, **23**, 933.
56. F. Kukurugya, P. Nielsen and L. Horckmans, *Minerals*, 2020, **20**, 906.
57. K. S. Wang, K. L. Lin and Z. Q. Huang, *Cem. Concr. Res.*, 2001, **31**, 97.
58. N. Menad, N. Kanari and M. Save, *Int. J. Miner. Process.*, 2014, **126**, 1.
59. J. Xiong, Z. He, Q. Mahmood, D. Liu, X. Yang and E. Islam, *J. Hazard. Mater.*, 2008, **152**, 211.
60. Z. Guo, D. Zhu, J. Pan, T. Wu and F. Zhang, *Metals*, 2016, **6**, 86.
61. D. Zhu, J. Xu, Z. Guo, J. Pan, S. Li, L. Pan and C. Yang, *J. Cleaner Prod.*, 2020, **250**, 119462.
62. N. Ma and J. B. Houser, *J. Cleaner Prod.*, 2014, **82**, 221.
63. S. Li, J. Pan, D. Zhu, Z. Guo, J. Xu and J. Chou, *Powder Technol.*, 2019, **347**, 159.
64. J. Xiang, Q. Huang, X. Lv and C. Bai, *J. Hazard. Mater.*, 2017, **336**, 1.
65. L. Meng, Z. Wang, Y. Zhong, L. Guo, J. Gao, K. Chen, H. Cheng and Z. Guo, *Chem. Eng. J.*, 2017, **326**, 540.
66. Y. Li, J. Gao, Y. Du and Z. Guo, *Miner. Eng.*, 2020, **155**, 106471.
67. X. Lan, J. Gao, Y. Li and Z. Guo, *J. Hazard. Mater.*, 2019, **367**, 473.
68. L. Lin, Y. ping Bao, M. Wang, W. Jiang and H. mei Zhou, *J. Iron Steel Res. Int.*, 2014, **21**, 496.
69. X. Lan, J. Gao, Z. Huang and Z. Guo, *Metall. Mater. Trans. B*, 2018, **49**, 1165.
70. Y. Du, J. Gao, X. Lan and Z. Guo, *Ceram. Int.*, 2020, **46**, 9885.
71. J. Gao, Y. Zhong and Z. Guo, *ISIJ Int.*, 2016, **56**, 1352.
72. S. Panda, S. Mishra, D. S. Rao, N. Pradhan, U. Mohapatra, S. Angadi and B. K. Mishra, *Korean J. Chem. Eng.*, 2015, **32**, 667.
73. K. Mikula, G. Izydorczyk, D. Skrzypczak, K. Moustakas, A. Witek-Krowiak and K. Chojnacka, *J. Hazard. Mater.*, 2021, **403**, 123602.
74. B. A. Wills and J. A. Finch, Wills' Mineral Processing Technology, Butterworth Heinemann Elsevier, Oxford United Kingdom, 2015.
75. A. Sarrafi, B. Rahmati, H. R. Hassani and H. H. A. Shirazi, *Miner. Eng.*, 2004, **17**, 457.
76. W. J. Bruckard, M. Somerville and F. Hao, *Miner. Eng.*, 2004, **17**, 495.
77. J. Liu, G. Liu, Y. Huang and Z. Zhang, *Appl. Surf. Sci.*, 2019, **491**, 624.
78. H. Zhou, G. Liu, L. Zhang and C. Zhou, *J. Hazard. Mater.*, 2021, **401**, 123293.
79. H. S. Altundoğan and F. Tümen, *Hydrometallurgy*, 1997, **44**, 261.
80. J. Zhang, W. Zhang, L. Zhang and S. Gu, *Int. J. Miner. Process.*, 2015, **138**, 20.
81. Y. Ji, S. Shen, J. Liu and Y. Xue, *J. Cleaner Prod.*, 2017, **149**, 1068.
82. T. Jiang, J. Wen, M. Zhou and X. Xue, *J. Alloys Compd.*, 2018, **742**, 402.
83. H. Tagawa, *Thermochim. Acta*, 1984, **80**, 23.

84. M. I. Muravyov and N. V. Fomchenko, *Appl. Biochem. Microbiol.*, 2013, **49**, 562.
85. Z. Guo, L. Zhang, Y. Cheng, X. Xiao, F. Pan and K. Jiang, *Hydrometallurgy*, 2010, **104**, 25.
86. A. Potysz, P. N. L. Lens, J. van de Vossenberg, E. R. Rene, M. Grybos, G. Guibaud, J. Kierczak and E. D. van Hullebusch, *Appl. Geochem.*, 2016, **68**, 39.
87. A. H. Kaksonen, S. Särkijärvi, E. Peuraniemi, S. Junnikkala, J. A. Puhakka and O. H. Tuovinen, *Hydrometallurgy*, 2017, **168**, 135.
88. A. H. Kaksonen, S. Särkijärvi, J. A. Puhakka, E. Peuraniemi, S. Junnikkala and O. H. Tuovinen, *Miner. Eng.*, 2017, **106**, 97.
89. A. van Zomeren, S. R. van der Laan, H. B. A. Kobesen, W. J. J. Huijgen and R. N. J. Comans, *Waste Manage.*, 2011, **31**, 2236.
90. M. Baghalha, V. G. Papangelakis and W. Curlook, *Hydrometallurgy*, 2007, **85**, 42.
91. A. Potysz and J. Kierczak, *Minerals*, 2019, **9**, 542.
92. P. Kinnunen, J. Mäkinen, M. Salo, R. Soth and K. Komnitsas, *Minerals*, 2020, **10**, 654.
93. Z. Yang, M. Rui-Lin, N. Wang-Dong and W. Hui, *Hydrometallurgy*, 2010, **103**, 25.
94. S. Anand, K. Sarveswara Rao and P. K. Jena, *Hydrometallurgy*, 1983, **10**, 305.
95. H. S. Altundoğan, M. Boyrazli and F. Tumen, *Miner. Eng.*, 2004, **17**, 465.
96. E. A. Gorrepati, P. Wongthahan, S. Raha and H. S. Fogler, *Langmuir*, 2010, **25**, 10467.
97. I. M. Ahmed, A. A. Nayl and J. A. Daoud, *J. Saudi Chem. Soc.*, 2016, **20**, 280.
98. A. Potysz, A. Pędziwiatr, S. Hedwig and M. Lenz, *J. Environ. Chem. Eng.*, 2020, **8**, 104450.
99. L. B. Sukla, R. N. Kar and V. V. Panchanadikar, *Int. J. Environ. Stud.*, 1995, **47**, 81.
100. D. E. Lazo, L. G. Dyer and R. D. Alorro, *Miner. Eng.*, 2017, **100**, 115.
101. R. Baciocchi, G. Costa, A. Polettini and R. Pomi, *Energy Procedia*, 2009, **1**, 4859.
102. H. Hocheng, C. Su and U. U. Jadhav, *Chemosphere*, 2014, **117**, 652.
103. C. S. Gahan, M. L. Cunha and Å. Sandström, *Hydrometallurgy*, 2009, **95**, 190.
104. M. Dopson, A. K. Halinen, N. Rahunen, D. Boström, J. E. Sundkvist, M. Riekkola-Vanhanen, A. H. Kaksonen and J. A. Puhakka, *Biotechnol. Bioeng.*, 2008, **99**, 811.
105. M. Dopson, L. Lövgren and D. Boström, *Hydrometallurgy*, 2009, **96**, 288.
106. A. N. Banza, E. Gock and K. Kongolo, *Hydrometallurgy*, 2002, **67**, 63.
107. F. Carranza, R. Romero, A. Mazuelos, N. Iglesias and O. Forcat, *Hydrometallurgy*, 2009, **97**, 39.
108. M. I. Muravyov, N. V. Fomchenko, A. V. Usoltsev, E. A. Vasilyev and T. F. Kondrat'Eva, *Hydrometallurgy*, 2012, **119–120**, 40.

109. N. Fomchenko and M. Muravyov, *Minerals*, 2020, **10**, 1097.
110. E. Ahmed and S. J. M. Holmström, *Microb. Biotechnol.*, 2014, 7, 196.
111. M. Saha, S. Sarkar, B. Sarkar, B. K. Sharma, S. Bhattacharjee and P. Tribedi, *Environ. Sci. Pollut. Res.*, 2016, **23**, 3984.
112. M. Hofmann, G. Retamal-Morales and D. Tischler, *Nat. Prod. Rep.*, 2020, **37**, 1262.
113. A. Potysz, M. Grybos, J. Kierczak, G. Guibaud, P. N. L. Lens and E. D. van Hullebusch, *Appl. Geochem.*, 2016, **64**, 92.
114. E. D. van Hullebusch, N. H. Yin, N. Seignez, J. Labanowski, A. Gauthier, P. N. L. Lens, C. Avril and Y. Sivry, *J. Environ. Manage.*, 2015, **147**, 297.
115. J. Petersen, *Hydrometallurgy*, 2016, **165**, 206.
116. S. Hedrich, R. Kermer, T. Aubel, M. Martin, A. Schippers, D. Barrie Johnson and E. Janneck, *Solid State Phenom.*, 2017, **262**, 107.
117. K. Vijayaraghavan and R. Balasubramanian, *J. Environ. Manage.*, 2015, **160**, 283.
118. A. Robalds, G. M. Naja and M. Klavins, *J. Hazard. Mater.*, 2016, **304**, 553.
119. M. K. Khalid, J. Hamuyuni, V. Agarwal, J. Pihlasalo, M. Haapalainen and M. Lundström, *J. Cleaner Prod.*, 2019, **215**, 1005.
120. J. J. Wang, Y. Bao, R. Ma, Y. Li, L. Gong, Y. Zhang, Z. Niu and B. Xin, *Hydrometallurgy*, 2018, **177**, 140.
121. F. Mathieux, F. Ardente, S. Bobba, P. Nuss, G. A. Blengini, P. A. Dias, D. Blagoeva, C. Torres De Matos, D. Wittmer, C. Pavel, T. Hamor, H. Saveyn, B. Gawlik, G. Orveillon, D. Huygens, E. Garbarino, E. Tzimas, F. Bouraoui and S. Solar, *Report EUR 28832 EN*, 2017.
122. S. Anand, P. K. Rao and P. K. Jena, *Hydrometallurgy*, 1980, **5**, 355.
123. L. B. Sukla, S. C. Panda and P. K. Jena, *Hydrometallurgy*, 1986, **16**, 153.
124. A. V. Beşe, O. N. Ata, C. Çelik and S. Çolak, *Chem. Eng. Process.*, 2003, **42**, 291.
125. A. V. Beşe, *Ultrason. Sonochem.*, 2007, **14**, 790.
126. S. L. Mishra, C. Vijayalakshmi, N. V. Thakur and K. S. Koppiker, *Miner. Process. Extr. Metall. Rev.*, 1992, **10**, 275.
127. C. Arslan and F. Arslan, *Hydrometallurgy*, 2002, **67**, 53.
128. O. Herreros, R. Quiroz, E. Manzano, C. Bou and J. Viñals, *Hydrometallurgy*, 1998, **49**, 87.
129. S. M. Abdel Basir and M. A. Rabah, *Hydrometallurgy*, 1999, **53**, 31.
130. R. K. Nadirov, L. I. Syzdykova, A. K. Zhussupova and M. T. Usserbaev, *Int. J. Miner. Process.*, 2013, **24**, 2799.
131. Ch. Zhang, B. Hu, H. Wang, M. Wang and X. Wang, *Min., Metall., Explor.*, 2020, **37**, 1241.
132. G. Shi, Y. Liao, B. Su, Y. Zhang, W. Wang and J. Xi, *Sep. Purif. Technol.*, 2020, **241**, 116699.
133. C. M. Du, X. Gao, S. Ueda and S. Y. Kitamura, *Hydrometallurgy*, 2019, **189**, 105109.
134. G. Bulut, *Waste Manage. Res.*, 2006, **24**, 1.
135. M. Dimitrijević, D. Urošević, S. Milić, M. Sokić and R. Marković, *J. Min. Metall., Sect. B*, 2017, **53**, 407.

136. J. Che, Y. Chen, B. Ma, C. Wang and W. Zhang, *Hydrometallurgy*, 2020, **193**, 105321.
137. A. Król, K. Mizerna and M. Bożym, *J. Hazard. Mater.*, 2020, **384**, 121502.
138. S. Song, W. Sun, L. Wang, R. Liu, H. Han, Y. Hu and Y. Yang, *J. Environ. Chem. Eng.*, 2019, 7, 102777.
139. K. Gargul, B. Boryczko, A. Bukowska, P. Jarosz and S. Małecki, *Arch. Civ. Mech. Eng.*, 2019, **19**, 648.
140. L. Tian, Z. Xu, L. Chen, Y. Liu and T. A. Zhang, *Hydrometallurgy*, 2019, **184**, 45.
141. M. D. Turan, Z. A. Sari and J. D. Miller, *Trans. Nonferrous Met. Soc. China*, 2017, **27**, 1404.
142. Y. Li, V. G. Papangelakis and I. Perederiy, *Hydrometallurgy*, 2009, **97**, 185.
143. Y. Li, I. Perederiy and V. G. Papangelakis, *J. Hazard. Mater.*, 2008, **152**, 607.
144. P. K. Gbor, I. B. Ahmed and C. Q. Jia, *Hydrometallurgy*, 2000, **57**, 13.
145. L. L. De Windt, P. Chaurand and J. Rose, *Waste Manage.*, 2011, **31**, 225.
146. A. M. Fällman, *Stud. Environ. Sci.*, 1997, **71**, 531.
147. E. Rudnik, L. Burzyńska and W. Gumowska, *Miner. Eng.*, 2009, **22**, 88.
148. L. B. Sukla, R. N. Kar and V. Panchanadikar, *BioMetals*, 1992, **5**, 169.
149. K. D. Mehta, B. D. Pandey and Premchand, *Mater. Trans., JIM*, 1999, **40**, 214.
150. E. A. Vestola, M. K. Kuusenaho, H. M. Närhi, O. H. Tuovinen, J. A. Puhakka, J. J. Plumb and A. H. Kaksonen, *Hydrometallurgy*, 2010, **103**, 74.
151. A. H. Kaksonen, S. Särkijärvi, J. A. Puhakka, E. Peuraniemi, S. Junnikkala and O. H. Tuovinen, *Hydrometallurgy*, 2016, **159**, 46.
152. A. Schippers, *Solid State Phenom.*, 2017, **262**, 61.
153. P. Georgiev, I. Spasova, V. Groudeva, M. Nicolova, A. Lazarova, M. Iliev, R. Ilieva and S. Groudev, *Solid State Phenom.*, 2017, **262**, 696.
154. Y. Cheng, Z. Guo, X. Liu, H. Yin, G. Qiu, F. Pan and H. Liu, *Bioresour. Technol.*, 2009, **100**, 2737.
155. K. Q. Jiang, Z. H. Guo, X. Y. Xiao and X. Y. Wei, *Trans. Nonferrous Met. Soc. China*, 2012, **22**, 3120.
156. J. Wang, Q. Huang, T. Li, B. Xin, S. Chen, X. Guo, C. Liu and Y. Li, *J. Environ. Manage.*, 2015, **159**, 11.
157. S. Ahmadi, M. Vafaie Sefti, M. M. Shadman, Z. Azimi Dijvejin and H. Hosseini, *J. Adv. Environ. Health Res.*, 2017, **5**, 154.
158. H. I. Gomes, V. Funari, W. M. Mayes, M. Rogerson and T. J. Prior, *J. Environ. Manage.*, 2018, **222**, 30.
159. O. Bayat, E. Sever, B. Bayat, V. Arslan and C. Poole, *Appl. Biochem. Biotechnol.*, 2009, **152**, 117.
160. S. Thallner, C. Hemmelmair, S. Martinek and W. Schnitzhofer, *Solid State Phenom.*, 2017, **262**, 79.

Subject Index